ELECTRICAL
& ELECTRONIC
SYSTEMS

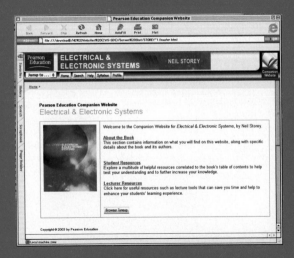

Visit the *Electrical & Electronic Systems* Companion Website at
www.booksites.net/storey_elec

to find valuable teaching and learning material including:

• Learning objectives for each chapter

• Multiple choice questions to help test your learning

Also: This site has a syllabus manager, search functions, and
email results functions.

NEIL STOREY **ELECTRICAL & ELECTRONIC SYSTEMS**

PEARSON

Prentice
Hall

Harlow, England • London • New York • Boston • San Francisco • Toronto
Sydney • Tokyo • Singapore • Hong Kong • Seoul • Taipei • New Delhi
Cape Town • Madrid • Mexico City • Amsterdam • Munich • Paris • Milan

Pearson Education Limited

Edinburgh Gate
Harlow
Essex CM20 2JE
England

and Associated Companies throughout the world

Visit us on the World Wide Web at:
www.pearsoned.co.uk

First published 2004

ISBN 0130 93046 6

British Library Cataloguing-in-Publication Data
A catalogue record for this book is available from the British Library

10 9 8 7 6 5 4 3 2 1
08 07 06 05 04

Typeset in 10/12 pt Times Roman by 35
Printed by Ashford Colour Press Ltd, Gosport

The publisher's policy is to use paper manufactured from sustainable forests.

Contents

Website Resources

For Students:
- Learning objectives for each chapter
- Multiple choice questions to help test your learning

For Lecturers:
- A secure, password protected site with teaching material
- Complete, downloadable Instructor's Manual Powerpoint slides that can be downloaded and used as OHTs

Also: This regularly maintained site has a syllabus manager, search functions, and email results functions.

Trademarks

1. PSpice is a registered trademark of Cadence Design Systems, Inc.

2. MATLAB is a registered trademark of The Maths Work, Inc.

3. Mathcad is a registered trademark of Mathsoft, Inc.

4. Intel, Pentium, i386 and i486 are trademarks or registered trademarks of Intel Corporation.

Preface

The area of *Electrical & Electronic Systems* represents one of the most important 'enabling technologies' of the modern age. Such systems are used at the heart of a wide range of products that extend from cars to aircraft, and from mobile phones to nuclear power stations. For this reason, *all* engineers, scientists and technologists need a basic understanding of such systems, while many will require a far more detailed knowledge of this area.

The importance of this topic has led to the creation of a large number of text books in this area, and readers may wonder how this book differs from the many others on offer. Perhaps the most significant characteristic of this text is that it adopts a 'top-down' approach to the subject which is not found in many of its competitors. It begins by outlining the uses and characteristics of electrical and electronic systems, before looking in detail at their analysis. This gives readers an insight into *why* topics are of importance before they are studied in detail, and greatly simplifies the learning process.

A great benefit of a top-down approach is that it makes the book more accessible for *all* its potential readers. For those who intend to specialise in electronic or electrical engineering the material is presented in a way that makes it easy to absorb, providing an excellent grounding for further study. For those intending to specialise in other areas of engineering or science, the order of presentation allows them to gain a good grounding in the *basics*, and to progress into the *detail* only as far as is appropriate for their needs.

Another unusual characteristic of this book is that it integrates both electrical engineering and electronic engineering within a single volume. This not only permits the purchase of one book rather than two, but also allows the two subjects to be integrated, to produce a unified approach to these closely related areas.

Who should read this book

This text is intended for undergraduate students in all fields of engineering and science. For students of electronics and electrical engineering it provides a first-level introduction to electrical and electronic systems that will provide a sound basis for further study. For students of other disciplines it includes most of the electrical and electronic material that they will need throughout their course.

Assumed knowledge

The book assumes very little by way of prior knowledge, except for a basic understanding of school level mathematics and a little physics. Most readers will have met electrical concepts such as voltage, current and Ohm's law during their time at school, but a summary of such topics is included for those who would like some revision of this material.

Companion website

The text is supported by a comprehensive companion website that will greatly increase both your understanding and·your enjoyment of the book. The site contains a range of support material, including computer-marked self-assessment exercises for each chapter. These exercises not only give you instant feedback on your understanding of the material, but also give useful guidance on areas of difficulty. To visit the site, go to www.booksites.net/storey_elec

Circuit simulation

Circuit simulation offers a powerful and simple means of gaining an insight into the operations of electronic circuits. Throughout the book there are numerous **Computer Simulation Exercises** that support the material in the text. These are marked by icons in the margin. The exercises may be performed using any circuit simulation package, although the use of **PSpice** is recommended. PSpice is one of the most widely used packages both within industry and within Universities and Colleges. It comes as part of a suite of programs that provides schematic capture of circuits and the graphical display of simulation results.

PSpice simulation

Many students will have access to PSpice within their University or College, but you may also obtain the software for use on your own computer if you wish. A student edition of the PSpice package can be downloaded **free of charge** from the manufacturer's website or obtained on a **free CD**. Details of how to get your free copy of the simulation software are given on the book's companion website.

To simplify the use of simulation as an aid to understanding the material within the book, a series of PSpice demonstration files can also be downloaded from the website. The name of the relevant demonstration file is given in the margin under the computer icon for the associated computer simulation exercise. Computer icons are also found next to some circuit diagrams where simulation files have been provided to aid understanding of the operation of the circuit. The demonstration files come with full details of how to carry out the various simulation exercises.

Problems using simulation have also been included within the exercises at the end of the various chapters. These exercises do *not* have demonstration files and are set to develop and test the reader's understanding of the use of simulation as well as the circuits concerned.

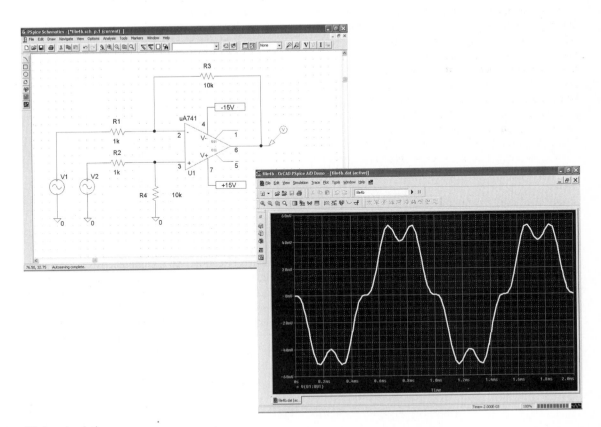

PSpice simulation

To the instructor

A comprehensive set of online support material is available for instructors using this book as a course text. This includes a set of editable **PowerPoint slides** to aid in the preparation of lectures, plus an **instructor's manual** that gives **fully worked solutions** to all the numerical problems and **sample answers** for the various non-numerical exercises. Guidance is also given on **course preparation** and on the selection of topics to meet the needs of students with different backgrounds and interests. This material, together with the various online study aids, simulation exercises and self-assessment tests, should greatly assist both the instructor and the student to gain maximum benefit from courses based on this text. Instructors adopting this book should visit the book's companion website at: www.booksites.net/storey_elec for details of how to gain access to the secure website that holds the instructor's support material.

Acknowledgements

I would like to express my gratitude to the various people who have assisted in the preparation of this book. In particular I would like to thank my colleagues at the University of Warwick who have provided useful feedback and suggestions.

I would also like to thank the companies who have given permission to reproduce their material. These are: RS Components for the photographs in Figures 3.2(a), 3.2(b), 3.3(a), 3.3(b), 3.4, 3.5(a), 3.5(b), 3.6, 3.7, 3.9(a), 3.9(b), 3.12, 4.1, 4.3, 4.4, 4.6, 11.14(a); Farnell Electronic Components Limited for the photographs in Figures 4.2, 11.12 and 11.13(a); and Cadence Design Systems Inc. for permission to use the PSpice images in this Preface.

Finally, I would like to give special thanks to my family for their help and support during the writing of this book. In particular I wish to thank my wife Jillian for her never-failing encouragement and understanding.

Chapter 1

Engineering Systems

Objectives

When you have studied the material in this chapter you should be able to:

- **discuss the relevance of electrical and electronic engineering, and its importance as an 'enabling technology';**
- **describe the characteristics and advantages of a 'systems approach' to engineering;**
- **list the main elements that form the basis of all electrical and electronic systems and be able to cite real-world examples of such components;**
- **identify the inputs and outputs of an engineering system and understand the significance of the choice of system boundary;**
- **explain the varied characteristics of physical quantities and the need to represent these quantities by electrical signals;**
- **use block diagrams to represent complex engineering systems.**

1.1   Introduction

As we begin our look at **electrical and electronic systems** it is perhaps appropriate to start by explaining what is meant by the terms 'electrical' and 'electronic' in this context. Both terms relate to the use of electrical energy but, unfortunately, different engineers use these terms in slightly different ways. There are also variations in the ways these words are used in different countries. 'Electrical' is often used to refer to applications that are concerned with the generation, transmission or use of large amounts of electrical energy. 'Electronic' applications often involve smaller amounts of power, and in many cases the electrical energy is used to convey information rather than as a source of power. Unfortunately, many situations do not fall neatly within either of these two camps but occupy the middle ground between them. For this reason, in this text we will assume the broad distinction outlined above but will be fairly liberal in our use of the two terms, since much of the material covered is relevant to all forms of electrical and electronic system.

As well as understanding the meanings of the title of this text, it is also important to understand the relevance of the material it contains. To do this we need to look initially at the very nature of engineering.

1.1.1　The interdisciplinary nature of engineering

All real engineering projects of any size are inherently interdisciplinary. One only has to look at a few examples of typical engineering projects, such as cars, aircraft, heating systems or dams, to see the interaction and interdependence of the various forms of expertise required.

A modern car, for example, is the result of many years of development involving a wide range of engineering disciplines. Although it is tempting to see this work simply as automotive engineering, in reality a majority of the engineers working for automotive companies have qualifications in other areas. Clearly, mechanical engineering plays a large part in the design of the structure of the car and of its various mechanical components, such as the body, the engine and the gearbox. However, modern vehicles also contain a large number of electrical components, such as the battery, the alternator and the ignition system. Electronic components are also numerous, such as the engine management system, the anti-lock braking system and any entertainment components. In modern cars it is not uncommon for more than 40 percent of the cost of the vehicle to be within its various electrical and electronic systems, many of which are computer-based. Therefore, in addition to automotive and mechanical engineering, car production also involves elements of electrical, electronic and computer systems engineering. The metal parts of the car are often subjected to various anti-corrosion treatments that fall within the area of chemical engineering, while the process of actually assembling the car would normally be considered to fall within manufacturing or industrial engineering.

It can be seen that the production of a modern car uses techniques and skills from a wide range of engineering disciplines to produce a single, high-quality product. Looking at any other large-scale project is likely to produce a similar result, since all real projects of any size are inherently multidisciplinary.

Since all real engineering projects require a wide range of engineering skills, it might seem that engineers should be expert in *all* fields of engineering. While this might be a utopian objective, the diversity of modern engineering makes this completely impractical. No single person can master all the engineering disciplines, and most would rather specialise and become proficient in a particular field than study a wide range of subjects at a lower level. However, it is essential that engineers can work effectively alongside others with complementary skills, and this requires that each member of the team has an appreciation of all aspects of the work. For this reason, all engineers should have a broad-based education that covers mathematics, basic science and what might be termed 'enabling technologies' before specialising in a particular area. They will also need to gain an understanding of the application itself by obtaining experience in the relevant industry or application domain.

1.1.2 Electrical and electronic systems as enabling technologies

One can identify a number of engineering topics that could be considered to be essential for all engineers. Without doubt the fundamentals of electrical and electronic systems must fall within this group.

Almost all engineering projects have elements of electrical engineering within them, and in recent years more and more applications have included electronic or computer-based elements. Many engineers will wish to be involved in the production of these electrical/electronic components and will therefore require a detailed knowledge of their design and construction. Other engineers, such as mechanical or civil engineers, are unlikely to be responsible for the detailed design of complex electronic systems, but they *are* likely to be involved in projects that include such systems. It is therefore essential that they understand the characteristics of such components in order that they can interact with specialists in these areas, and to enable them to take full advantage of the possibilities within these technologies. Thus *all* engineers need a basic understanding of the characteristics and use of electrical and electronic systems, while specialists in these areas also require skills in design, analysis and construction.

1.1.3 A systems approach to engineering

Several areas of human endeavour are associated with the solution of problems that involve great complexity. These include topics as diverse as the comprehension of biological organisms, the proof of complex mathematical relationships and the rationalisation of philosophical arguments. Over the years several distinct approaches have evolved for tackling such problems, and many of these are of direct relevance to the production of complex systems within engineering.

One method is to adopt what might be termed a **systematic** approach, in which a complex problem or system is simplified by dividing it into a number of smaller elements. These elements are then themselves subdivided, the process being repeated until the various constituents have been devolved into elements that are sufficiently simple to be easily understood. This approach is widely used within engineering in what is termed **top-down design**, where a complex system is progressively divided into simpler and simpler subsystems. This results in a series of modules that are of a manageable level of complexity and size to allow direct implementation. This approach is based, to some extent, on what might be seen as a 'reductionist' view, implying that a complex system is no more than the sum of its parts.

A problem with the reductionist view is that it ignores characteristics that are features of the 'whole' rather than of individual components. These **systemic** properties are often complex in nature and may relate to several diverse aspects of the system. For example, the 'ride' and 'feel' of a car are not determined by a single module or subsystem but by the interaction of a vast number of individual components.

In recent years, modern engineering practice has evolved a more 'holistic' approach that combines the best elements of a *systematic* approach together with considerations of *systemic* issues. This results in what is called a **systems approach** to engineering.

The systems approach has its origins back in the 1960s but has gained favour within many engineering disciplines only recently. It is categorised by a number of underlying principles, which include a strong emphasis on the application of scientific methods, the use of systematic project management techniques and, perhaps most importantly, the adoption of a broad-based interdisciplinary or team approach. In addition to specialists from a range of engineering disciplines, a project might also involve experts in other fields such as artistic design, ergonomics, sociology, psychology or law. A key feature of a systems approach is that it places as much importance on identifying the relationships *between* components and events as it does on identifying the characteristics of the components and events themselves.

This text does not claim to provide a true 'systems approach' to the material it covers. In many ways such a broad-based treatment would be inappropriate for an introductory text of this kind. However, it does attempt to present information on components and techniques within the context of the systems in which they are used. For this reason, rather unusually, the book starts by looking at electrical and electronic systems before describing in detail the components in them. This allows the reader to understand *why* these components are required to have the characteristics that they have and *how* the techniques relate to the applications in which they are used.

1.2   Systems

Before looking at the nature of electrical and electronic systems, it is perhaps appropriate to make sure that we understand what we mean by the word **system**.

In an engineering context, a system can be defined as any closed volume for which all the inputs and outputs are known. This definition allows us to consider an infinite number of 'systems' depending on the volume of space that we decide to select. However, in practice, we normally select our 'closed volume' to enclose a component, or group of components, that are of interest to us. Thus we could select a volume that includes the components that control the engine of a car and call this an 'engine management system'. Alternatively, we might select a larger volume that includes the complete car and call this an 'automotive system'. A larger volume might include a complete 'transportation system', while a volume that contains the Earth might be described as an 'ecosystem'. Since we can freely select the boundaries of our systems, we can use this approach to subdivide large systems into smaller, more manageable blocks. For example, a car might be considered as a large number of smaller systems (or subsystems), each responsible for a different function.

As we change the elements within our 'system' we also change the inputs and outputs. The signals going into and out from an engine management system relate to the status, or condition, of various parts of the engine and the car. If we consider the complete car as our system, then the inputs include petrol, water and commands from the driver, while the outputs include work (in the form of movement), heat and exhaust gases. If we look at our planet as a single system, then the inputs and outputs are primarily different forms of radiation. From outside a system, only the inputs and outputs are visible. However, it may be possible to learn something of the nature of the system by observing the relationship between these inputs and outputs. One way of describing the characteristics of any system is in terms of the nature of the inputs and the outputs and the relationship between them.

In some situations only particular inputs and outputs to a system are of interest, and others may be completely ignored. For example, one input to a mobile phone might be air entering or leaving its case. An electronic engineer designing such a system might decide to ignore this form of input and to concentrate only on those inputs related to the operation of the unit. This general principle can be extended so that only particular kinds of input and output are considered. Thus a company's accounting system might consider only the flow of money into and out of the company. This concept can be extended so that the 'volume' of the system becomes nebulous, and the system is effectively defined solely by its inputs, its outputs and the relationship between them.

1.3   Electrical and electronic systems

From the discussion above it is clear that the term 'system' could be applied to any arrangement that has identifiable inputs and outputs. However, the term is normally reserved for arrangements that perform some useful function, such as telephone systems or power generation systems.

If we look at a large number of electrical and electronic systems, it is possible to identify a number of basic functions, or processes, that appear again and again. These processes, each of which is concerned with electrical energy in one form or another, can be divided into five main groups, which are:

1. generation
2. transmission or communication

3. control or processing
4. utilisation
5. storage.

Most electrical or electronic systems can be seen as a combination of elements within these groups, although the forms of these elements may vary considerably.

1.3.1   Generation

In *electrical* systems, generation is often concerned with the production of electricity for use as a power source. Elements within this group include the electrical generators in power stations or hydroelectric dams. Smaller units, such as the alternators used in cars, also fall within this group. In *electronic* systems, generation is more often concerned with the production of small electrical signals that are used to represent physical quantities, rather than as a source of power. Many sensors produce electrical signals in response to changes in their environment. These signals are usually relatively small and would not represent a useful source of electrical energy. However, the information that they convey may be extremely useful. An example of such a sensor is a photodiode, which can produce an electrical signal related to the amount of light falling on it. We will look at various sensors in more detail in Chapter 3.

1.3.2   Transmission or communication

In *electrical* systems, transmission is often concerned with the distribution of electrical power. This might involve overhead or underground power cables or, in smaller applications, the use of thick copper wires. In *electronic* systems, communication is more often concerned with the transmission of information than it is with the transmission of power. This might be achieved using electrical wires or fibre-optic cables. Alternatively, information might be transmitted using radio (waves) or microwaves.

1.3.3   Control or processing

Control is related to command and regulation functions. The simplest form of control is that of a switch, which allows or inhibits the flow of electrical energy. Switches can take an almost limitless number of forms from simple mechanical arrangements, such as domestic light switches, to high-power switchgear, such as that used to control the flow of electricity on the national power grid.

In addition to simple ON/OFF functions, elements within this group can also be concerned with more complex forms of control. In *electrical* systems it is common to need to regulate or vary the power supplied to some piece of apparatus, while in *electronic* systems it is often necessary to perform extremely complex processing functions, perhaps using computer-based techniques.

1.3.4   Utilisation

Utilisation is concerned with the use of electrical energy to perform some useful function. In *electrical* systems this might involve the production of heat, light or motion, while in *electronic* systems this might involve the production of sound or the display of visual information. Elements within this group are extremely varied, and it is these components that produce the required output of a system.

1.3.5   Storage

Although electrical energy can be stored directly, as in a capacitor, in many applications storage is achieved by converting electrical energy into another form. Perhaps the most common form of electrical storage is the use of rechargeable batteries, such as those used in cars. Here electrical energy from the alternator is stored chemically in the battery in a way that allows it to be converted back to an electrical form when required. Other examples within the *electrical* area include the use of flywheels and pumped water systems. In the former, an electric motor is used to spin a flywheel, thus converting electrical energy into kinetic energy. At a later time, the flywheel can be used to drive a generator, thus converting the stored energy back into an electrical form. Some power stations use excess output power to pump water to a raised reservoir. When required, this stored potential energy can be converted to electrical energy by discharging the water through a turbine generator.

Electronic applications also employ methods of storing electrical energy, though normally on a smaller scale. Here we are often more interested in storing information, as in the case of the random-access memory (RAM) used in computers. Here the presence or absence of a small amount of electrical charge is used to store a single element of information. By combining a large number of such elements, it is possible to store the vast amounts of data used by modern computers. These information storage techniques differ from the electrical storage methods discussed earlier in that the energy used cannot normally be recovered – it is the information that is being stored, not the electrical energy itself. As in the electrical examples, it is common for storage to be achieved by converting electrical energy into another form. For example, in a videocassette information is stored by converting an electrical signal into a magnetic pattern on the tape.

Figure 1.1 A power
distribution system

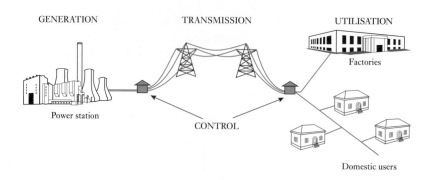

1.3.6   System examples

Elements of these five basic functions may be combined to form a wide
range of different systems. However, if one looks at a large number of
such systems it is apparent that many fall within a few distinct categories.
These include arrangements that are responsible for:

- power generation and distribution;
- monitoring of some equipment or process;
- control of some equipment or process;
- signal processing;
- communication.

Figure 1.1 shows an example of the first of these groups – a power distribution system. Here a power station is the generation element, switching
equipment corresponds to the control aspects, and high-voltage power lines
are responsible for transmission. In this example, the utilisation elements
are the domestic or industrial users of the distributed electricity. This
particular system does not include any storage elements, although these are
present in many power distribution networks. This might take the form of
a pumped storage system as discussed earlier.

You might like to identify other electrical or electronic systems and to
see if these fall within one of the categories listed above. You might also
like to identify the basic elements within these systems and decide which
of the five basic functions they represent.

1.4   System inputs and outputs

In Section 1.2 we noted that a system may be described solely by its inputs,
its outputs and the relationship between them. This concept prompts us to
look at the nature of the inputs and outputs of a system and the way that
these interact with the system itself.

Figure 1.2 represents a generalised system, together with its inputs and
outputs. This diagram makes no assumptions about the form of any of its

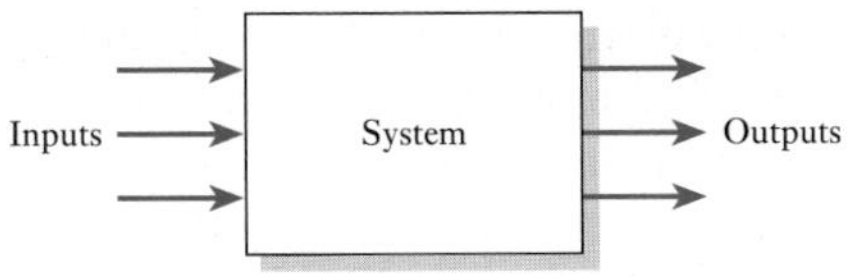

Figure 1.2 A generalised system

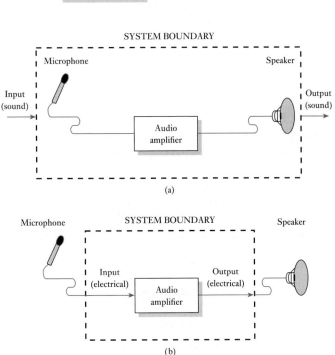

Figure 1.3 The effects of choosing system boundaries

components, and this could represent a mechanical or biological system just as easily as an electrical or electronic arrangement. Thus the inputs and outputs in this case could be forces, temperatures, velocities or any other physical quantities. Alternatively, they could be electrical quantities such as voltages or currents.

When considering electrical and electronic systems, we are concerned with arrangements that generate, or manipulate, electrical energy in one form or another. However, the nature of the inputs and outputs to such systems may depend on where we choose to draw the system's boundaries. This is illustrated in Figure 1.3, which shows an arrangement involving an audio amplifier, a microphone and a speaker. In Figure 1.3(a), we have chosen to consider the microphone and speaker as parts of our system. Here the input and output are in the form of sound waves. In Figure 1.3(b), we are considering the system to consist only of the audio amplifier itself. Now the microphone and speaker are external to the system, and the input and output are in an electrical form.

In Figure 1.3, the microphone senses variations in the external environment (in this case sound waves) and represents them electrically for processing within the electronic parts of the system. Conversely, the speaker takes the electrical outputs produced by the system and uses them to affect the external environment (again, in this case, by creating sound waves). Components that interact with the outside world in this way are referred to as *sensors* and *actuators,* and without such devices our electrical and electronic systems would be useless. We will therefore look at a range of such devices in Chapters 3 and 4. For the moment we will simply note that such elements exist and that they can be used to enable a system to interact with the world around it.

1.5	Physical quantities and electrical signals

The electrical fluctuations produced by a sensor convey information about some varying physical quantity. Such a representation is termed an electrical *signal*. In Figure 1.3, the output of the microphone is an electrical signal that represents the sounds that it detects. Similarly, the output from the amplifier is an electrical signal that represents the sounds to be produced by the speaker. Signals may take a number of forms, but before discussing these it is perhaps appropriate to look at the nature of the physical quantities that they may represent.

1.5.1 Physical quantities

The world about us may be characterised by a number of physical properties or quantities, many of which vary with time. Examples of these include temperature, humidity, pressure, altitude, position and velocity. The time-varying nature of such physical quantities allows them to be categorised into those that vary in a *continuous* manner and those that exhibit a *discontinuous* or *discrete* nature.

The vast majority of real-world physical quantities (such as temperature, pressure and humidity) vary in a continuous manner. This means that they change smoothly from one value to another, taking an infinite number of values. In contrast, discrete quantities do not change smoothly but instead switch abruptly between distinct values. Few natural quantities exhibit this characteristic (although there are some examples, such as population), but many man-made quantities are discrete.

1.5.2 Electrical signals

It is often convenient to represent a varying physical quantity by an electrical signal. This is because the processing, communication and storage of information is often much easier when it is represented electrically.

Having noted that physical quantities may be either continuous or discrete in nature, it is not surprising that the electrical signals that represent them may also be either continuous or discrete. However, there is not

Figure 1.4   Examples of
analogue and digital signals

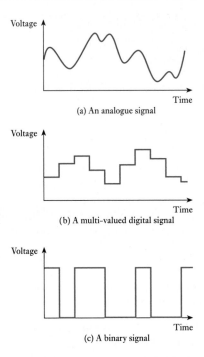

(a) An analogue signal

(b) A multi-valued digital signal

(c) A binary signal

necessarily a direct correspondence between these forms, since it may be convenient to represent a continuous quantity by a discrete signal, or vice versa. For reasons that are largely historical, continuous signals are normally referred to as *analogue*, while discrete signals are described as *digital*.

Both analogue and digital signals can take many forms. Perhaps one of the simplest is where the voltage of a signal corresponds directly to the magnitude of the physical quantity being represented. This format is used for both the input and the output signals in Figure 1.3(b), where the voltages of the signals correspond directly to fluctuations in the input and output air pressure (sound). Most people will have seen the output of a microphone displayed on an oscilloscope and noted the relationship between the sound level and the magnitude of the displayed waveform. Figure 1.4(a) shows an example of a typical analogue signal waveform.

Although it is very common to represent the magnitude of a continuous quantity by the voltage of an electrical signal, many other forms are also used. For example, it might be more convenient to represent the value of a physical quantity by the magnitude of the *current* flowing in a wire (rather than by the voltage on it), or by the frequency of a sinusoidal waveform. These and other formats are used in certain situations, these being chosen to suit the application.

Digital signals may also vary in form. Figure 1.4(b) shows a signal that takes a number of discrete levels. It could be that this signal represents numerical information, such as the number of people in a building. Since the signal changes abruptly from one value to another it is digital in nature, and in many cases such signals have a limited number of allowable values.

The most common forms of digital signals are those that have only two possible values, as shown in Figure 1.4(c). Such signals, which are described as *binary* signals, are widely used since they are produced by many simple sensors and can be used to control many forms of actuator. For example, a simple domestic light switch has two possible states (ON and OFF), and therefore the voltage controlled by such a switch can be seen as a binary signal representing the required state of the lights. In such an arrangement the light bulb represents the 'utilisation' element of the arrangement (see Section 1.3), which in this case is used in only two possible states (again, ON and OFF). Many electrical and electronic systems are based on the use of this form of ON/OFF control and therefore all make use of binary signals of one form or another. However, binary signals are also used in more sophisticated systems (such as those based on computers), since such signals are very easy to process, store and communicate. We will return to look at these issues in later chapters.

1.6	**System block diagrams**

It is often convenient to represent a complex arrangement by a simplified diagram that shows the system as a set of modules or blocks. This modular approach hides unnecessary detail and often aids comprehension. An example of a typical block diagram is shown in Figure 1.5. This shows a simplified representation of an engine control unit (ECU) that might be found in a car. This diagram shows the major components of the system and indicates the flow of energy or information between the various parts. The arrows in the diagram indicate the direction of flow.

When energy or information flows from a component we often refer to that component as the *source* of that energy or information. Similarly, when energy or information flows into a component, we often say that the component represents a *load* on the arrangement. Thus in Figure 1.5 we could consider the various sensors and the power supply to be sources for the ECU and the ignition coil to be a load.

In electrical systems a flow of energy requires an electrical *circuit*. Figure 1.6 shows a simple system with a single source and a single load. In

Figure 1.5   An automotive engine control unit (ECU)

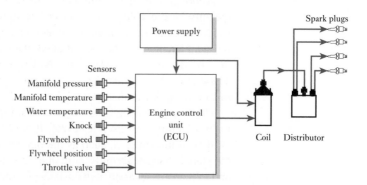

Figure 1.6 Sources and loads

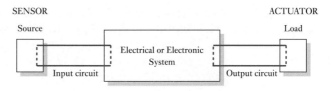

Figure 1.7 System partitioning

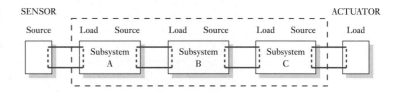

this figure the source of energy is some form of sensor, and the load is some form of actuator, perhaps representing a part of the ECU of Figure 1.5. In other systems the source could be a generator or power storage device, and the load could be any form of 'utilisation' element (see Section 1.3). In any event the source is linked to the system by an *input circuit*, and the load is connected by an *output circuit*.

We noted earlier that we are free to choose the boundaries of our system to suit our needs. We might therefore choose to divide the system of Figure 1.6 into a number of subsystems, or modules, as shown in Figure 1.7. This process is referred to as *partitioning* and can greatly simplify the design of complex systems. It can be seen that the output of each of these subsystems represents the input of the next. Thus the output of each module represents a *source*, while the input of each module represents a *load*. In the arrangements of Figures 1.6 and 1.7, each of the various modules has a single input and a single output. In practice, modules may have multiple inputs and outputs depending on the function of that part of the system.

We noted in the early stages of this chapter that a system may be defined solely by its inputs, its outputs and the relationship between them. It follows that each of the modules in our system can be defined solely in terms of the characteristics of the sources and loads that it represents and the relationship between its input and output signals. The design of such systems therefore involves the production of modules that take signals from appropriate input devices (be they sensors, generators or other modules) and produce from them appropriate signals to drive the relevant output devices (such as actuators). Therefore, before looking at the design of these circuits, we must first know something of the nature of the signals associated with these sensors and actuators. For this reason, we will look at these devices in Chapters 3 and 4.

In order to look at signals, sensors and actuators in a meaningful way, we need to understand something about basic electrical components and circuits. Fortunately, almost all readers of this text will have covered this material before progressing to this level of study. However, for completeness, this material is reviewed in Chapter 2.

Key points

- Engineering is inherently interdisciplinary, and all engineers should have an understanding of the basic principles of electrical and electronic engineering, if only so that they can talk intelligently to others who are specialists in this area.

- Engineers often adopt a 'systems approach' to design, which combines top-down *systematic* techniques with multidisciplinary *systemic* methods.

- As far as its appearance from outside is concerned, a system can be defined solely by its inputs, its outputs and the relationship between them.

- Most electrical or electronic systems can be seen as combinations of elements that are responsible for the *generation*, *transmission* or *communication*, *control* or *processing*, *utilisation*, or *storage* of electrical energy.

- Systems interact with the outside world through the use of *sensors* and *actuators*.

- Physical quantities may be either continuous or discrete. It is often convenient to represent physical quantities by electrical signals. These may also be either continuous or discrete. Continuous signals are normally referred to as *analogue*, while discrete signals are usually described as *digital*.

- Complex systems are often represented by block diagrams. These hide unnecessary detail and can aid comprehension.

- Energy or information flows from a *source* and flows into a *load*. Any module presents a *load* to whatever is connected to its input and represents a *source* to whatever is connected to its output.

- In order to design electrical or electronic systems, we need to understand the nature of the signals produced by the sensors and actuators that form their input and output devices.

Exercises

1.1 Discuss the meanings of the terms 'electrical engineering' and 'electronic engineering'.

1.2 List ten fields of engineering that might be associated with the construction of a railway system.

1.3 Explain the distinction between a systematic approach and a systemic approach to design. Which of these methods is associated with a systems approach?

1.4 Describe briefly what is meant by a system.

1.5 List five basic functions, or processes, that appear in electrical and electronic systems.

1.6 Categorise each of the following into one of the five groups described in the previous exercise: a car battery; a computer monitor; a solar cell; a light bulb; high-voltage power lines; a compact disc (CD); and a traffic light sequencer.

1.7 Identify examples of systems that are electrical, mechanical, hydraulic, pneumatic and biological, and in each case describe the nature of the inputs and outputs.

1.8 Explain why the choice of a system's boundaries affects the form of its inputs and outputs.

1.9 Identify five naturally occurring physical quantities not mentioned in the text that are continuous in nature.

1.10 Identify five naturally occurring physical quantities not mentioned in the text that are discrete in nature.

1.11 Give an example of a situation where a continuous physical quantity is represented by a digital signal.

1.12 Give an example of a situation where a discrete physical quantity is represented by an analogue signal.

1.13 Describe what is meant by 'partitioning' with respect to the design of an electronic system.

1.14 Explain how a module in an electrical system may be described in terms of 'sources' and 'loads'.

Chapter 2

Basic Electric Circuits and Components

Objectives

When you have studied the material in this chapter you should be able to:

- give the Système International (SI) units for a range of electrical quantities;
- use a range of common prefixes to represent multiples of these units;
- describe the basic characteristics of resistors, capacitors and inductors;
- apply Ohm's law, and Kirchhoff's voltage and current laws, to simple electrical circuits;
- calculate the effective resistance of resistors in series or in parallel, and analyse simple resistive potential divider circuits;
- define the terms 'frequency' and 'period' as they apply to sinusoidal quantities;
- draw the circuit symbols for a range of common electrical components.

2.1 Introduction

Most books on electrical and electronic engineering begin with a fairly lengthy treatment of basic material such as systems of measurement (units), Ohm's law and circuit analysis. For reasons that are explained in some detail in the Preface, this text adopts a different approach and attempts to explain the nature and characteristics of engineering systems before looking in detail at their analysis.

Most readers of this book will have met the basic concepts of electrical circuits long before embarking on study at this level. Therefore, the early chapters of the book assume only that the reader is familiar with this elementary material. Later, we will look at these basic concepts in some detail and extend them to give a greater understanding of the behaviour of the circuits and systems that we will have met by that time.

The list below gives an indication of the topics that you should be familiar with before reading the following chapters.

- The Système International (SI) units for quantities such as energy, power, temperature, frequency, charge, potential, resistance, capacitance and inductance. You should also know the symbols used for these units.
- The prefixes used to represent common multiples of these units and their symbols (for example, 1 kilometre = 1 km = 1000 metres).
- Electrical circuits and quantities such as charge, e.m.f. and potential difference.
- Direct and alternating currents.
- The basic characteristics of resistors, capacitors and inductors.
- Ohm's law, Kirchhoff's laws and power dissipation in resistors.
- The effective resistance of resistors in series and parallel.
- The operation of resistive potential dividers.
- The terms used to describe sinusoidal quantities.
- The circuit symbols used for resistors, capacitors, inductors, voltage sources and other common components.

If, having read through the list above, you are confident that you are familiar with all these topics you can move on immediately to Chapter 3. However, just in case there are a few areas that might need some reinforcement, the remainder of this chapter provides what might be seen as a *revision* section on this material. This does not aim to give a detailed treatment of these topics (where appropriate this will be given in later chapters) but simply explains them in sufficient detail to allow an understanding of the early parts of the book.

In this chapter, worked examples are used to illustrate several of the concepts involved. One way of assessing your understanding of the various topics is to look quickly through these examples to see if you can perform the calculations involved, before looking at the worked solutions. Most readers will find the early examples trivial, but experience shows that many will feel less confident on those related to *potential dividers*. This is a very important topic, and a clear understanding of these circuits will make it much easier to understand the remainder of the book.

The exercises at the end of this chapter are included to allow you to test your understanding of the 'assumed knowledge' listed above. If you can perform these exercises easily you should have no problems with the technical content of the next few chapters. If not, you would be well advised to invest a little time in looking at the relevant sections of this chapter before continuing.

2.2 Système International units

The Système International (SI) d'Unités (International System of Units) defines units for a large number of physical quantities but, fortunately for our current studies, we need very few of them. These are shown in Table 2.1. In later chapters, we will introduce additional units as necessary, and Appendix B gives a more comprehensive list of units relevant to electrical and electronic engineering.

Table 2.1 Some important units

Quantity	Quantity symbol	Unit	Unit symbol
Capacitance	C	farad	F
Charge	Q	coulomb	C
Current	I	ampere	A
Electromotive force	E	volt	V
Frequency	f	hertz	Hz
Inductance (self)	L	henry	H
Period	T	second	s
Potential difference	V	volt	V
Power	P	watt	W
Resistance	R	ohm	Ω
Temperature	T	kelvin	K
Time	t	second	s

2.3 Common prefixes

Table 2.2 lists the most commonly used unit prefixes. These will suffice for most purposes although a more extensive list is given in Appendix B.

Table 2.2 Common unit prefixes

Prefix	Name	Meaning (multiply by)
T	tera	10^{12}
G	giga	10^{9}
M	mega	10^{6}
k	kilo	10^{3}
m	milli	10^{-3}
μ	micro	10^{-6}
n	nano	10^{-9}
p	pico	10^{-12}

2.4 Electrical circuits

2.4.1 Electric charge

Charge is an amount of electrical energy and can be either positive or negative. In atoms, protons have a positive charge and electrons have an equal negative charge. While protons are fixed within the atomic nucleus, electrons are often weakly bound and may therefore be able to move. If a body or region develops an excess of electrons it will have an overall negative charge, while a region with a deficit of electrons will have a positive charge.

2.4.2   Electric current

An electric current is a flow of electric charge, which in most cases is a flow of electrons. Conventional current is defined as a flow of electricity from a positive to a negative region. This conventional current is in the opposite direction to the flow of the negatively charged electrons. The unit of current is the **ampere** or **amp** (A).

2.4.3   Current flow in a circuit

A sustained electric current requires a complete circuit for the recirculation of electrons. It also requires some stimulus to cause the electrons to flow around this circuit.

2.4.4   Electromotive force and potential difference

The stimulus that causes an electric current to flow around a circuit is termed an electromotive force or e.m.f. The e.m.f. represents the energy introduced into the circuit by a source such as a battery or a generator.

The energy transferred from the source to the load results in a change in the electrical potential at each point in the load. Between any two points in the load there will exist a certain potential difference, which represents the energy associated with the passage of a unit of charge from one point to the other.

Both e.m.f. and potential difference are expressed in units of **volts**, and clearly these two quantities are related. Figure 2.1 illustrates the relationship between them: e.m.f. is the quantity that produces an electric current, while a potential difference is the effect on the circuit of this passage of energy.

Some students have difficulty in visualising e.m.f., potential difference, resistance and current, and it is sometimes useful to use an analogy. Consider, for example, the arrangement shown in Figure 2.2. Here a water pump forces water to flow around a series of pipes and through some form of restriction. While no analogy is perfect, this model illustrates the basic properties of the circuit of Figure 2.1. In the water-based diagram, the

Figure 2.1 Electromotive force and potential difference

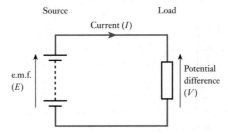

Figure 2.2   A water-based
analogy of an electrical circuit

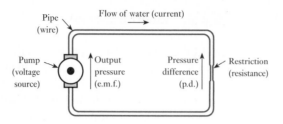

water pump forces water around the arrangement and is equivalent to the
voltage source (or battery), which pushes electrical charge around the
corresponding electrical circuit. The flow of water through the pipe corresponds to the flow of charge around the circuit and therefore the *flow rate*
represents the *current* in the circuit. The *restriction* within the pipe opposes
the flow of water and is equivalent to the *resistance* of the electrical circuit.
As water flows through the restriction the pressure will fall, creating a
pressure difference across it. This is equivalent to the *potential difference*
across the resistance within the electrical circuit. The flow rate of the water
will increase with the output pressure of the pump and decrease with the
level of restriction present. This is analogous to the behaviour of the electrical circuit, where the current increases with the e.m.f. of the voltage
source and decreases with the magnitude of the resistance.

2.4.5   Voltage reference points

Electromotive forces and potential differences in circuits produce different
potentials (or voltages) at different points in the circuit. It is normal to describe the voltages throughout a circuit by giving the potential at particular
points with respect to a single reference point. This reference is often called
the **ground** or **earth** of the circuit. Since voltages at points in the circuit are
measured with respect to ground, it follows that the voltage on the ground
itself is zero. Therefore, ground is also called the **zero volt line** of the circuit.

In a circuit, a particular point or junction may be taken as the zero volt
reference and this may then be labelled as 0 V, as shown in Figure 2.3(a).
Alternatively, the ground point of the circuit may be indicated using the
ground symbol, as shown in Figure 2.3(b).

Figure 2.3   Indicating voltage
reference points

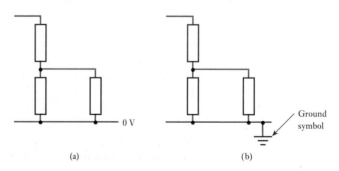

Figure 2.4 Indicating voltages in circuit diagrams

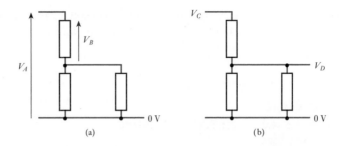

2.4.6 Representing voltages in circuit diagrams

Conventions for representing voltages in circuit diagrams vary considerably between countries. In the UK, and in this text, it is common to indicate a potential difference by an arrow, which is taken to represent the voltage at the head of the arrow with respect to that at the tail. This is illustrated in Figure 2.4(a). In many cases, the tail of the arrow will correspond to the zero volt line of the circuit (as shown in V_A in the figure). However, it can indicate a voltage difference between any two points in the circuit (as shown by V_B).

In some cases, it is inconvenient to use arrows to indicate voltages in circuits and simple labels are used instead, as shown in Figure 2.4(b). Here the labels V_C and V_D represent the voltage at the corresponding points *with respect to ground* (that is, with respect to the zero volt reference).

2.4.7 Representing currents in circuit diagrams

Currents in circuit diagrams are conventionally indicated by an arrow in the direction of the *conventional* current flow (that is, in the opposite direction to the flow of electrons). This was illustrated in Figure 2.1. This figure also shows that for positive voltages and currents the arrow for the current flowing out of a voltage source is in the *same direction* as the arrow representing its e.m.f. However, the arrow representing the current in a resistor is in the *opposite direction* to the arrow representing the potential difference across it.

2.5 Direct current and alternating current

The currents associated with electrical circuits may be constant or may vary with time. Where currents vary with time they may also be unidirectional or alternating.

When the current in a conductor always flows in the same direction this is described as a direct current (DC). Such currents will often be associated with voltages of a single polarity. Where the direction of the current periodically changes, this is referred to as alternating current (AC), and such currents will often be associated with alternating voltages. One of the most common forms of alternating waveform is the sine wave, as discussed in Section 2.13.

2.6   Resistors, capacitors and inductors

2.6.1   Resistors

Resistors are components whose main characteristic is that they provide resistance between their two electrical terminals. The **resistance** of a circuit represents its opposition to the flow of electric current. The unit of resistance is the **ohm (Ω)**. One may also define the **conductance** of a circuit as its ability to *allow* the flow of electricity. The conductance of a circuit is equal to the reciprocal of its resistance and has the units of **siemens (S)**. We will look at resistance in some detail in Chapter 12.

2.6.2   Capacitors

Capacitors are components whose main characteristic is that they exhibit capacitance between their two terminals. **Capacitance** is a property of two conductors that are electrically insulated from each other, whereby electrical energy is stored when a potential difference exists between them. This energy is stored in an electric field that is created between the two conductors. Capacitance is measured in **farads (F)**, and we will return to look at capacitance in more detail in Chapter 13.

2.6.3   Inductors

Inductors are components whose main characteristic is that they exhibit inductance between their two terminals. **Inductance** is the property of a coil that results in an e.m.f. being induced in the coil as a result of a change in the current in the coil. Like capacitors, inductors can store electrical energy and in this case it is stored in a magnetic field. The unit of inductance is the **henry (H)**, and we will look at inductance in Chapter 14.

2.7   Ohm's law

Ohm's law states that the current I flowing in a conductor is directly proportional to the applied voltage V and inversely proportional to its resistance R. This determines the relationship between the units for current, voltage and resistance, and the **ohm** is defined as the resistance of a circuit in which a current of 1 **amp** produces a potential difference of 1 **volt**.

The relationship between voltage, current and resistance can be represented in a number of ways, including:

$$V = IR \tag{2.1}$$

$$I = \frac{V}{R} \tag{2.2}$$

$$R = \frac{V}{I} \tag{2.3}$$

Figure 2.5 The relationship between V, I and R

A simple way of remembering these three equations is to use the 'virtual triangle' of Figure 2.5. The triangle is referred to as 'virtual' simply as a way of remembering the order of the letters. Taking the first three letters of VIRtual and writing them in a triangle (starting at the top) gives the arrangement shown in the figure. If you place your finger on one of the letters, the remaining two show the expression for the selected quantity. For example, to find the expression for 'V' put your finger on the V and you see I next to R, so V = IR. Alternatively, to find the expression for 'I' put your finger on the I and you are left with V above R, so I = V/R. Similarly, covering 'R' leaves V over I, so R = V/I.

Example 2.1

Voltage measurements (with respect to ground) on part of an electrical circuit give the values shown in the diagram below. If the resistance of R_2 is 220 Ω, what is the current I flowing through this resistor?

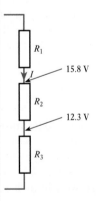

From the two voltage measurements, it is clear that the voltage difference across the resistor is 15.8 − 12.3 = 3.5 V. Therefore using the relationship

$$I = \frac{V}{R}$$

we have

$$I = \frac{3.5}{220} = 15.9 \text{ mA}$$

2.8   Kirchhoff's laws

2.8.1   Current law

At any instant, the algebraic sum of all the currents flowing into any junction in a circuit is zero.

$$\Sigma I = 0 \tag{2.4}$$

A **junction** is any point where electrical paths meet. The law comes about from consideration of conservation of charge – the charge flowing into a point must equal that flowing out.

2.8.2   Voltage law

At any instant, the algebraic sum of all the voltages around any loop in a circuit is zero.

$$\Sigma V = 0 \tag{2.5}$$

The term **loop** refers to any continuous path around the circuit, and the law comes about from consideration of conservation of energy.

With both laws, it is important that the various quantities are assigned the correct sign. When summing currents, those flowing *into* a junction are given the opposite polarity to those flowing *out* from it. Similarly, when summing the voltages around a loop, *clockwise* voltages will be assigned the opposite polarity to *anticlockwise* ones.

Example 2.2   Use Kirchhoff's current law to determine the current I_2 in the following circuit.

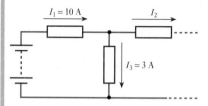

From Kirchhoff's current law

$$I_2 = I_1 - I_3$$
$$= 10 - 3$$
$$= 7 \text{ A}$$

Example 2.3 Use Kirchhoff's voltage law to determine the magnitude of V_1 in the following circuit.

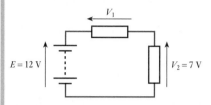

From Kirchhoff's voltage law (summing the voltages clockwise around the loop)

$$E - V_1 - V_2 = 0$$

or, rearranging

$$V_1 = E - V_2$$

$$= 12 - 7$$

$$= 5 \text{ V}$$

2.9 Power dissipation in resistors

The instantaneous power dissipation P of a resistor is given by the product of the voltage across the resistor and the current passing through it. Combining this result with Ohm's law gives a range of expressions for P. These are:

$$P = VI \tag{2.6}$$

$$P = I^2 R \tag{2.7}$$

$$P = \frac{V^2}{R} \tag{2.8}$$

Example 2.4 Determine the power dissipation in the resistor R_3 in the following circuit.

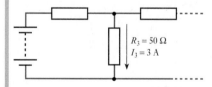

From Equation 2.7

$$P = I^2 R$$

$$= 3^2 \times 50$$

$$= 450 \text{ W}$$

2.10 Resistors in series

The effective resistance of a number of resistors in series is equal to the sum of their individual resistances.

$$R = R_1 + R_2 + R_3 + \ldots + R_n \tag{2.9}$$

For example, for the three resistors shown in Figure 2.6 the total resistance R is given by

$$R = R_1 + R_2 + R_3$$

Figure 2.6 Three resistors in series

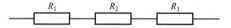

Example 2.5 | Determine the equivalent resistance of this combination.

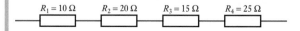

From above

$$R = R_1 + R_2 + R_3 + R_4$$
$$= 10 + 20 + 15 + 25$$
$$= 70 \ \Omega$$

2.11 Resistors in parallel

The effective resistance of a number of resistors in parallel is given by the following expression:

$$\frac{1}{R} = \frac{1}{R_1} + \frac{1}{R_2} + \frac{1}{R_3} + \cdots \frac{1}{R_n} \tag{2.10}$$

For example, for the three resistors shown in Figure 2.7 the total resistance R is given by

$$\frac{1}{R} = \frac{1}{R_1} + \frac{1}{R_2} + \frac{1}{R_3}$$

Figure 2.7 Three resistors in parallel

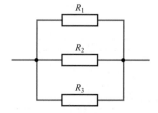

Example 2.6 | **Determine the equivalent resistance of this combination.**

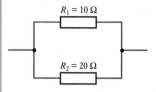

From above

$$\frac{1}{R} = \frac{1}{R_1} + \frac{1}{R_2}$$

$$= \frac{1}{10} + \frac{1}{20}$$

$$= \frac{3}{20}$$

$$\therefore R = \frac{20}{3} = 6.67 \ \Omega$$

Note that the effective resistance of a number of resistors in parallel will always be less than that of the lowest-value resistor.

2.12 Resistive potential dividers

When several resistors are connected in series the current flowing through each resistor is the same. The magnitude of this current is given by the voltage divided by the total resistance. For example, if we connect three resistors in series, as shown in Figure 2.8, the current is given by:

$$I = \frac{V}{R_1 + R_2 + R_3}$$

The voltage across each resistor is then given by this current multiplied by its resistance. For example, the voltage V_1 across resistor R_1 will be given by

$$V_1 = IR_1 = \left(\frac{V}{R_1 + R_2 + R_3} \right) R_1$$

Figure 2.8 A resistive potential divider

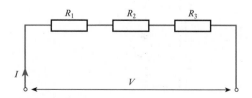

Figure 2.9   The division of
voltages in a potential divider

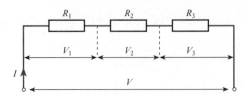

Therefore, the *fraction* of the total voltage across each resistor is equal to
its *fraction* of the total resistance, as shown in Figure 2.9, where

$$\frac{V_1}{V} = \frac{R_1}{R_1 + R_2 + R_3} \quad \frac{V_2}{V} = \frac{R_2}{R_1 + R_2 + R_3} \quad \frac{V_3}{V} = \frac{R_3}{R_1 + R_2 + R_3}$$

or, rearranging

$$V_1 = V\frac{R_1}{R_1 + R_2 + R_3} \quad V_2 = V\frac{R_2}{R_1 + R_2 + R_3}$$

$$V_3 = V\frac{R_3}{R_1 + R_2 + R_3}$$

To calculate the voltage at a point in a chain of resistors, one must determine the voltage across the complete chain, calculate the voltage across
those resistors between that point and one end of the chain and add this to
the voltage at that end of the chain. For example, in Figure 2.10:

$$V = V_2 + (V_1 - V_2)\frac{R_2}{R_1 + R_2} \tag{2.11}$$

Figure 2.10   A simple potential
divider

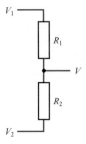

| **Example 2.7** | Determine the voltage *V* in the following circuit. |

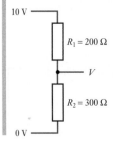

As described above, we first determine the voltage across the chain (by subtracting the voltages at either end of the chain). Then we calculate the voltage across the relevant resistor and add this to the voltage at the appropriate end of the chain.

In this case one end of the chain of resistors is at zero volts, so the calculation is very straightforward. The voltage across the chain is 10 V, and V is simply the voltage across R_2, which is given by

$$V = 10\frac{R_2}{R_1 + R_2}$$

$$= 10\frac{300}{200 + 300}$$

$$= 6 \text{ V}$$

Note that a common mistake in such calculations is to calculate $R_1/(R_1 + R_2)$, rather than $R_2/(R_1 + R_2)$. The value used as the numerator in this expression represents the resistor across which the voltage is to be calculated.

Potentiometer calculations are slightly more complicated where neither end of the chain of resistors is at zero volts.

Example 2.8 | Determine the voltage V in the following circuit.

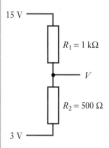

Again, we first determine the voltage across the chain (by subtracting the voltages at either end of the chain). Then we calculate the voltage across the relevant resistor and add this to the voltage at the appropriate end of the chain. Therefore

$$V = 3 + (15 - 3)\frac{R_2}{R_1 + R_2}$$

$$= 3 + 12\frac{500}{1000 + 500}$$

$$= 3 + 4$$

$$= 7 \text{ V}$$

In this case we pick one end of the chain of resistors as our reference point (we picked the lower end) and calculate the voltage on the output with respect to this point. We then add to this calculated value the voltage at the reference point.

2.13   Sinusoidal quantities

Sinusoidal quantities have a magnitude that varies with time in a manner described by the **sine** function. The variation of any quantity with time can be described by drawing its **waveform**. The waveform of a sinusoidal quantity is shown in Figure 2.11. The length of time between corresponding points in successive cycles of the waveform is termed its **period**, which is given the symbol T. The number of cycles of the waveform within one second is termed its **frequency**, which is usually given the symbol f.

The frequency of a waveform is related to its period by the expression

$$f = \frac{1}{T} \tag{2.12}$$

Figure 2.11   A sine wave

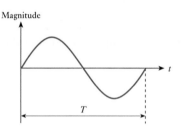

Example 2.9   What is the period of a sinusoidal quantity with a frequency of 50 Hz?

From above we know that

$$f = \frac{1}{T}$$

and therefore the period is given by

$$T = \frac{1}{f} = \frac{1}{50} = 0.02\,\text{s} = 20\,\text{ms}$$

2.14   Circuit symbols

The following are circuit symbols for a few basic electrical components.

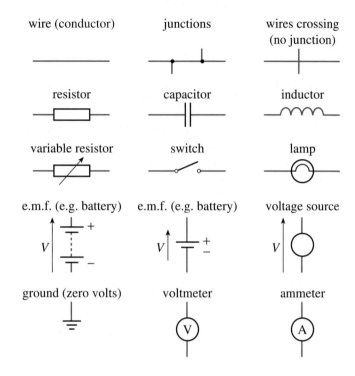

In later chapters we will meet a number of additional component symbols, but these are sufficient for our current needs.

Key points

Since this chapter introduces no new material there are very few key points. However, the importance of a good understanding of this 'assumed knowledge' encourages me to emphasise the following:

- Understanding the next few chapters of the book relies on understanding the various topics covered in this chapter.

- A clear concept of voltage and current is essential for all readers.

- Ohm's law and Kirchhoff's voltage and current laws are used extensively in later chapters.

- Experience shows that students have most problems with potential dividers – a topic that is widely used in these early chapters. You are therefore advised to make very sure that you are happy with this material before continuing.

Exercises

2.1 Give the prefixes used to denote the following powers: 10^{-12}; 10^{-9}; 10^{-6}; 10^{-3}; 10^3; 10^6; 10^9; 10^{12}.

2.2 Explain the difference between 1 ms, 1 m/s and 1 mS.

2.3 Explain the difference between 1 mΩ and 1 MΩ.

2.4 If a resistor of 1 kΩ has a voltage of 5 V across it, what is the current flowing through it?

2.5 A resistor has 9 V across it and a current of 1.5 mA flowing through it. What is its resistance?

2.6 A resistor of 25 Ω has a voltage of 25 V across it. What power is being dissipated by the resistor?

2.7 If a 400 Ω resistor has a current of 5 µA flowing through it, what power is being dissipated by the resistor?

2.8 What is the effective resistance of a 20 Ω resistor in series with a 30 Ω resistor?

2.9 What is the effective resistance of a 20 Ω resistor in parallel with a 30 Ω resistor?

2.10 What is the effective resistance of a series combination of a 1 kΩ resistor, a 2.2 kΩ resistor and a 4.7 kΩ resistor?

2.11 What is the effective resistance of a parallel combination of a 1 kΩ resistor, a 2.2 kΩ resistor and a 4.7 kΩ resistor?

2.12 Calculate the effective resistance between the terminals A and B in the following arrangements.

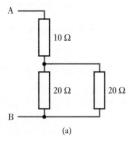

(a)

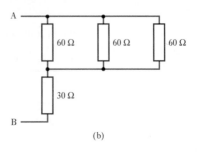

(b)

2.13 Calculate the effective resistance between the terminals A and B in the following arrangements.

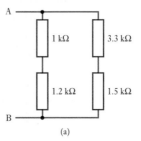

(a)

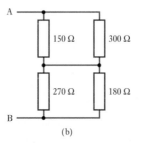

(b)

2.14 Calculate the voltages V_1, V_2 and V_3 in the following arrangements.

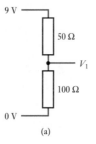

(a)

Exercises continued

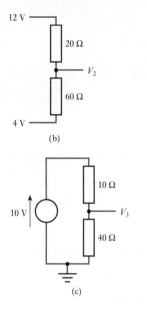

(b)

(c)

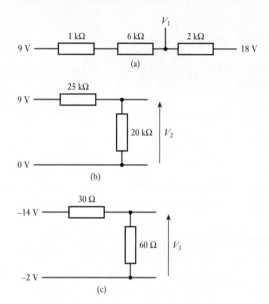

(a)

(b)

(c)

2.15 Calculate the voltages V_1, V_2 and V_3 in the following arrangements.

2.16 A sinusoidal quantity has a frequency of 1 kHz. What is its period?

2.17 A sinusoidal quantity has a period of 20 μs. What is its frequency?

Chapter 3

Sensors

Objectives

When you have studied the material in this chapter you should be able to:

- discuss the role of sensors in electrical and electronic systems;
- outline the requirement for a range of sensors of different types to meet the needs of varied applications;
- explain the meaning of terms such as range, resolution, accuracy, precision, linearity and sensitivity, as they apply to sensors;
- describe the operation and characteristics of a variety of devices for sensing various physical quantities;
- give examples from the diversity of sensing devices available and outline the different characteristics of these components;
- discuss the need for interfacing circuitry to make the signals produced by sensors compatible with the systems to which they are connected.

3.1 Introduction

In order to perform useful tasks electrical and electronic systems must interact with the world about them. To do this they use **sensors** to sense external physical quantities and **actuators** to affect or control them.

Sensors and actuators are often referred to as transducers. A **transducer** is a device that converts one physical quantity into another, and different transducers convert between a wide range of physical quantities. Examples include a mercury-in-glass thermometer, which converts variations in temperature into variations in the length of a mercury column, and a microphone, which converts sound into electrical signals.

In this text we are primarily interested in transducers that are used in electrical or electronic systems, so we are mainly interested in devices that produce or use electrical signals of some form. Transducers that convert physical quantities into electrical signals will normally be used to produce inputs for our system and will therefore be referred to as *sensors*. Transducers that take electrical input signals and control or affect an external physical quantity will be referred to as *actuators*. In this chapter we will look at the characteristics of sensors and in the next we will consider actuators.

Thermometers and microphones are both examples of sensors that convert one form of analogue quantity into another. Other sensors can be used with digital quantities, converting one digital quantity into another. Such systems include all forms of event counter, such as those used to count the number of people going through a turnstile.

A third class of sensors take an analogue quantity and represent it in a digital form. In some instances the output is a simple binary representation of the input, as in a thermostat, which produces one of two output values depending on whether a temperature is above or below a certain threshold. In other devices the analogue quantity at the input is represented by a multi-valued output, as in the case of a digital voltmeter, where an analogue input quantity is represented by a numerical (and therefore discrete) output. Representing an analogue quantity by a digital quantity is, by necessity, an approximation. However, if the number of allowable discrete states is sufficient, the representation may be adequate for a given application. Indeed, in many cases the error caused by this approximation is small compared with the noise or other errors within the system and can therefore be ignored.

For completeness one should say that there is a final group of sensors that take a digital input quantity and use this to produce an analogue output. However, such components are less common, and there are very few widely used examples of such devices.

Almost any physical property of a material that varies in response to some excitation can be used to produce a sensor. Commonly used devices include those whose operation is:

- resistive
- inductive
- capacitive
- piezoelectric
- photoelectric
- elastic
- thermal.

The range of sensing devices available is vast, and in this chapter we will restrict ourselves to a few examples that are widely used in electrical and electronic systems. The examples chosen have been selected to show the diversity of devices available and to illustrate some of their characteristics. These examples include sensors for a variety of physical quantities and devices that are both analogue and digital in nature. However, before we start looking at individual devices it is appropriate to consider how we quantify the performance of such components.

3.2   **Describing sensor performance**

When describing sensors and instrumentation systems we make use of a range of terms to quantify their characteristics and performance. It is important to have a clear understanding of this terminology, so we will look briefly at some of the more important terms.

3.2.1   Range

This defines the maximum and minimum values of the quantity that the sensor or instrument is designed to measure.

3.2.2   Resolution or discrimination

This is the smallest discernible change in the measured quantity that the sensor is able to detect. This is usually expressed as a percentage of the range of the device; for example, the resolution might be given as 0.1 percent of the full-scale value (that is, one-thousandth of the range).

3.2.3   Error

This is the difference between a measured value and its true value. Errors may be divided into random errors and systematic errors. **Random errors** produce *scatter* within repeated readings. The effects of such errors may be quantified by comparing multiple readings and noting the amount of scatter present. The effects of random errors may also be reduced by taking the average of these repeated readings. **Systematic errors** affect all readings in a similar manner and are caused by factors such as mis-calibration. Since all readings are affected in the same way, taking multiple readings does not allow quantification or reduction of such errors.

3.2.4   Accuracy, inaccuracy and uncertainty

The term *accuracy* describes the maximum expected error associated with a measurement (or a sensor) and may be expressed as an absolute value or as a percentage of the range of the system. For example, the accuracy of a vehicle speed sensor might be given as ±1 mph or as ±0.5 percent of the full-scale reading. Strictly speaking, this is actually a measure of its *inaccuracy*, and for this reason the term *uncertainty* is sometimes used.

3.2.5   Precision

This is a measure of the lack of random errors (scatter) produced by a sensor or instrument. Devices with high precision will produce repeated

Figure 3.1 Accuracy and precision

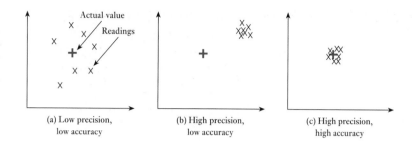

(a) Low precision, low accuracy

(b) High precision, low accuracy

(c) High precision, high accuracy

readings with very little spread. It should be noted that precision is very often confused with accuracy, which has a very different meaning. A sensor might produce a range of readings that are very consistent but that are all very inaccurate. This is illustrated in Figure 3.1, which shows the performance of three sensor systems.

3.2.6 Linearity

In most situations it is convenient to have a sensor where the output is linearly proportional to the quantity being measured. If one were to plot a graph of the output of a sensor against the measured quantity, a perfectly linear device would produce a straight line going through the origin. In practice real sensors will have some non-linearity, which is defined as the maximum deviation of any reading from this straight line. Non-linearity is normally expressed as a percentage of the full-scale value.

3.2.7 Sensitivity

This is a measure of the change produced at the output for a given change in the quantity being measured. A sensor that has high sensitivity will produce a large change in its output for a given input change. The units of this measure reflect the nature of the measured quantity. For example, for a temperature sensor the sensitivity might be given as 10 mV/°C, meaning that the output would change by 10 mV for every 1 °C change in temperature.

3.3 Temperature sensors

The measurement of temperature is a fundamental part of a large number of control and monitoring systems, ranging from simple temperature-regulating systems for buildings to complex industrial process-control plants.

Temperature sensors may be divided into those that give a simple binary output to indicate that the temperature is above or below some threshold value and those that allow temperature measurements to be made.

Binary output devices are effectively temperature-operated switches, an example being the **thermostat**, which is often based on a **bimetallic strip**. This is formed by bonding together two materials with different thermal expansion properties. As the temperature of the bimetallic strip increases it bends, and this deflection is used to operate a mechanical switch.

A large number of different techniques are used for temperature measurement, but here we will consider just three forms.

3.3.1   Resistive thermometers

The electrical resistance of all conducting materials changes with temperature. The resistance of a piece of metal varies linearly with its absolute temperature. This allows temperature to be measured by determining the resistance of a sample of the metal and comparing it with its resistance at a known temperature. Typical devices use platinum wire; such devices are known as **platinum resistance thermometers** or **PRT**s.

PRTs can produce very accurate measurements at temperatures from less than −150 °C to nearly 1000 °C to an accuracy of about 0.1 °C, or 0.1 percent. However, they have poor **sensitivity**. That is, a given change in the input temperature produces only a small change in the output signal. A typical PRT might have a resistance of 100 Ω at 0 °C, which increases to about 140 Ω at 100 °C. Figure 3.2(a) shows a typical PRT element. PRTs are also available in other forms, such as the probe shown in Figure 3.2(b).

Figure 3.2   Platinum resistance thermometers (PRTs)

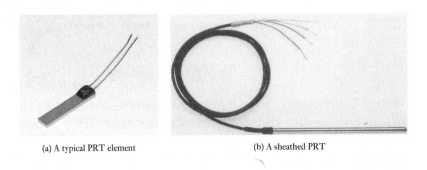

(a) A typical PRT element   (b) A sheathed PRT

3.3.2   Thermistors

Like PRTs, these devices also change their resistance with temperature. However, they use materials with high thermal coefficients of resistance to give much improved sensitivity. A typical device might have a resistance of 5 kΩ at 0 °C and 100 Ω at 100 °C. Thermistors are inexpensive and robust but are very non-linear and often suffer from great variability in their nominal value between devices. Figure 3.3(a) shows a typical disc thermistor, while Figure 3.3(b) shows a device incorporating a threaded section for easy attachment.

Figure 3.3 Thermistors

(a) A typical disc thermistor   (b) A threaded thermistor

Figure 3.4 A *pn* junction temperature sensor

3.3.3 *pn* junctions

A *pn* junction is a semiconductor device that has the properties of a **diode**. That is, it conducts electricity in one direction (when the device is said to be *forward-biased*) but opposes the flow of electricity in the other direction (when the device is said to be *reverse-biased*). The properties and uses of semiconductor devices will be discussed in more detail in Chapter 19.

At a fixed current, the voltage across a typical forward-biased semiconductor diode changes by about 2 mV per °C. Devices based on this property use additional circuitry to produce an output voltage or current that is directly proportional to the junction temperature. Typical devices might produce an output voltage of 1 mV per °C, or an output current of 1 μA per °C, for temperatures above 0 °C. These devices are inexpensive, linear and easy to use but are limited to a temperature range from about −50 °C to about 150 °C by the semiconductor materials used. Such a device is shown in Figure 3.4.

3.4 Light sensors

Sensors for measuring light intensity fall into two main categories: those that generate electricity when illuminated and those whose properties (for example, their resistance) change under the influence of light. We will consider examples of both of these classes of device.

3.4.1 Photovoltaic

Light falling on a *pn* junction produces a voltage and can therefore be used to generate power from light energy. This principle is used in solar cells. On a smaller scale, **photodiodes** can be used to measure light intensity,

Figure 3.5   Light sensors

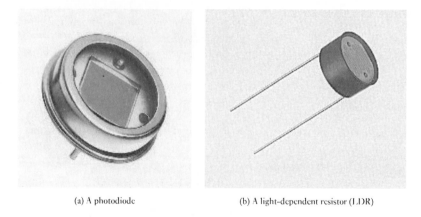

(a) A photodiode   (b) A light-dependent resistor (LDR)

since they produce an output voltage that depends on the amount of light falling on them. A disadvantage of this method of measurement is that the voltage produced is not related linearly to the incident light intensity. Figure 3.5(a) shows examples of typical photodiode light sensors.

3.4.2   Photoconductive

Photoconductive sensors do not generate electricity, but their conduction of electricity changes with illumination. The photodiode described above as a photovoltaic device may also be used as a photoconductive device. If a photodiode is reverse-biased by an external voltage source, in the absence of light it will behave like any other diode and conduct only a negligible leakage current. However, if light is allowed to fall on the device, charge carriers will be formed in the junction region and a current will flow. The magnitude of this current is proportional to the intensity of the incident light, making it more suitable for measurement than the photovoltaic arrangement described earlier.

The currents produced by photodiodes in their photoconductive mode are very small. An alternative is to use a **phototransistor**, which combines the photoconductive properties of the photodiode with the current amplification of a transistor to form a device with much greater sensitivity. The operation of transistors will be discussed in later chapters.

A third class of photoconductive device is the **light-dependent resistor** or **LDR**. As its name implies, this is a resistive device that changes its resistance when illuminated. Typical devices are made from materials such as cadmium sulphide (CdS) which have a much lower resistance when illuminated. One advantage of these devices in some applications is that they respond to different wavelengths of light in a manner similar to the human eye. Unfortunately, their response is very slow, taking perhaps 100 ms to respond to a change in illumination compared with a few microseconds, or less, for the semiconductor junction devices. A typical LDR is shown in Figure 3.5(b).

In addition to sensors that measure light intensity there are also a large number of sensors that use light to measure other quantities, such as position, motion and temperature. We will look at an example of such a sensor in Section 3.6 when we consider opto-switches.

3.5 Force sensors

3.5.1 Strain gauge

The resistance between opposite faces of a rectangular piece of uniform electrically conducting material is proportional to the distance between the faces and inversely proportional to its cross-sectional area. The shape of such an object may be changed by applying an external force to it. The term **stress** is used to define the force per unit area applied to the object, and the term **strain** refers to the deformation produced. In a strain gauge, an applied force deforms the sensor, increasing or decreasing its length (and its cross-section) and therefore changing its resistance. Figure 3.6 shows the construction of a typical device.

The gauge is in the form of a thin layer of resistive material arranged to be sensitive to deformation in only one direction. The long thin lines of the sensor are largely responsible for the overall resistance of the device. Stretching or compressing the gauge in the direction shown will extend or contract these lines and will have a marked effect on the total resistance. The comparatively thick sections joining these lines contribute little to the overall resistance of the unit. Consequently, deforming the gauge perpendicular to the direction shown will have little effect on the total resistance of the device.

In use, the gauge is bonded to the surface in which strain is to be measured. The fractional change in resistance is linearly related to the applied strain. If it is bonded to a structure with a known stress-to-strain characteristic, the gauge can be used to measure force. Thus it is often found at the heart of many force transducers or **load cells**. Similarly, strain gauges may be connected to diaphragms to produce **pressure sensors**.

Figure 3.6 A strain gauge

Direction of sensitivity

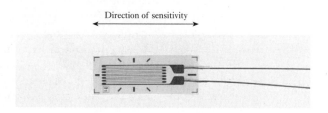

3.6 Displacement sensors

Displacement or position may be sensed using a very wide range of methods, including resistive, inductive, mechanical and optical techniques. As with many classes of sensor, both analogue and digital types are used.

3.6.1 Potentiometers

Resistive potentiometers are among the most common position transducers, and most people will have encountered them as the controls used in radios and other electronic equipment. Potentiometers may be angular or linear, consisting of a length of resistive material with an electrical terminal at each end and a third terminal connected to a sliding contact on to the resistive *track*. When used as a position transducer, a potential is placed across the two end terminals and the output is taken from the terminal connected to the sliding contact. As the sliding contact moves the output voltage changes between the potentials on each end of the track. Generally, there is a linear relationship between the position of the slider and the output voltage.

3.6.2 Inductive sensors

The inductance of a coil is affected by the proximity of ferromagnetic materials, an effect that is used in a number of position sensors. One of the simplest of these is the inductive **proximity sensor**, in which the proximity of a ferromagnetic plate is determined by measuring the inductance of a coil. Figure 3.7 shows examples of typical proximity sensors. We will return to consider the operation of such devices in Chapter 14 after we have looked at electromagnetism. In that chapter, we will also look at another inductive displacement sensor – the **linear variable differential transformer** or **LVDT**.

Figure 3.7 Inductive displacement or proximity sensors

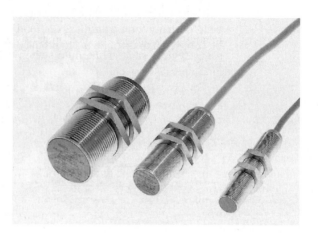

3.6.3 Switches

The simplest digital displacement sensors are mechanical switches. These are used in many forms and may be manually operated or connected to a mechanism of some form. Manually operated switches include toggle switches, which are often used as power ON/OFF switches on electrical equipment, and momentary-action pushbutton switches as used in computer keyboards. It may not be immediately apparent that switches of this type are position sensors, but clearly they output a value dependent on the position of the input lever or surface and are therefore binary sensors.

When a switch is connected to some form of mechanism, its action as a position sensor becomes more obvious. A common form of such a device is the **microswitch**, which consists of a small switch mechanism attached to a lever or push-rod, allowing it to be operated by some external force. Microswitches are often used as **limit switches**, which signal that a mechanism has reached the end of its safe travel. Such an arrangement is shown in Figure 3.8(a). Switches are also used in a number of specialised position-measuring applications, such as liquid level sensors. One form of such a sensor is shown in Figure 3.8(b). Here the switch is operated by a float, which rises with the liquid until it reaches some specific level.

Figure 3.8 Switch position sensors

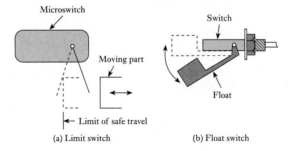

(a) Limit switch   (b) Float switch

3.6.4 Opto-switches

In addition to the use of mechanical switches, position can also be sensed using devices such as the **opto-switch**, which, as its name suggests, is a light-operated switch.

The opto-switch consists of a light sensor, usually a phototransistor, and a light source, usually a light-emitting diode (LEDs will be described in the next chapter), housed within a single package. Two physical arrangements are widely used, as illustrated in Figure 3.9.

Figure 3.9(a) shows a reflective device in which the light source and sensor are mounted adjacent to each other on one face of the unit. The presence of a reflective object close to this face will cause light from the source to reach the sensor, causing current to flow in the output circuit. Figure 3.9(b) shows a slotted opto-switch in which the source and sensor

Figure 3.9 Reflective and slotted opto-switches

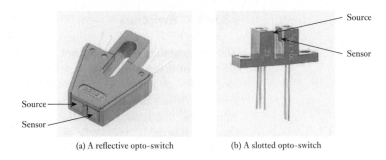

(a) A reflective opto-switch   (b) A slotted opto-switch

are arranged to face each other on either side of a slot in the device. In the absence of any object in the slot, light from the source will reach the sensor, and this will produce a current in the output circuit. If the slot is obstructed, the light path will be broken and the output current will be reduced.

Although opto-switches may be used with external circuitry to measure the current flowing and thus to determine the magnitude of the light reaching the sensor, it is more common to use them in a binary mode. In this arrangement, the current is compared with some threshold value to decide whether the opto-switch is ON or OFF. In this way, the switch detects the presence or absence of objects, the threshold value being adjusted to vary the sensitivity of the arrangement. We will consider some applications of the opto-switch later in this section.

3.6.5 Absolute position encoders

Figure 3.10 illustrates the principle of a simple linear absolute position encoder. A pattern of light and dark areas is printed on to a strip and is detected by a sensor that moves along it. The pattern takes the form of a series of lines that alternate between light and dark. It is arranged so that the combination of light and dark areas on the various lines is unique at each point along the strip. The sensor, which may be a linear array of phototransistors or photodiodes, one per line, picks up the pattern and produces an appropriate electrical signal, which can be decoded to determine the sensor's position. The combination of light and dark lines at each point represents a **code** for that position.

Figure 3.10 An absolute position encoder

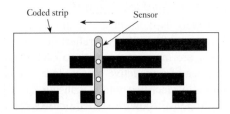

Since each point on the strip must have a unique code, the number of distinct positions along the strip that can be detected is determined by the number of lines in the pattern. For a sensor of a given length, increasing the number of lines in the pattern increases the resolution of the device but also increases the complexity of the detecting array and the accuracy with which the lines must be printed.

Although linear absolute encoders are available, the technique is more commonly applied to angular devices. These often resemble rotary potentiometers, but they have a coded pattern in a series of concentric rings in place of the conducting track and an array of optical sensors in place of the wiper. Position encoders have excellent linearity and a long life, but they generally have poorer resolution than potentiometers and are usually more expensive.

3.6.6 Incremental position encoders

The incremental encoder differs from the absolute encoder in that it has only a single detector, which scans a pattern consisting of a regular series of stripes perpendicular to the direction of travel. As the sensor moves over the pattern, the sensor will detect a series of light and dark regions. The distance moved can be determined by counting the number of transitions. One problem with this arrangement is that the direction of motion cannot be ascertained, as motion in either direction generates similar transitions between light and dark. This problem is overcome by the use of a second sensor, slightly offset from the first. The direction of motion may now be determined by noting which sensor is first to detect a particular transition. This arrangement is shown in Figure 3.11, which also illustrates the signals produced by the two sensors for motion in each direction.

In comparison with the absolute encoder, the incremental encoder has the disadvantage that external circuitry is required to count the transitions, and that some method of resetting this must be provided to give a reference point or datum. However, the device is simple in construction and can provide high resolution. Again, both linear and angular devices are available. Figure 3.12 shows a small angular incremental position encoder.

Figure 3.11 An incremental position encoder

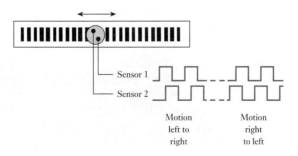

Sensor 1

Sensor 2

Motion left to right

Motion right to left

Figure 3.12   An angular
incremental position encoder

Figure 3.12   An angular
incremental position encoder

3.6.7   Other counting techniques

Incremental encoders employ event counting to determine displacement. Several other techniques use this method, and Figure 3.13 shows two examples.

Figure 3.13(a) shows a technique that uses an inductive proximity sensor, as described earlier in this section. Here a ferromagnetic gear wheel is placed near the sensor; as the wheel rotates the teeth pass close to the sensor, increasing its inductance. The sensor can therefore detect the passage of each tooth and thus determine the distance travelled. A great advantage of this sensor is its tolerance to dirty environments.

Figure 3.13(b) shows a sensor that uses the slotted opto-switch discussed earlier. This method uses a disc that has a number of holes or slots spaced equally around its perimeter. The disc and opto-switch are mounted such that the edge of the disc is within the slot of the switch. As the disc rotates the holes or slots cause the opto-switch to be periodically opened and closed, producing a train of pulses with a frequency determined by the speed of rotation. The angle of rotation can be measured by counting the number of pulses. A similar method uses an inductive proximity sensor in place of the opto-switch, and a ferromagnetic disc.

Figure 3.13   Examples of
displacement sensors using
counting

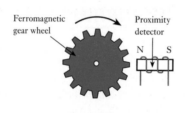

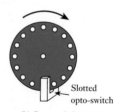

(a) Inductive sensor

(b) Opto-switch sensor

3.6.8 Rangefinders

Measurement of large distances usually requires a non-contact method. Both passive systems (which simply observe their environment) and active systems (which send signals out into the environment) are available. Passive techniques include optical triangulation methods, in which two slightly displaced sights are aligned on a common target. The angular difference between the two sights can then be measured using one of the angular sensors described above. Trigonometry is then used to calculate the distance between the sights and the target. This method is employed in rangefinding equipment used for surveying. Active systems transmit either sound or electromagnetic energy and detect the energy reflected from a distant object. By measuring the time taken for the energy to travel to the object and back to the transmitter, the distance between them may be determined. Because the speed of light is so great, some optical systems use the phase difference between the transmitted and received signals, rather than time of travel, to determine the distance (phase difference, and its measurement, will be discussed in Chapter 11).

3.7 Motion sensors

In addition to the measurement of displacement, it is often necessary to determine information concerning the motion of an object, such as its velocity or acceleration. These quantities may be obtained by differentiation of a position signal with respect to time, although such techniques often suffer from noise, since differentiation tends to amplify high-frequency noise present in the signal. Alternatively, velocity and acceleration can be measured directly using a number of sensors.

The counting techniques described earlier for the measurement of displacement can also be used for velocity measurement. This is achieved by measuring the frequency of the waveforms produced instead of counting the number of pulses. This gives a direct indication of speed. In fact, many of the counting techniques outlined earlier are more commonly used for speed measurement than for measurement of position. In many applications the direction of motion is either known or is unimportant, and these techniques often provide a simple and inexpensive solution.

Direct measurement of acceleration is made using an **accelerometer**. Most accelerometers make use of the relationship between force, mass and acceleration:

$$\text{force} = \text{mass} \times \text{acceleration}$$

A mass is enclosed within the accelerometer. When the device is subjected to acceleration the mass experiences a force, which can be detected in a number of ways. In some devices a force transducer, such as a strain gauge, is incorporated to measure the force directly. In others, springs are used to convert the force into a corresponding displacement, which is then measured with a displacement transducer. Because of the different modes of operation of the devices, the form of the output signal also varies.

3.8   Sound sensors

A number of techniques are used to detect sound. These include carbon (resistive) microphones, capacitive microphones, piezoelectric microphones and moving-coil microphones. Of these, the last are probably the most common. A moving-coil microphone consists of a permanent magnet and a coil connected to a diaphragm. Sound waves move the diaphragm, which causes the coil to move with respect to the magnet, thus generating an electrical signal. This process is illustrated in Figure 3.14. The generation of electrical currents using electromagnetism is discussed in later chapters.

Figure 3.14   A microphone

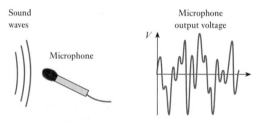

3.9   Sensor interfacing

Many electrical and electronic systems require their inputs to be in the form of electrical signals in which the voltage or current of the signal is related to the physical quantity being sensed. While some sensors produce an output voltage or current that is directly related to the physical quantity being measured, others require additional circuitry to generate such signals. The process of making the output of one device compatible with the input of another is often referred to as **interfacing**. Fortunately, the circuitry required is usually relatively simple, and this section gives a few examples of such techniques.

3.9.1   Resistive devices

In a potentiometer, the resistance between the central moving contact and the two end terminals changes as the contact is moved. This arrangement can easily be used to produce an output voltage that is directly related to the position of the central contact. If a constant voltage is placed across the outer terminals of a potentiometer, the voltage produced on the centre contact varies with its position. If the resistance of the track varies linearly,

Figure 3.15 Using a resistive sensor in a potential divider

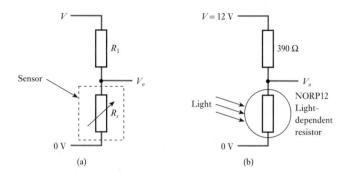

(a)   (b)

then the output voltage will be directly proportional to the position of the centre contact, and hence to the input displacement.

Many sensors represent changes in a physical quantity by changes in resistance. Examples include platinum resistance thermometers, photoconductive sensors and some forms of microphone. One way of converting a changing resistance into a changing voltage is to use the sensor in a potential divider circuit, as illustrated in Figure 3.15(a), where R_s represents the variable resistance of the sensor.

The output voltage V_o of this arrangement is given by the expression

$$V_o = V \frac{R_s}{R_1 + R_s}$$

and clearly the output voltage V_o varies with the sensor resistance R_s. An example of the use of this arrangement is shown in Figure 3.15(b), which depicts a simple light meter based on a light-dependent resistor (LDR). Light falling on the resistor affects its resistance (as discussed in Section 3.4), which in turn determines the output voltage of the circuit. The LDR shown changes its resistance from about 400 Ω (at 1000 lux) to about 9 kΩ (at 10 lux), which will cause the output voltage V_o to change from about 6 V to about 11.5 V in response to such changes in the light level.

While the arrangement of Figure 3.15(a) produces an output voltage that varies with the sensor resistance R_s, this is not a linear relationship. One way of producing a voltage that *is* linearly related to the resistance of a sensor is to pass a constant current through the device, as shown in Figure 3.16. From Ohm's law, the output of the circuit is given by

$$V_o = I R_s$$

and since I is constant, the output is clearly linearly related to the sensor voltage. The constant current I in the figure comes from some external circuitry – not surprisingly, such circuits are called **constant current sources**.

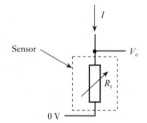

Figure 3.16 Using a resistive sensor with a constant current source

3.9.2 Switches

Most switches have two contacts, which are connected electrically when the switch is in one state (the closed state) and disconnected (or open

Figure 3.17   Generating a
binary signal using a switch

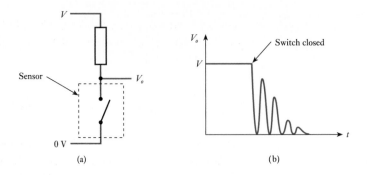

circuit) when the switch is in the other state (the open state). This arrangement can be used to generate binary electrical signals simply by adding a
voltage source and a resistance, as shown in Figure 3.17(a). When the
switch is closed, the output is connected to the zero volts line and therefore the output voltage V_o is zero. When the switch is open, the output is no
longer connected to the zero volts line but is connected through the resistance R to the voltage supply V. The output voltage will therefore be equal
to the supply voltage minus any voltage drop across the resistance. This
voltage drop will be determined by the value of the resistance R and the
current flowing into the output circuit. If the value of R is chosen such that
this voltage drop is small compared with V, we can use the approximation that the output voltage is zero when the switch is closed and V when it
is open. The value chosen for R clearly affects the accuracy of this approximation; we will be looking at this in later chapters when we consider
equivalent circuits.

One problem experienced by all mechanical switches is that of **switch
bounce**. When the moving contacts in the switch come together, they have
a tendency to bounce rather than to meet cleanly. Consequently, the electric circuit is first made, then broken, then made again, sometimes several
times. This is illustrated in Figure 3.17(b), which shows the output voltage
of the circuit of Figure 3.17(a) as the switch is closed. The length of the
oscillation will depend on the nature of the switch but might be a few
milliseconds in a small switch and perhaps tens of milliseconds in a large
circuit breaker. Switch bounce can cause severe problems, particularly if
contact closures are being counted. Although good mechanical design can
reduce this problem it cannot be eliminated, making it necessary to overcome this problem in other ways. Several electronic solutions are possible,
and it is also possible to tackle this problem using computer software techniques in systems that incorporate microcomputers.

Although the above discussion assumes the use of a mechanical switch,
the circuit of Figure 3.17(a) can also be used with an opto-switch sensor.
Optical switches do not produce a perfect 'closed circuit' when activated,
but the effective resistance of the device does change dramatically between
its ON and OFF states. Therefore, by choosing an appropriate value for the
external resistance R, it can be arranged that the circuit produces a binary

voltage signal that changes from approximately zero to approximately V volts, depending on the state of the switch. Optical switches do not suffer from switch bounce.

3.9.3 Capacitive and inductive devices

Sensors that change their capacitance or inductance in response to external influences normally require the use of alternating current (AC) circuitry. Such circuits need not be complicated, but they do involve techniques that are yet to be discussed in this text. We will therefore leave further consideration of such circuits until later.

3.10 Sensors – a summary

It is not the purpose of this section to provide an exhaustive list of all possible sensors. Rather, it sets out to illustrate some of the important classes of sensor that are available and to show the ways in which they provide information. It will be seen that some sensors generate output currents or voltages related to changes in the quantity being measured. In doing so they extract power from the environment and can deliver power to external circuitry (although usually the power available is small). Examples of such sensors are photovoltaic sensors and moving-coil microphones.

Other devices do not deliver power to external circuits but simply change their physical attributes, such as resistance, capacitance or inductance, in response to variations in the quantity being measured. Examples include resistive thermometers, photoconductive sensors, potentiometers, inductive position transducers and strain gauges. When using such sensors, external circuitry must be provided to convert the variation in the sensor into a useful signal. Often this circuitry is very simple, as illustrated in the last section.

Unfortunately, some sensors do not produce an output that is linearly related to the quantity being measured (for example, a thermistor). In these cases, it may be necessary to overcome the problem by using electronic circuitry or processing to compensate for any non-linearity. This process is called **linearisation.** The ease or difficulty of linearisation depends on the characteristics of the sensor and the accuracy required.

Example 3.1

Selecting an appropriate sensor for a computer mouse.

In this chapter, we have looked at a number of displacement and motion sensors. Armed with this information, we will select a suitable method of determining the displacement of a mouse for use as a computer pointing device. The resolution of the sensing arrangement should be such that the user can select an individual pixel (the smallest definable point in the display). A typical screen might have a 1024×768 pixel display, although displays may have a resolution several times greater than this. Movement of the cursor from one side of the screen to the other should require a movement of the mouse of a few centimetres (the sensitivity of the mouse is often selectable using software in the computer).

A mouse normally senses motion using a small rubber ball that projects from its base. As the mouse is moved over a horizontal surface, the ball rotates about two perpendicular axes, and this motion is used to determine the position of a cursor on a computer screen.

We have looked at several sensors that may be used to measure angular position. These include simple potentiometers and position encoders. Sensing the *absolute position* of the ball (and hence the mouse) could represent a problem, since for high-performance displays this could require a resolution of better than one part in 2000. Sensors with such a high resolution are often expensive and physically large. In this application, it is probably more appropriate to sense *relative* motion of the mouse. This reduces the complexity of the sensing mechanism and also means that the mouse is not tied to a fixed absolute position.

Measurement of the relative motion of the rubber ball suggests the use of some form of incremental sensor. This could use a proprietary incremental encoder, but because this is a very high-volume application, it is likely that a more cost-effective solution could be found. The diagram below shows a possible arrangement based on the use of slotted wheels and optical sensors (as described in Section 3.6).

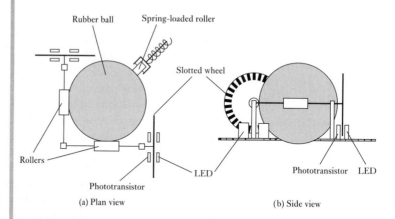

(a) Plan view   (b) Side view

In order to resolve rotation of the ball into two perpendicular components, the ball is pressed against two perpendicular rollers by a third, spring-loaded roller. Rotation of the ball in a particular direction causes one or both of the sensing rollers to turn. Each of these rollers is connected to a slotted wheel that is placed between two slotted optical switches. The switches are positioned to allow the direction of rotation to be detected in a manner similar to that shown in Figure 3.11. The signals from the sensors are fed to the computer, which keeps track of the movement of the ball and hence determines the appropriate cursor position. This arrangement has a range limited only by the method used to count the moving slots. The sensitivity is determined by the relative sizes of the ball and the pulleys, and by the number of slots in the wheels.

Key points

- A wide range of sensors is available to meet the needs of a spectrum of possible applications.

- Some sensors produce an output voltage or current that is related to the quantity being measured. They therefore supply power (albeit in small quantities).

- Other devices simply change their physical attributes, such as resistance, capacitance or inductance, in response to changes in the measured quantity.

- Interfacing circuitry may be required with some sensors to produce a signal in the desired form.

- Some sensors produce an output that is linearly related to the quantity being measured.

- Other devices are non-linear in operation.

- In some applications linearity is unimportant. For example, a proximity detector may simply be used to detect the presence or absence of an object.

- In other applications, particularly where an accurate measurement is required, linearity is of more importance. In such applications, we will use either a sensor that has a linear characteristic or some form of linearisation to overcome non-linearities in the measuring device.

Exercises

3.1 Explain the meanings of the terms 'sensor', 'actuator' and 'transducer'.

3.2 What is meant by the resolution of a sensor?

3.3 Explain the difference between random and systematic errors.

3.4 Define the terms 'accuracy' and 'precision'.

3.5 Give an example of a digital temperature sensor.

3.6 What is the principal advantage and disadvantage of platinum resistance thermometers (PRTs) when making accurate temperature measurements?

3.7 A PRT has a resistance of $100\,\Omega$ at $0\,°C$ and a temperature coefficient of $+0.385\,\Omega$ per $°C$. What would be its resistance at $100\,°C$?

The PRT is connected to an external circuit that measures the resistance of the sensor by passing a constant current of $10\,mA$ through it and measuring the voltage across it. What would this voltage be at $100\,°C$?

Exercises continued

3.8 The PRT described in the last exercise is connected as shown in the diagram below to form an arrangement where the output voltage V_o is determined by the temperature of the PRT.

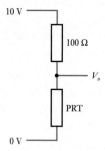

Derive an expression for V_o in terms of the temperature of the PRT.

The resistance of the PRT is linearly related to its absolute temperature. Is V_o linearly related to temperature?

3.9 How do thermistors compare with PRTs?

3.10 *pn* junction temperature sensors are inexpensive, linear and easy to use. However, they do have certain limitations, which restricts their use. What are these limitations?

3.11 Why might one choose to use a photodiode as a photoconductive light sensor rather than in a photovoltaic mode?

3.12 What is the advantage of a phototransistor light sensor in comparison with a photodiode sensor?

3.13 In what situations might one use a slow light-dependent resistor sensor in preference to a much faster photodiode or phototransistor sensor?

3.14 Explain the meanings of the terms 'stress' and 'strain'.

3.15 Suggest a suitable method for using a strain gauge to measure the vertical force applied to the end of a beam that is supported at one end.

3.16 Suggest a suitable method of employing two strain gauge to measure the vertical force applied to the end of a beam that is supported at one end. Why might this approach be used in preference to that described in the last exercise?

3.17 Describe two methods of measurement that would be suitable for a non-contact, automatic rangefinder for distances up to 10 m.

3.18 In an earlier exercise, we considered a PRT that has a resistance of 100 Ω at 0 °C and a temperature coefficient of +0.385 Ω per °C. If such a device is connected to a constant current source of 10 mA, in an arrangement as shown in Figure 3.16, what would be the output voltage of the arrangement at 0 °C?

What would be the sensitivity of this arrangement (in mV/°C) at temperatures above 0 °C?

3.19 The arrangement of Figure 3.17 produces an output of 0 V if the switch is closed and *V* if the switch is open. Devise a similar circuit that reverses these two voltages.

3.20 Suggest ten physical quantities, not discussed in this chapter, that are measured regularly, giving in each case an application where this measurement is required.

Chapter 4

Actuators

Objectives

When you have studied the material in this chapter you should be able to:
- discuss the need for actuators in electrical and electronic systems;
- describe a range of actuators, both analogue and digital, for controlling various physical quantities;
- explain the requirement for actuators with different properties for use in different situations;
- describe the use of interface circuitry to match a particular actuator to the system that drives it.

4.1 Introduction

Sensors provide only half of the interaction required between an electrical system and its surroundings. In addition to being able to sense physical quantities in their environment, systems must also be able to affect the outside world in some way so that their various functions can be performed. This might require the system to move something, change its temperature or simply provide information via some form of display. All these functions are performed by **actuators**.

As with the sensors discussed in the last chapter, actuators are **transducers** since they convert one physical quantity into another. Here we are interested in actuators that take electrical signals from our system and use them to vary some external physical quantity. As one would expect, there are a large number of different forms of actuator, and it would not be appropriate to attempt to provide a comprehensive list of such devices. Rather, this chapter sets out to show the diversity of such devices and to illustrate some of their characteristics.

4.2   Heat actuators

Most heating elements may be considered as simple **resistive heaters**, which output the power that they absorb as heat. For applications requiring only a few watts of heat output, ordinary resistors of the appropriate power rating may be used. Special heating cables and elements are available for larger applications, which may dissipate many kilowatts.

4.3   Light actuators

Most lighting for general illumination is generated using conventional incandescent or fluorescent lamps. The power requirements of such devices can range from a fraction of a watt to hundreds or perhaps thousands of watts.

For signalling and communication applications, the relatively low speed of response of conventional lamps makes them unsuitable, and other techniques are required.

4.3.1   Light-emitting diodes

One of the most common light sources used in electronic circuits is the **light-emitting diode** or **LED**. This is a semiconductor diode constructed in such a way that it produces light when a current passes through it. A range of semiconductor materials can be used to produce infrared or visible light of various colours. Typical devices use materials such as gallium arsenide, gallium phosphide or gallium arsenide phosphide.

The characteristics of these devices are similar to those of other semiconductor diodes (which will be discussed in Chapter 19) but with different operating voltages. The light output from an LED is approximately proportional to the current passing through it; a typical small device might have an operating voltage of 2.0 V and a maximum current of 30 mA.

LEDs can be used individually or in multiple-element devices. One example of the latter is the LED **seven-segment display** shown in Figure 4.1. This consists of seven LEDs, which can be switched ON or OFF individually to display a range of patterns.

Figure 4.1 LED seven-segment displays

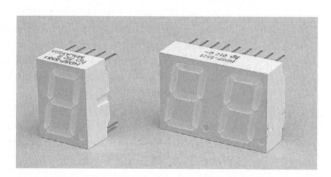

Infrared LEDs are widely used with photodiodes or phototransistors to enable short-range wireless communication. Variations in the current applied to the LED are converted into light with a fluctuating intensity, which is then converted back into a corresponding electrical signal by the receiving device. This technique is widely used in **remote control** applications for televisions and video recorders. In these cases, the information transmitted is generally in a digital form. Because there is no electrical connection between the transmitter and the receiver, this technique can also be used to couple digital signals between two circuits that must be electrically isolated. This is called **opto-isolation**. Small self-contained **opto-isolators** are available that combine the light source and sensor in a single package. The input and output sections of these devices are linked only by light, enabling them to produce electrical isolation between the two circuits. This is particularly useful when the two circuits are operating at very different voltage levels. Typical devices will provide isolation of up to a few kilovolts.

4.3.2 Liquid crystal displays

Liquid crystal displays (**LCDs**) consist of two sheets of polarised glass with a thin layer of oily liquid sandwiched between them. An electric field is used to rotate the plane of polarisation of the liquid in certain regions, making some parts of the display opaque while others are transparent. The display segments can be arranged to create specific patterns (such as those of seven-segment displays) or in a matrix to allow any characters or images to be displayed.

A great advantage of LCDs (compared with LEDs) is that they are able to use ambient light, greatly reducing power consumption and allowing them to be used in a wide range of low-power applications. When insufficient ambient light is available they can also be backlit, although this increases their power consumption considerably. LCDs are widely used in wristwatches, mobile phones and many forms of battery-operated electronic equipment. They are also used in computer displays and other high-resolution applications. An example of a small LCD module is shown in Figure 4.2.

4.3.3 Fibre-optic communication

For long-distance communication, the simple techniques used in television remote-control units are not suitable as they are greatly affected by ambient light, that is, light present in the environment. This problem can be overcome by the use of a **fibre-optic cable**, which captures the light from the transmitter and passes it along the cable to the receiver without interference from external light sources. Fibres are usually made of either an optical polymer or glass. The former are inexpensive and robust, but their

Figure 4.2 A liquid crystal display module

high attenuation makes them suitable for only short-range communications of up to about 20 metres.

Glass fibres have a much lower attenuation and can be used over several hundred kilometres, but they are more expensive that polymer fibres. For long-range communications, the power available from a conventional infrared LED is insufficient. In such applications **laser diodes** may be used. These combine the light-emitting properties of an LED with the light amplification of a laser to produce a high-power, coherent light source.

4.4 Force, displacement and motion actuators

In practice, actuators for producing force, displacement and motion are often closely related. A simple DC permanent magnet motor, for example, if opposed by an immovable object, will apply a force to that object determined by the current in the motor. Alternatively, if resisted by a spring, the motor will produce a displacement that is determined by its current and, if able to move freely, it will produce a motion related to the current. We will therefore look at several actuators that can be used to produce each of these outputs, as well as some that are designed for more specific applications.

4.4.1 Solenoids

A solenoid consists of an electrical coil and a ferromagnetic slug that can move into, or out of, the coil. When a current is passed through the solenoid, the slug is attracted towards the centre of the coil with a force determined by the current in the coil. The motion of the slug may be opposed by a spring to produce a displacement output, or the slug may simply be free to move. Most solenoids are linear devices, the electric current producing a linear force/displacement/motion. However, rotational solenoids are also available that produce an angular output. Both forms may be used with a continuous analogue input, or with a simple ON/OFF

Figure 4.3 Small linear
solenoids

(digital) input. In the latter case, the device is generally arranged so that
when it is energised (that is, turned ON) it moves in one direction until it
reaches an end stop. When de-energised (turned OFF) a return spring forces
it to the other end of its range of travel, where it again reaches an end
stop. This produces a binary position output in response to a binary input.
Figure 4.3 shows examples of small linear solenoids.

4.4.2 Meters

Panel meters are important output devices in many electronic systems providing a visual indication of physical quantities. Although there are various
forms of panel meter, one of the simplest is the **moving-iron meter**, which
is an example of the rotary solenoid described above. Here a solenoid
produces a rotary motion, which is opposed by a spring. This produces an
output displacement that is proportional to the current flowing through the
coil. A needle attached to the moving rotor moves over a fixed scale to indicate the magnitude of the displacement. Moving-iron meters can be used
for measuring AC or DC quantities. They produce a displacement that is
related to the magnitude of the current and is independent of its polarity.

Although moving-iron meters are used in some applications, a more
common arrangement is the **moving-coil meter**. Here, as the name implies,
it is the coil that moves with respect to a fixed magnet, producing a meter
that can be used to determine the polarity of a signal as well as its magnitude. The deflection of a moving-coil meter is proportional to the average
value of the current. AC quantities can be measured by incorporating a
rectifier and applying suitable calibration. However, it should be noted
that the calibration usually assumes that the quantity being measured is
sinusoidal, and incorrect readings will result if other waveforms are used
(this issue will be discussed in Chapter 11). Examples of typical moving-
coil meters are shown in Figure 4.4.

Typical panel meters will produce a full-scale deflection for currents of
50 μA to 1 mA. Using suitable series and shunt resistances, it is possible to
produce meters that will display either voltages or currents with almost any
desired range.

Figure 4.4   Moving-coil meters

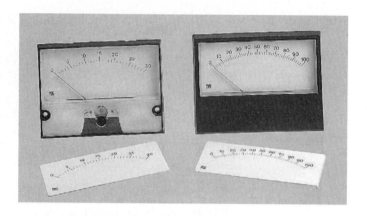

4.4.3   Motors

Electric motors fall into three broad types: AC motors, DC motors and stepper motors. **AC motors** are primarily used in high-power applications and situations where great precision is not required. Control of these motors is often by simple ON/OFF techniques, although variable-power drives are also used. We will return to look at the operation of AC motors in more detail in Chapter 23.

DC motors are extensively used in precision position-control systems and other electronic systems, particularly in low-power applications. These motors have very straightforward characteristics, with their speed being determined by the applied voltage and their torque being related to their current. The speed range of DC motors can be very wide, with some devices being capable of speeds from tens of thousands of revolutions per minute down to a few revolutions per day. Some motors, in particular DC permanent-magnet motors, have an almost linear relationship between speed and voltage and between torque and current. This makes them particularly easy to use. DC motors will be considered in more detail in Chapter 23.

Stepper motors, as their name implies, move in discrete steps. The motor consists of a central rotor surrounded by a number of coils (or windings). The form of a simple stepper motor is shown in Figure 4.5, and a typical motor is shown in Figure 4.6.

Figure 4.5   A simple stepper motor

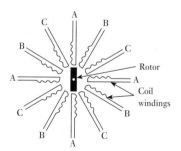

Figure 4.6   A typical stepper
motor

Diametrically opposite pairs of coils are connected together so that
energising any one pair of coils will cause the rotor to align itself with that
pair. By applying power to each set of windings in turn, the rotor is made
to 'step' from one position to another and thus generate rotary motion. In
order to reduce the number of external connections to the motor, groups of
coils are connected together in sequence. In the example shown, every third
coil is joined to give three coil sets, which have been labelled A, B and C.
If initially winding A is energised, the rotor will take up a position aligned
with the nearest winding in the A set. If now A is de-energised and B is
activated, the motor will 'step' around to align itself with the next coil. If
now B is de-energised and C is activated, the rotor will again step to the
adjacent coil. If the activated coil now reverts to A, the rotor will move on
in the same direction to the next coil, since this is the closest coil in the A
set. In this way, the rotor can be made to rotate by activating the coils in
the sequence 'ABCABCA . . .'. If the sequence in which the windings are
activated is reversed (CBACBAC . . .), the direction of rotation will also
reverse. Each element in the sequence produces a single step that results in
an incremental movement of the rotor.

The waveforms used to activate the stepper motor are binary in nature,
as shown in Figure 4.7. The motor shown in Figure 4.5 has twelve coils,
and consequently twelve steps would be required to produce a complete
rotation of the rotor. Typical small stepper motors have more than twelve
coils and might require 48 or 200 steps to perform one complete revolution.
The voltage and current requirements of the coils will vary with the size
and nature of the motor.

The speed of rotation of the motor is directly controlled by the frequency
of the waveforms used. Some stepper motors will operate at speeds of

Figure 4.7 Stepper motor current waveforms

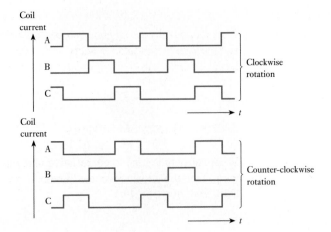

several tens of thousands of revolutions per minute, but all have a limited rate of acceleration determined by the inertia of the rotor. All motors have a 'maximum pull-in speed', which is the maximum speed at which they can be started from rest without losing steps. To operate at speeds above this rate, they must be accelerated by gradually increasing the frequency of the applied waveforms. Since the movement of the rotor is directly controlled by the waveforms applied to the coils, the motor can be made to move through a prescribed angle by applying an appropriate number of transitions to the coils. This is not possible using a DC motor, since the speed of rotation is greatly affected by the applied load.

4.5   Sound actuators

4.5.1   Speakers

Most speakers (or loudspeakers) have a fixed permanent magnet and a movable coil connected to a diaphragm. Input to the speaker generates a current in the coil, which causes it to move with respect to the magnet, thereby moving the diaphragm and generating sound. The nominal impedance of the coil in the speaker is typically in the range 4 to 15 Ω, and the power-handling capacity may vary from a few watts for a small domestic speaker to several hundreds of watts for speakers used in public address systems.

4.5.2   Ultrasonic transducers

At very high frequencies, the permanent-magnet speakers described earlier are often replaced by **piezoelectric actuators**. Such transducers are usually designed to operate over a narrow range of frequencies.

4.6 Actuator interfacing

The actuators discussed above all consume electrical power in order to vary some external physical quantity. Therefore the process of interfacing is largely concerned with the problem of enabling an electrical or electronic system to control the power in such as device.

4.6.1 Resistive devices

Where an actuator is largely resistive in nature, as in a resistive heating element, then the power dissipated in the device will be related to the voltage applied to it by the relationship

$$P = \frac{V^2}{R}$$

Here the power supplied to the actuator is simply related to the voltage applied across it. In such cases, the problems of interfacing are largely related to the task of supplying sufficient power to drive the actuator. In the case of devices requiring just a few watts (or less), this is relatively simple. However, as the power requirements increase, the task of supplying this power becomes more difficult. We will consider methods of driving high power loads when we look at power electronic circuits in Chapter 22.

One way of simplifying the control of high-power devices is to operate them in an ON/OFF manner. Where a device can be turned ON or OFF manually, this can be achieved using a simple mechanical switch. Alternatively, this function can be achieved under system control using an **electrically operated switch** (we will look at the operation of such circuits at a later stage).

In many cases, it is necessary to vary the power dissipated in an actuator rather than just to turn it ON and OFF. This may also be achieved using switching techniques in some cases. By repeatedly turning a device ON and OFF at high speed, it is possible to control the power dissipated in the component by altering the fraction of time for which the device is ON. Such techniques are used in conventional domestic **light dimmers**.

4.6.2 Capacitive and inductive devices

Capacitive and inductive actuators, such as motors and solenoids, create particular interfacing problems. This is particularly true when using the switching techniques described above. We will leave discussion of these problems until we have looked at capacitance and inductance in more detail.

4.7 Actuators – a summary

All the actuators we have discussed take an electrical input signal and from it generate a non-electrical output. In each case, power is taken from the input in order to apply power at the output. The power requirements are quite small in some cases, such as an LED or a panel meter, which consume only a fraction of a watt. In other cases the power required may be considerable. Heaters and motors, for example, may consume hundreds or even thousands of watts.

The **efficiency** of conversion also varies from device to device. In a heater, effectively all the power supplied by the input is converted to heat. We could say that the conversion efficiency is 100 percent. LEDs, however, despite being one of the more efficient methods of converting electrical power into light, have an efficiency of only a few percent, the remaining power being dissipated as heat.

Some actuators can be considered as simple resistive loads in which the current will vary in direct proportion to the applied voltage. Most heaters and panel meters would come into this category. Other devices, such as motors and solenoids, have a large amount of inductance as well as resistance, while others possess a large capacitance. Such devices behave very differently from simple resistive loads, particularly when a rapidly changing signal is applied. A third group of devices are non-linear and cannot be represented by simple combinations of passive components. LEDs and semiconductor laser diodes come into this third group. When designing electronic systems, it is essential to know the characteristics of the various actuators to be used so that appropriate interfacing circuitry can be produced.

Example 4.1

Selecting an appropriate actuator for a wristwatch.

In this chapter, we have looked at a number of forms of electric motor. Here we will use this information to illustrate the selection of an actuator for a given application. In this case, we need to select an electric motor to drive the hands in a conventional **quartz wristwatch**.

At the heart of a modern wristwatch is a quartz crystal, vibrating at about 32,000 times per second. This is used with appropriate electronic circuitry to produce a train of pulses with a frequency of precisely 1 Hz. This signal is used to drive the second hand of the watch, which then drives the minute and hour hands through a series of gears. Our task, therefore, is to select a motor to produce a rotation of precisely one revolution per minute from a train of pulses.

While it is possible to produce DC motors that will rotate at such low speeds, such motors are usually physically too large for this application. A more attractive option would be to use a DC motor at a higher speed and to use some form of reduction gearbox to reduce the speed appropriately. However, the primary problem with the use of a DC motor (or indeed an AC motor) in this application relates to the difficulty of accurately controlling the speed.

Since in this application our time reference comes from a train of digital pulses, it makes sense to use a *digital* actuator in the form of a stepper motor. A typical wristwatch would incorporate a stepper motor that rotates once for every sixty input pulses. This is connected directly to the second hand, which therefore 'steps' around by one second for each pulse. A 60:1 reduction gear arrangement then drives the minute hand, and a similar arrangement drives the hour hand. Fortunately, it is possible to produce very small stepper motors that consume very little power, making them ideal for this application.

Key points

- All useful systems need to affect their environment in order to perform their intended functions.

- Systems affect their environment using actuators.

- Most actuators take power from their inputs in order to deliver power at their outputs. The power required varies tremendously between devices.

- Some devices consume only a fraction of a watt. Others may consume hundreds or perhaps thousands of watts.

- In most cases, the energy conversion efficiency of an actuator is less than 100 percent, and sometimes it is much less.

- Some actuators resemble resistive loads, while others have considerable capacitance or inductance. Others still are highly non-linear in their characteristics.

- The ease or difficulty of driving actuators varies with their characteristics.

Exercises

4.1 Explain the difference between a transducer and an actuator.

4.2 What form of device would normally be used as a heat actuator when the required output power is a few watts?

4.3 What form of heat actuator would be used in applications requiring a power output of several kilowatts?

4.4 Estimate the efficiency of a typical heat actuator.

4.5 What forms of light actuator would typically be used for general illumination? What would be a typical range for the output power for such devices?

4.6 Why are conventional light bulbs unsuitable for signalling and communication applications? What forms of transducer are used in such applications?

4.7 How do light-emitting diodes (LEDs) differ from conventional semiconductor diodes?

4.8 What would be a typical value for the operating voltage of an LED, and what would be a typical value for its maximum current?

4.9 From the information given for the last exercise, what would be a typical value for the maximum power dissipation of an LED?

4.10 In addition to displaying the digits 0–9, the seven-segment display of Figure 4.1 can be used to indicate some alphabetic characters (albeit in a rather crude manner). List the upper and lower case letters that can be shown in this way and give examples of simple status messages (such as 'Start' and 'Stop') that can be displayed using an array of these devices.

4.11 Briefly describe the operation and function of an opto-isolator.

4.12 What environmental factor causes problems for optical communication systems using conventional LEDs and photo-detectors? How may this problem be reduced?

4.13 What form of optical fibre would be preferred for communication over a distance of several kilometres? What form of light source would normally be used in such an arrangement?

4.14 Explain how a single form of transducer might be used as a force actuator, a displacement actuator or a motion actuator.

4.15 Describe the operation of a simple solenoid.

4.16 Explain how a solenoid may be used as a binary position actuator.

4.17 Explain why a simple panel meter may be thought of as a rotary solenoid.

4.18 What is the most common form of analogue panel meter? What would be typical operating currents for such devices?

4.19 List three basic forms of electric motor.

4.20 What form of motor would typically be used in high-power applications?

4.21 What form of motor might be used in an application requiring precise position control?

4.22 Briefly describe the operation of a stepper motor.

4.23 How are the speed and the direction of rotation of a stepper motor controlled?

4.24 What would be a typical value for the impedance of the coil of a loudspeaker?

4.25 Explain how the power dissipated in an actuator may be varied using an electrically operated switch.

Chapter 5

Signals and Data Transmission

Objectives

When you have studied the material in this chapter you should be able to:

- list a range of analogue and digital signal types that are widely used in modern electronic systems;
- characterise signals in terms of a range of physical attributes and properties;
- describe the limitations imposed on electrical signals by electronic circuits and communications channels;
- explain the need to change the frequency range of signals in some situations and outline several modulation techniques for achieving this;
- discuss the problems associated with distortion and noise in electrical and electronic circuits.

5.1 Introduction

In Chapter 1 we looked at the nature of **physical quantities**, such as temperature, pressure and humidity, which characterise the environment in which we live. We also noted that it is often convenient to represent these quantities by electrical **signals**, since this can simplify the processing, communication and storage of information relating to these quantities. In Chapter 3, we looked at examples of sensors that perform the task of producing an electrical representation of a physical quantity, and in Chapter 4 we considered the reverse operation of taking an electrical signal and using this to vary some external physical quantity. In this chapter, we will complete this investigation by looking in more detail at the nature and characteristics of electrical signals of different types.

Electrical signals can take many forms and in particular they can be either analogue or digital in nature. In the next few sections, we will look at some widely used signal formats and at examples of their use.

5.2 Analogue signals

We noted in Chapter 1 that analogue signals are free from discontinuities and can take an infinite number of values.

The use of analogue electrical signals dates back to the early nineteenth century to the work of scientists such as Michael Faraday, Luigi Galvani and Alessandro Volta. In 1876, the Scottish inventor Alexander Graham Bell demonstrated the first telephone system and later, in 1899, the Italian Guglielmo Marconi used analogue signals for wireless communications between England and France. Over the following decades a range of new technologies were developed, including television, sound recording and mobile communications. In each case these technologies were initially based on the use of analogue signals of one form or another, although digital techniques are also widely used in these areas. Analogue signals can take an almost limitless number of forms and are used in almost all forms of electrical or electronic system.

Perhaps the simplest form of analogue signal to understand is where a physical quantity is represented by a *voltage* (or sometimes a *current*) that is directly proportional to that quantity. This is a very common signal format in electronic systems, and in Chapter 3 we met several sensors that are designed to produce such signals. Examples include a *pn*-junction temperature sensor, which produces an output voltage that is directly proportional to temperature, and a photovoltaic light sensor, which produces an output voltage directly proportional to light intensity. Many systems use simple voltage signals of this type, and most readers will have come across these in various applications. For example, many readers will have used an oscilloscope to look at voltage signals representing music or speech in an audio system. Here the voltage of the waveform represents the variation in air pressure associated with the sound. This is illustrated in Figure 5.1.

Figure 5.1 Voltage representation of a sound waveform

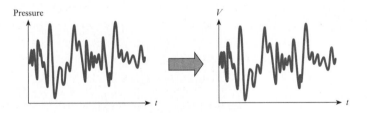

5.3 Digital signals

In Chapter 1, we saw that digital signals can take only certain discrete values, with discontinuities marking the transitions between them. In most applications only two values are used, producing a **binary** signal. In some cases these two values are represented by the presence or absence of a voltage on a wire or conductor, so it is common to refer to the two values as ON and OFF. It is also common to give the values the labels 1 and 0.

Figure 5.2 A simple digital signal

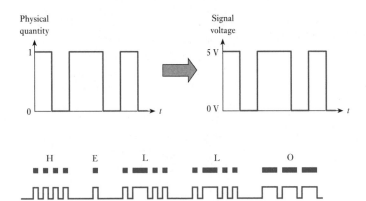

Figure 5.3 Morse code

Perhaps the most common form of binary digital signal is where the two values are represented by different voltages. An example of such a signal is shown in Figure 5.2.

As with their analogue counterparts, the use of digital signals has a long and distinguished history. In 1837, two Englishmen, William Cooke and Charles Wheatstone, produced the first practical telegraph instrument, which was used for railway communications. At around the same time, the American Samuel Morse developed his famous signalling code for sending text over such a channel. Morse's code represents individual characters by combinations of short and long pulses, as illustrated in Figure 5.3.

For several decades, long-distance communication was dominated by the digital technology of the telegraph. Then, forty years after its invention, it was largely overshadowed by the development of the telephone. This took analogue speech signals and transmitted them directly, without the need for coding and decoding of the information at either end. The domination of analogue techniques in communications was to last for nearly a hundred years until the development of advanced digital communication methods during the second half of the twentieth century. This, together with the development of powerful computers, has revolutionised all aspects of engineering and, in turn, all aspects of our lives. Digital technology now dominates many areas of electronics, including computing, communications and control. The reasons for this dominance are numerous. Some will be discussed in later sections of this chapter, while others will become apparent in later chapters.

The simple digital signal of Figure 5.2 represents the state of a single binary quantity. It could, for example, represent the output of a binary sensor such as a switch. In Chapter 3, we looked at a simple interfacing circuit that could be used to produce a binary voltage signal from a switch, and this is shown symbolically in Figure 5.4. The information represented by such a signal could be represented by a single binary variable or **binary digit**. A binary digit is usually referred to as a **bit** (this being an abbreviation for *b*inary dig*it*), and the signals of Figures 5.2 and 5.4 therefore represent single-bit quantities.

Figure 5.4 A switch as a single-bit information source

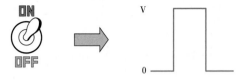

Figure 5.5 A multiple-bit information source

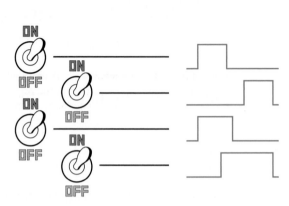

While there are many applications for single-bit signals there is also a need to represent more complex digital information. It could be, for example, that we wish to represent the state of not one switch, but a group of many switches as shown in Figure 5.5.

Rather than represent the state of many switches by a group of separate signals, it is sometimes more convenient to group the information together into a single **digital word**. Logically, a word that represents n independent binary digits is referred to as an n-bit word. We could therefore represent the state of the four switches of Figure 5.5 by a single 4-bit word.

While multiple-bit words can be used to represent arrays of switches, they can also be used to represent other forms of information. A 4-bit word can represent the sixteen possible combinations of the states of four switches, but it could also represent the sixteen different states of another form of digital variable. By choosing a digital word of an appropriate length, we can create digital variables with any desired number of states.

A digital variable of n bits can take 2^n different values and can therefore represent 2^n different values or states. For example, an 8-bit variable can take 256 values. We could therefore use an 8-bit variable to represent the state of eight switches, a number in the range 0 to 255 or perhaps a single printable character (there being more than enough scope to represent all the alphabetic, numeric and punctuation characters in a single 8-bit word). For reasons that will become apparent in later chapters, we often select particular word lengths in preference to others. 8-bit words are particularly common and are given the special name of **bytes**.

Example 5.1 How many different values can be represented by binary words of 8, 16 and 32 bits?

An n-bit word can take 2^n different values. Therefore:

An 8-bit word can take $2^8 = 256$ values

A 16-bit word can take $2^{16} = 65,536$ values

A 32-bit word can take $2^{32} = 4,294,967,296$ values

If a quantity is represented by a digital word, the number of bits in the word determines the resolution of the representation. This in turn limits the accuracy that can be associated with such quantities.

Example 5.2 Determine the resolution of quantities represented by words of 8, 16 and 32 bits.

An 8-bit word can take 256 values. This gives a resolution of 1 part in 256, or about 0.39%.

A 16-bit word can take 65,536 values. This gives a resolution of 1 part in 65,536 or about 0.0015%.

A 32-bit word can take 4,294,967,296 values. This gives a resolution of 1 part in 4,294,967,296 or about 0.000000023%.

Having arranged our information into digital words, we are now faced with the problem of how to represent these words by appropriate signals, for example when we wish to transmit this information from one place to another. One approach is simply to generate a series of voltage signals on an appropriate number of separate signal lines. Clearly, since an n-bit word represents the states of n binary variables, we would need n separate lines to represent it (plus an earth connection). This would be referred to as a **parallel** data format and is illustrated in Figure 5.6(a). Alternatively, we could opt to transmit the data on a single signal line by sending the digits one after another. This is referred to as a **serial** data format and is shown in Figure 5.6(b). Each approach has its advantages and disadvantages. The parallel format is faster, since all the data is sent in one go, but this

Figure 5.6 Parallel and serial data formats

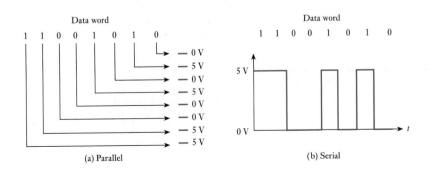

(a) Parallel

(b) Serial

approach requires many separate lines. The serial format is much slower but has the advantage of requiring only a single line (plus an associated earth line). We will return to this topic in later chapters when we look at data communication in more detail.

5.4 Signal properties

Electrical signals may be characterised in a number of ways. First, we could describe the range of the amplitude of the signal in terms of its maximum and minimum values. In the case of binary voltage signals, this would require us to define the voltages used to represent the two possible states of the signal. Second, we could say something about the polarity of the signal. Some signals are **unipolar** (that is, they have voltages or currents of only one polarity), while others are **bipolar** (that is, they have both positive and negative values). Examples of unipolar and bipolar waveforms are shown in Figure 5.7. Electrical signals that repeatedly change from positive to negative are also referred to as **alternating signals**, and we will learn more about this important class of signals in Chapter 11.

Another important characteristic of a signal is the range of frequencies present within it. For example, a human voice might contain frequencies from about 50 Hz to about 7 kHz (depending on the person concerned), and a high-quality microphone might convert this information into an electrical signal with a similar range of frequencies. One way of describing the frequencies present in a physical quantity or a signal is to draw its **frequency spectrum**, which shows the magnitude of its various frequency components. A signal that is a simple sinusoid of a fixed frequency has a spectrum that consists of a single line at that frequency, the height of the line representing its magnitude. A signal that is formed by the addition of two sinusoidal signals has a spectrum with two lines, and this principle can be extended to represent signals formed from various sinusoidal components.

Figure 5.7 Unipolar and bipolar signals

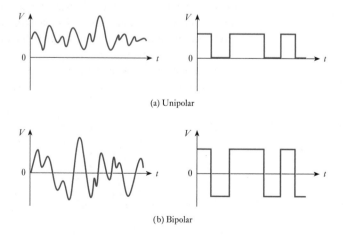

(a) Unipolar

(b) Bipolar

Example 5.3 **A signal consists of a sinusoidal voltage with a frequency of 1 kHz and a magnitude of 2 V. Sketch its frequency spectrum.**

Since the signal consists of a single frequency, its spectrum is a single line, its length indicating its magnitude.

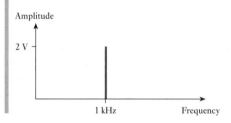

Example 5.4 **A signal is formed by adding two sinusoidal voltages. The first has a frequency of 1 kHz and a magnitude of 2 V. The second has a frequency of 1.5 kHz and a magnitude of 1 V. Sketch the frequency spectrum of this signal.**

This signal has two frequency components and therefore has two lines in its spectrum.

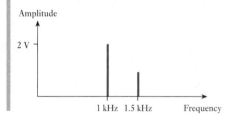

Complex waveforms might contain components of several discrete frequencies, resulting in a **line spectrum** as illustrated in Figure 5.8. However, it is more common for signals to have a broad spread of frequency components, resulting in a continuous distribution of frequencies. This results in a **continuous spectrum**. An example of such a spectrum is shown in Figure 5.9, which shows a possible frequency spectrum for a speech waveform.

Frequency spectra can be plotted in a number of ways. The vertical axis, representing the magnitude of the quantity, might be used to represent the

Figure 5.8   A line spectrum

Figure 5.9   A frequency
spectrum for a speech waveform

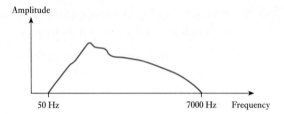

voltage of a frequency component of an electrical signal, but it could
alternatively represent the magnitude of some physical quantity. It is also
common to plot the power associated with a particular component in a
signal rather than its magnitude. In electrical signals, power is proportional
to the square of the voltage of the component. In such cases, the spectrum
is sometimes referred to as a **power spectrum**. It is also common to plot
spectra against a logarithmic frequency axis.

The frequencies present in a spectrum relate to the rate at which the
quantity changes. Rapidly changing events correspond to high frequencies,
while slow fluctuations relate to low-frequency components.

While a signal may have either a continuous or a discrete spectrum, all
signals will tend to occupy a certain **frequency range**. The difference
between the highest and lowest frequencies present in a signal is termed its
bandwidth. For example, a signal with a frequency range from 1 kHz to
4 kHz would have a bandwidth of 3 kHz.

| **Example 5.5** | A typical human speech might have a frequency range from about 50 Hz to about 7 kHz. Estimate the bandwidth of such speech. |

The bandwith is simply the difference between the maximum and minimum
frequencies present. In this case

$$\text{bandwidth} = 7 \text{ kHz} - 50 \text{ Hz}$$

$$= 6.95 \text{ kHz}$$

$$\approx 7 \text{ kHz}$$

5.5   **System
limitations**

In practice, all physical systems impose restrictions on the signals that can
be used with them. For example, all systems will have limits to the magnitude of the input signals that they can accept and to the output signals that
they can produce.

Systems also have limits to the range of frequencies over which they
will operate, and signal components outside this range may be distorted or
even removed. For example, while typical speech might have a frequency
range from about 50 Hz to 7 kHz, a conventional telephone channel only
transmits frequencies from about 300 Hz to about 3400 Hz. This removes

Figure 5.10 The effects of a restricted frequency range

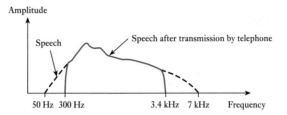

high- and low-frequency components from the signal, distorting its spectrum as shown in Figure 5.10. The result is a signal with reasonable intelligibility but poor overall quality.

The range of frequencies over which a system or component operates is related to its **frequency response**. In the next chapter, we will consider this topic in some detail and will see how this term is defined. For the moment, it is adequate to consider that it simply represents the limits of a system's frequency range. We have already noted that the bandwidth of a signal is given by the difference between the highest and lowest frequencies present in it. We can adopt a similar notation to describe the bandwidth of a communications channel or other system, as the difference between the highest and lowest frequencies over which it will operate. Again, we will look at a more precise definition of this term in the next chapter.

In Figure 5.10, we saw the effects of transmitting a signal over a channel (in this case a telephone line) that has a smaller bandwidth than the signal itself. Problems may also arise when the channel, though of adequate bandwidth, has a frequency response that does not correspond to the frequency range of the signal to be transmitted. This is illustrated in Figure 5.11. Here the channel has sufficient bandwidth to transmit the signal, but the frequency ranges of the signal and channel are different. The problems of mismatched frequency ranges can be tackled through the use of **modulation**, which allows us to change the frequency range of a signal.

Figure 5.11 The effects of an ill-matched frequency response

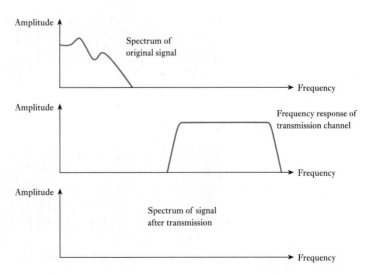

5.6   Modulation

Most readers will be aware of at least two commonly used forms of modulation, since these are used for radio broadcasting. These are **amplitude modulation (AM)** and **frequency modulation (FM)**, which are used to convert signals with frequencies in the normal audio range (from a few tens of hertz to a few kilohertz) into signals with frequencies appropriate for radio transmission (perhaps from hundreds of kilohertz up to hundreds of megahertz).

These techniques are illustrated in Figure 5.12. Here a low-frequency input signal, as shown in Figure 5.12(a), is to be transmitted over some form of communications channel. Modulation is performed using a **carrier signal**, as shown in Figure 5.12(b), of a frequency appropriate to the channel

Figure 5.12 Amplitude and frequency modulation

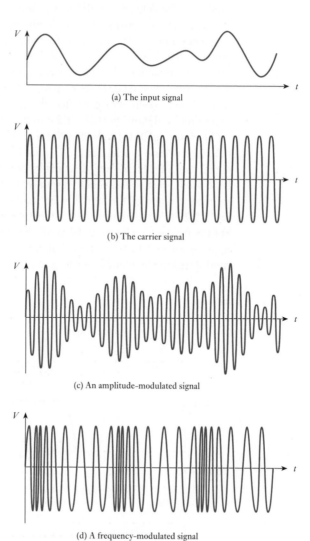

(a) The input signal

(b) The carrier signal

(c) An amplitude-modulated signal

(d) A frequency-modulated signal

to be used. In practice, the carrier would often have a frequency several orders of magnitude greater that of the input signal. In the figure, this has been reduced so that the effects of modulation are more apparent. Amplitude modulation can be performed in many ways, an example being shown in Figure 5.12(c). Here the amplitude of the carrier is varied (or *modulated*) to represent the instantaneous magnitude of the input signal. You will observe that the envelope of the resultant waveform mirrors the shape of the original waveform. Figure 5.12(d) shows an example of frequency modulation, where the amplitude of the carrier is kept constant but its frequency is modulated to represent the input signal.

Amplitude modulation and frequency modulation each have their own characteristics, but both have the effect of shifting the frequency range of the signal. In each case, the carrier wave is being varied in some way, and therefore the resulting modulated signal is no longer a pure sinusoid. It therefore contains components at frequencies other than that of the original carrier. However, in each case the resultant spectrum is centred on the carrier frequency. Therefore, by appropriate choice of this carrier frequency, the input signal can be shifted to any appropriate frequency band. This process is illustrated in Figure 5.13, which shows how the frequency range of a signal might be changed by modulation. Note that in this example there is not a simple shift of the frequency components of the original signal. The spectral content of a modulated signal is dependent on the modulation method used.

Amplitude and frequency modulation can also be used with digital signals, as shown in Figure 5.14. Such techniques are used in a range of applications. For example, such techniques are used by **computer modems** to convert the digital information produced by computers into a form appropriate for transmission over a telephone line. Here the carrier frequency would be chosen to lie within the frequency range of the telephone line, which we have previously noted is from about 300 Hz to 3.4 kHz. When applied to digital signals, these techniques are often given particular

Figure 5.13   Modulation spectra

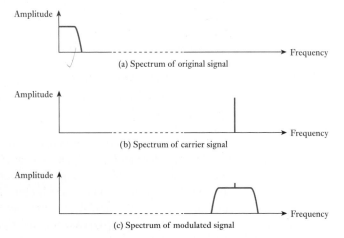

(a) Spectrum of original signal

(b) Spectrum of carrier signal

(c) Spectrum of modulated signal

Figure 5.14 Digital modulation techniques

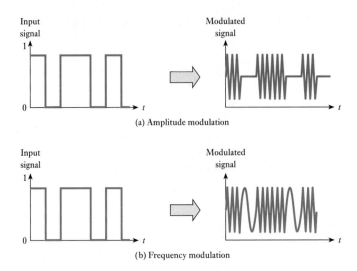

(a) Amplitude modulation

(b) Frequency modulation

names. The technique illustrated in Figure 5.14(a) is known as **amplitude shift keying** (or ASK), while that shown in Figure 5.14(b) is called **frequency shift keying** (or FSK). A third form of modulation, **phase shift keying** (PSK), is also widely used, including in high-speed computer modems. Phase and phase shift will be discussed in Chapter 11.

5.7   Demodulation

The process of modulation would be of little benefit if we could not at some later stage recover our original signal. This is achieved through **demodulation**.

The circuitry required to demodulate a signal will depend on the modulation method that has been used. In some cases, it is necessary for the demodulator to generate a carrier signal similar to that used during modulation and to use this to reverse the frequency-shifting operation to reproduce the original waveform. In other cases a simpler approach is possible. For example, with the waveform of Figure 5.12(c) the original signal is represented by the 'envelope' of the modulate signal. This can be recovered easily using a circuit known as an **envelope detector**. Such circuitry is very simple and is found, for example, in even rudimentary radio receivers such as 'crystal sets'. We will look at a circuit for such a demodulator in Chapter 19.

The basic principles of modulation and demodulation are shown in Figure 5.15. Here an input signal (which could be analogue or digital) is first modulated to make it compatible with a particular communications channel. This channel could be a radio link, a fibre-optic cable or any other form of channel. After modulation, the resultant signal is transmitted and is then demodulated to recover the original signal.

Figure 5.15 Modulation and demodulation

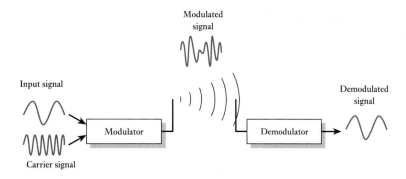

5.8   Multiplexing

The ability to shift the frequency range of a signal through modulation also allows us to make more effective use of the bandwidth of a communication channel by combining several signals together. This process is known as **multiplexing** and is illustrated in Figure 5.16. Here four independent signals are combined to form a single multiplexed signal that can be transmitted down a single communications channel. The signals are combined in this way by using a different carrier frequency when modulating each signal. In this way, a very large number of signals can be transmitted simultaneously on a single pair of copper wires, a single optical fibre, or a single radio channel. At the receiving end, the signals are separated by the reverse process of **demultiplexing** the transmitted signal.

The process described here is often known as **frequency-division multiplexing**, since the frequency range of the channel is divided into a number of separate sub-channels that can each carry an individual signal. All readers will be familiar with the use of frequency-division multiplexing

Figure 5.16 Frequency multiplexing

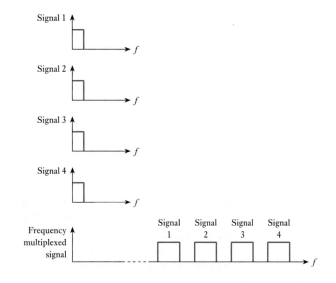

in the transmission of radio signals. Here the wide frequency spectrum available to radio signals is used to transmit a large number of separate radio stations at different frequencies. Here the demultiplexing is performed by the radio receiver, which is tuned to select one station from the many available. Frequency-division multiplexing is also widely used to transmit multiple signals over other forms of communication channel.

Multiplexing can also be performed by dividing the *time* available on a channel between a number of signals. This is known as **time-division multiplexing** and is also widely used, particularly for the transmission of digital signals.

<table>
<tr><td>**5.9**</td><td>**Distortion and noise**</td></tr>
</table>

No electrical or electronic circuits are perfect, and one of the ways in which circuit imperfections manifest themselves is in the **distortion** of signals that pass through them. Distortion can take many forms, and Figure 5.17 shows some examples of different types of distortion, as they might effect a sinusoidal signal.

Figure 5.17 Examples of the effects of distortion on a sine wave

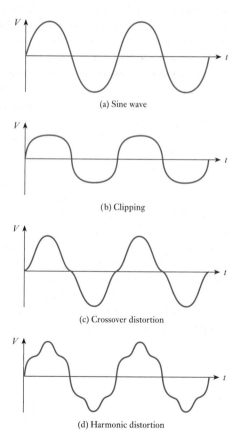

(a) Sine wave

(b) Clipping

(c) Crossover distortion

(d) Harmonic distortion

Figure 5.17(a) shows a sinusoidal signal that represents the input to a circuit. Figures 5.17(b) and 5.17(c) show the effects of different forms of non-linearity in the circuit, the first resulting in **clipping** of the waveform and the second resulting in **crossover distortion**. Figure 5.17(d) shows the effects of **harmonic distortion**, where a circuit adds additional components to a signal at frequencies that are multiples of the signal frequency. In practice, all circuits alter the signals that pass through them in a range of ways, including elements of each of these and other forms of distortion.

The importance of distortion will depend very much on the application. In an audio amplifier, large amounts of distortion would be audible and would degrade the quality of the sound produced. In a temperature-measuring system, distortion might result in a reduction in the accuracy of the measurements. Although some distortion is always present, its magnitude will depend on the circuit techniques used. Good design must ensure that the distortion is at an acceptable level for a given application.

Signals are also affected by **noise**. This is a random fluctuation of a signal that is produced either by variations in the system or by external effects of the environment. It has a number of causes and is always present in electronic systems.

Since the amount of noise produced in a system is not related to the input signal, its effects will be most marked when the input signal is small. This effect is readily experienced when listening to a radio or tape player. Noise, experienced as a background 'hiss', is much more apparent during quiet, rather than loud, passages.

Figure 5.18 illustrates the effects of noise on both analogue and digital signals. It can be seen that the addition of noise to the signals changes the waveforms in both cases. However, in the digital case it is still apparent which parts of the signal represent the higher voltage and which the lower, and it is therefore possible to extract the original information from the signal. This is not the case with the analogue signal, where separating the noise from the original signal is not usually possible. This illustrates an important distinction between analogue and digital signals, which will be developed further in later chapters.

An important distinction between distortion and noise is that distortion is *systematic* and is repeatable, while noise is *random* and is therefore not repeatable. If a signal is passed repeatedly through a particular circuit, then the distortion produced on each occasion will be the same, while the noise added to it will be different each time.

In electronic circuits, the presence of distortion and noise often limits the ultimate performance of a system. One of the major tasks of electronic design is to reduce the magnitude of these effects.

Figure 5.18 The effects of noise on analogue and digital signals

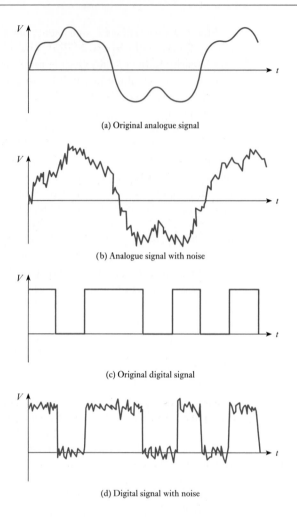

(a) Original analogue signal

(b) Analogue signal with noise

(c) Original digital signal

(d) Digital signal with noise

Key points

- It is often convenient to represent physical quantities by electrical signals, since this can simplify processing, transmission and storage.

- Electrical signals can take many forms and can be either analogue or digital.

- Both analogue and digital signals have been used since the early nineteenth century, and both play a vital role in modern electronics.

- In many cases, analogue signals take the form of a varying voltage, where the magnitude of the voltage is directly proportional to some physical quantity.

- Digital signals often take the form of a varying voltage, where the magnitude of the voltage alternates between two specific voltages to represent the 1's and 0's of a binary quantity.

- In some cases, digital information may be combined to form digital words, which represent several separate bits of information. These words may be transmitted or processed in parallel or in series.

- Signals may be unipolar or bipolar in nature. The frequencies present in the signal are described by the signal's frequency spectrum.

- All physical systems impose restrictions on the amplitude and frequency range of the signals that can be used with them.

- Modulation can be used to alter the characteristics of a signal to make it compatible with a particular transmission channel or system.

- Multiplexing, and the reverse process of demultiplexing, allows us to combine several signals on to a single communications channel to make more efficient use of the channel's capacity.

- All electrical circuits add distortion and noise to the signals that pass through them.

Exercises

5.1 Why is it often attractive to represent physical quantities by electrical signals, rather than by some other means?

5.2 Give an example of the use of analogue electrical signals in the nineteenth century.

5.3 Give an example of a sensor that produces a voltage that is directly related to the physical quantity being sensed.

5.4 Give an example of an actuator where the output physical quantity is directly related to the voltage of the signal applied to it.

5.5 Describe a common signal format for representing binary quantities.

5.6 Give an example of the use of digital electrical signals in the nineteenth century.

5.7 How many different values can be represented by a single bit of information?

5.8 How many different values can be represented by a single 12-bit digital word?

5.9 What would be the resolution of a quantity represented by a 12-bit word?

5.10 A system requires that a temperature be measured and stored to an accuracy of better than 1 percent. What is the minimum word length required to achieve this accuracy?

5.11 What are the advantages and disadvantages of transmitting information in a serial format rather than a parallel format?

5.12 Define the terms 'unipolar' and 'bipolar' as they apply to electrical signals.

5.13 What characteristics of a signal are described by its frequency spectrum?

5.14 Sketch the frequency spectrum of a signal that is formed by the addition of a sinusoidal signal with a magnitude of 3 V and a frequency of 100 Hz, and a sinusoidal signal with a magnitude of 2 V with a frequency of 50 Hz.

5.15 Explain, in general terms, what is meant by the frequency range and the bandwidth of a signal.

Exercises continued

5.16 A signal has a frequency range that extends from 1 kHz to 10 kHz. What is the bandwidth of this signal?

5.17 What is meant by the frequency response of a system?

5.18 Why is it important to match the frequency range of a system to the range of the signals with which it is to be used?

5.19 Explain what is meant by the term 'modulation'.

5.20 Modulation has the effect of 'shifting' the frequency range of a signal. For what reason might one wish to perform this operation?

5.21 Explain briefly the relationship between an amplitude-modulated signal and the waveform that it represents.

5.22 Explain briefly the relationship between a frequency-modulated signal and the waveform that it represents.

5.23 Explain the role of a carrier signal in amplitude modulation.

5.24 How is the frequency of the carrier signal selected?

5.25 Explain what is meant by ASK, FSK and PSK.

5.26 What is meant by the term 'envelope detection'?

5.27 How does multiplexing allow more effective use of channel bandwidth?

5.28 Explain the difference between distortion and noise.

5.29 How do the characteristics of analogue and digital techniques differ with respect to noise?

Chapter 6

Amplification

Objectives

When you have studied the material in this chapter you should be able to:

- **explain the concept of amplification;**
- **give examples of both active and passive amplifiers;**
- **use simple 'equivalent circuits' to determine the gain of an amplifier;**
- **discuss the effects of 'input resistance' and 'output resistance' on the voltage gain of an amplifier and use these quantities to calculate loading effects;**
- **define terms such as output power, power gain, voltage gain and frequency response;**
- **determine the effects of connecting several amplifiers in series;**
- **describe several common forms of amplifier, including differential and operational amplifiers.**

6.1 Introduction

In earlier chapters, we noted that many electrical and electronic systems comprise one or more *sensors*, which take information from the 'real world'; one or more *actuators*, which allow the system to output information to the real world; and some form of *processing*, which makes signals associated with the former appropriate for use with the latter. Although the form of the processing that is required varies greatly from one application to another, one element that is often required is *amplification*.

Simplistically, **amplification** means making things bigger, and the converse operation, **attenuation**, means making things smaller. These basic operations are fundamental to many systems, including both electrical and non-electrical applications.

Examples of non-electronic amplification are shown in Figure 6.1. The first shows a lever arrangement. Here the force applied at the output is greater than that applied at the input, and we have therefore amplified the input force. However, the distance moved by the output is less that that

Figure 6.1 Examples of mechanical amplifiers

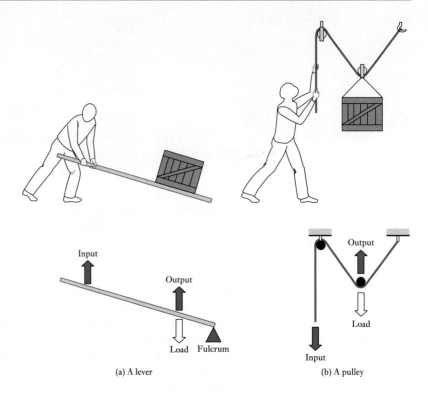

(a) A lever　　　　　　　　　　　　　　　　(b) A pulley

moved at the input. Thus, although the force has been amplified, the displacement has been reduced or attenuated. Note that, if the positions of the input and the output of the lever were reversed, we would produce an arrangement that amplified movement but attenuated force.

The second example in Figure 6.1 shows a pulley arrangement. As in the first example this produces a greater force at the output than is applied at the input, but the distance moved at the output is less than that at the input. Thus again we have a force amplifier but a movement attenuator.

In the lever arrangement shown in Figure 6.1(a), the direction of the output force is the same as that of the input. Such an amplifier is referred to as a **non-inverting amplifier**. In the pulley arrangement of Figure 6.1(b), a downward force at the input results in an upward force at the output and we have what is termed an **inverting amplifier**. Different arrangements of levers and pulleys may be either inverting or non-inverting.

Both the examples of Figure 6.1 are **passive systems**. That is, they have no external energy source other than the inputs. For such systems, the **output power** (that is, the power delivered at the output) can never be greater than the **input power** (that is, the power absorbed by the input) and in general it will be less because of losses. In our examples, losses would be caused by friction at the fulcrum and at the pulleys.

In order to be able to provide power gain, some amplifiers are not passive but **active**. This means that they have some form of external energy source that can be harnessed to produce an output that has more power than

Figure 6.2 A torque amplifier

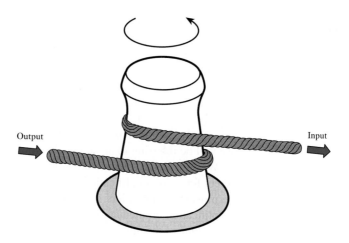

Output   Input

the input. Figure 6.2 shows an example of such an amplifier, called a **torque amplifier**. This consists of a rotating shaft with a rope or cable wound around it. The amplifier can be used as a power winch and is often found in boats and ships. One end of the rope (the output) is attached to a load, and a control force is applied to the other end (the input). If no force is applied to the input, the rope will hang loosely around the rotating shaft and little force will be applied at the output. The application of a force to the input tightens the rope around the shaft and increases the friction between them. This frictional force is applied to the rope and results in a force being exerted at the output. The greater the force applied to the input, the greater the frictional force experienced by the rope and the greater the force exerted at the output. We therefore have an amplifier where a small force applied at the input generates a larger force at the output. The magnitude of the amplification may be increased or decreased by changing the number of turns of the rope around the drum.

It should be noted that since the rope is continuous the distance moved by the load at the output is equal to the distance moved by the rope at the input. However, the force applied at the output is greater than that at the input, and the arrangement therefore delivers more power at the output than it absorbs at the input. It therefore provides not only force amplification but also power amplification. The extra power available at the output is supplied by the rotating shaft and will result in an increased drag being experienced by whatever force is causing it to rotate.

6.2   **Electronic amplifiers**

In electronics there are also examples of both passive and active amplifiers. Examples of the former include a step-up transformer, where an alternating voltage signal applied to the input will generate a larger voltage signal at the output. Although the voltage at the output is increased, the ability of the output to provide current to an external load is reduced. The power supplied

module that does not attempt to describe the internal construction of the unit but simply to model its external characteristics. Since an amplifier, or any other module, can be characterised by the nature of its inputs, its outputs and the relationship between them, we clearly need some way of describing these module attributes.

6.3.1 Modelling the input of an amplifier

To model the characteristics of the input of an amplifier, we need to be able to describe the way in which it appears to circuits that are connected to it. In other words, we need to model how it appears when it represents the *load* of another circuit.

Fortunately, in most cases the input circuitry of an amplifier can be modelled adequately by a single fixed resistance, which is termed its **input resistance** and which is given the symbol R_i. What this means is that, when an external voltage source applies an input voltage to the amplifier, the current that flows into the amplifier is related to the voltage as if the input were a simple resistance of value R_i. Therefore, an adequate model of the input of an amplifier is that shown in Figure 6.5. Such a model is termed the equivalent circuit of the input.

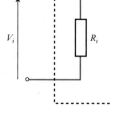

Figure 6.5 An equivalent circuit of the input of an amplifier

6.3.2 Modelling the output of an amplifier

To model the characteristics of the output of an amplifier, we need to be able to describe the way in which it appears to circuits that are connected to it. In other words, we need to model how it appears when it represents a *source* to another circuit.

If one takes a voltage source, such as a battery, and connects it across a resistance R, the current that flows I is related to the battery voltage V by Ohm's law:

$$I = \frac{V}{R}$$

If we connect the terminals of the source together, $R = 0$, so the current should be infinite. Of course, in practice the current is not infinite, because any real voltage source has some resistance associated with it. We can represent a real voltage source by an ideal **voltage generator** (that is, a voltage generator that has no internal resistance) in series with a resistance. This arrangement is shown in Figure 6.6(a). An ideal voltage generator is normally represented by the circular symbol shown in Figure 6.6(b). This form of representation is an example of a **Thévenin equivalent circuit**. We will return to look in detail at the use of such circuits in Chapter 12.

Since we now have an equivalent circuit of a voltage source, we can use this to model the output of our amplifier. This is shown in Figure 6.7. Here the voltage to be produced by the amplifier is represented by the voltage V,

Figure 6.6 An equivalent circuit of a voltage source

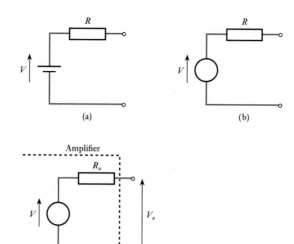

Figure 6.7 An equivalent circuit of the output of an amplifier

and R_o represents the resistance associated with the output circuitry; this is termed the **output resistance** of the circuit. The actual voltage that is produced at the output of the circuit V_o is equal to V minus the voltage drop across R_o. The magnitude of this voltage drop is given by Ohm's law and is equal to the product of R_o and the current flowing though this resistance.

6.3.3   Modelling the gain of an amplifier

Having successfully modelled the input and the output of our amplifier, we now need to turn our attention to the relationship between them. This is clearly the *gain* of the amplifier.

The gain of an amplifier can be modelled through the use of a **controlled voltage source**: that is, a source in which the voltage is controlled by some other quantity in the circuit. Such sources are also known as **dependent voltage sources**. In this case, the controlled voltage source produces a voltage that is equal to the input voltage of the amplifier multiplied by the circuit's gain. This is illustrated in Figure 6.8, where a voltage source produces a voltage equal to A_vV_i.

It should be noted that we are currently considering an equivalent circuit of an amplifier and *not* its physical implementation. You should not concern yourself with how a controlled voltage source operates, since it is simply a convenient way of modelling the operation of the circuit.

Figure 6.8 Representing gain in an equivalent circuit

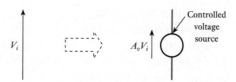

Figure 6.9 An equivalent circuit of an amplifier

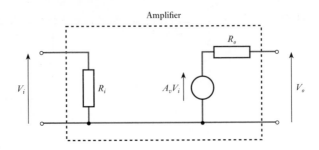

6.4 Equivalent circuit of an amplifier

Since we now have equivalent circuits for the input, the output and the gain of a module, we are in a position to draw an equivalent circuit for the amplifier of Figure 6.3. This is shown in Figure 6.9. The **input voltage** of the circuit V_i is applied across the input resistance R_i, which models the relationship between the input voltage and the corresponding input current. The voltage gain of the circuit A_v is represented by the controlled voltage source, which produces a voltage of $A_v V_i$. The way in which the output voltage of the circuit changes with the current taken from the circuit is modelled by placing a resistor in series with the voltage source. This is the output resistance R_o of the circuit. You will notice that the equivalent circuit has two input terminals and two output terminals as in the amplifier shown in Figure 6.3. The lower terminal in each case forms a reference point, since all voltages must be measured with respect to some reference voltage. In this circuit, the input and the output references are joined together. This common reference point is often joined to the earth or **chassis** of the system (as in Figure 6.3), and its potential is then taken as the 0 V reference.

Note that the equivalent circuit does not show any connection to a power source. The voltage source in the equivalent circuit would in practice take power from an external power supply of some form, but in this circuit we are only interested in its functional properties, not its physical implementation.

The advantage of an equivalent circuit is that it allows us easily to calculate the effects that external circuits (which also have input and output resistance) will have on the amplifier. This is illustrated in the following example.

Example 6.1

An amplifier has a voltage gain of 10, an input resistance of 1 kΩ and an output resistance of 10 Ω. The amplifier is connected to a sensor that produces a voltage of 2 V and has an output resistance of 100 Ω, and to a load resistance of 50 Ω. What will be the output voltage of the amplifier (that is, the voltage across the load resistance)?

First we draw an equivalent circuit of the amplifier, the sensor and the load.

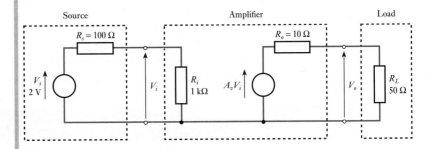

The input voltage is applied across a potential divider formed by R_s and R_i. From Chapter 2, we know how to calculate the resulting voltage, V_i, which is given by

$$V_i = \frac{R_i}{R_s + R_i} V_s$$

$$= \frac{1\,k\Omega}{100\,\Omega + 1\,k\Omega} 2\,V$$

$$= 1.82\,V$$

This voltage is amplified by the voltage source, and the resultant voltage is applied across a second potential divider formed by R_o and R_L, therefore the output voltage is given by

$$V_o = A_v V_i \frac{R_L}{R_o + R_L}$$

$$= 10 V_i \frac{50\,\Omega}{10\,\Omega + 50\,\Omega}$$

$$= 10 \times 1.82 \frac{50\,\Omega}{10\,\Omega + 50\,\Omega}$$

$$= 15.2\,V$$

As defined in Equation 6.1, the voltage gain of an amplifier is the ratio of the output voltage to the input voltage.

Example 6.2   **Calculate the voltage gain of the circuit of Example 6.1.**

$$\text{voltage gain } (A_v) = \frac{V_o}{V_i} = \frac{15.2}{1.82} = 8.35$$

Examples 6.1 and 6.2 show that, when an amplifier is connected to a source and a load, the resulting output voltage may be considerably less than one might have expected given the source voltage and the voltage gain of the amplifier in isolation. The input voltage to the amplifier is less than the source voltage due to the effects of the source resistance. Similarly, the output voltage of the circuit is affected by the load resistance as well as the characteristics of the amplifier. This effect is known as **loading**. The *lower* the resistance that is applied across the output of a circuit the more *heavily* it is loaded, since more current is drawn from it.

File 06A

Computer Simulation Exercise 6.1

Simulate the circuit of Example 6.1 and experiment with different values for the various resistors, noting the effect on the voltage gain of the circuit. How should the resistors be chosen to maximise the voltage gain? In other words, under what conditions are the loading effects at a minimum?

From the above example, it is clear that the gain produced by an amplifier is greatly affected by circuitry that is connected to it. For this reason, we should perhaps refer to the gain of the amplifier in isolation as the **unloaded amplifier gain**, since it is the ratio of the output voltage to the input voltage in the absence of any loading effects. When we connect a source and a load to the amplifier we produce **potential dividers** at the input and output, which reduce the effective gain of the circuit. You will remember that we noted the importance of potential dividers in Chapter 2 when we discussed their behaviour.

Earlier in this chapter, we noted that an ideal voltage amplifier would not affect the circuit to which it was connected and would give an output that was independent of the load. This implies that it would draw no current from the source and would be unaffected by loading at the output. Consideration of the analysis of Example 6.1 shows that this requires the input resistance R_i to be infinite and the output resistance R_o to be zero. Under these circumstances, the effects of the two potential dividers are removed and the voltage gain of the complete circuit becomes equal to the voltage gain of the unloaded amplifier, irrespective of the source and load resistance. This is illustrated in Example 6.3.

Example 6.3

An amplifier has a voltage gain of 10, an infinite input resistance and zero output resistance. The amplifier is connected, as in Example 6.1, to a sensor that produces a voltage of 2 V and has an output resistance of 100 Ω, and to a load resistance of 50 Ω. What will be the output voltage of the amplifier?

As before, we first draw an equivalent circuit of the amplifier, the sensor and the load.

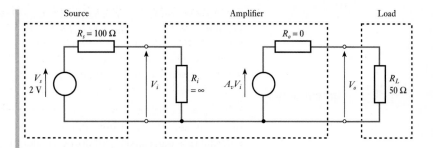

From the diagram

$$V_i = \frac{R_i}{R_s + R_i} V_s$$

When R_i is much larger than R_s, this approximates to

$$V_i = \frac{R_i}{R_s + R_i} V_s$$

$$\approx \frac{R_i}{R_i} V_s$$

In this case, R_i is infinite so we can say

$$V_i = \frac{R_i}{R_i} V_s$$

$$= V_s$$

$$= 2 \text{ V}$$

and therefore the output voltage is given by

$$V_o = A_v V_i \frac{R_L}{R_o + R_L}$$

$$= 10 \, V_i \frac{50 \text{ }\Omega}{0 + 50 \text{ }\Omega}$$

$$= 10 \, V_i$$

$$= 10 \times 2$$

$$= 20 \text{ V}$$

The arrangement now has a voltage gain of 10, and there are no loading effects.

No real amplifier can have an infinite input resistance or zero output resistance. However, if the input resistance is *large* compared with the source resistance and the output resistance is *small* compared with the load resistance, the effects of these resistances will be small and may often be neglected. This will produce the maximum voltage gain from the circuit.

For these reasons, a good voltage amplifier is characterised by a high input resistance and a low output resistance.

6.5   Output power

In the last section, we looked at the performance of an amplifier in terms of its voltage gain and how this performance is affected by internal and external resistances. We will now consider the performance of the amplifier in terms of the **power** that it can deliver to an external load.

The power dissipated in the load resistor (the **output power** P_o) of a circuit is simply

$$P_o = \frac{V_o^2}{R_L} \tag{6.4}$$

Example 6.4

Calculate the output power of the circuit of Example 6.1.

From the earlier analysis, the output voltage V_o of the circuit is 15.2 V, and the load resistance is 50 Ω. Therefore, the output power is

$$P_o = \frac{V_o^2}{R_L} = \frac{(15.2)^2}{50} = 4.6 \text{ W}$$

In the previous section, we noted that, unless the output resistance of an amplifier is zero, the voltage gain achieved is affected by the ratio of the load resistance to the output resistance. Therefore, for a given amplifier (with a particular value of output resistance) and a given input voltage, the output voltage will vary with the resistance of the load connected to it. Since the output power is related to the output voltage, it is clear that the output power will also vary with the load resistance. These two dependencies are illustrated in Table 6.1, which shows the output voltage and output power produced by the circuit of Example 6.1 for different values of load resistance.

You will notice that the output voltage increases steadily as the resistance of the load is increased. This is because the output rises as the amplifier is less heavily loaded.

Table 6.1 Variation of output voltage and output power with load resistance in the circuit of Example 6.1

Load resistance, $R_L(\Omega)$	Output voltage, $V_o(V)$	Output power, $P_o(W)$
1	1.65	2.7
2	3.03	4.6
3	4.20	5.9
10	9.10	8.3
33	14.0	5.9
50	15.2	4.6
100	16.5	2.7

However, the power output initially rises as the resistance of the load is increased from $1\ \Omega$ until a maximum is reached. It then drops as the load is increased further. To investigate this effect, we need to look at the expression for the output voltage of the circuit. From the analysis of Example 6.1, we know that this is

$$V_o = A_v V_i \frac{R_L}{R_o + R_L}$$

Since the power dissipated in a resistance is given by V^2/R, the power dissipated in the load resistance (the output power P_o) is given by

$$P_o = \frac{V_o^2}{R_L} = \frac{\left(A_v V_i \dfrac{R_L}{R_L + R_o}\right)^2}{R_L} = A_v^2 V_i^2 \frac{R_L}{(R_L + R_o)^2}$$

By differentiating this expression for P_o with respect to R_L it is easy to show that this expression has a maximum value when

$$R_L = R_o$$

This is left as an exercise for the reader.

Substituting for the component values used in Example 6.1 shows that maximum power should be dissipated in the load when its resistance is equal to R_o, which is $10\ \Omega$. The tabulated data in Table 6.1 confirms this result.

Thus in circuits in which the output characteristics can be adequately represented by a simple resistance, maximum power is transferred to the load when the load resistance is equal to the output resistance. This result holds for transfers between any two circuits and is a simplified statement of the **maximum power theorem**. We will return to this theorem when we look at power transfer in more detail in Chapter 16.

The process of choosing a load to maximise **power transfer** is called **matching** and is a very important aspect of circuit design in certain areas. However, it should be remembered that, since maximum power transfer occurs when the load and output resistances are equal, the voltage gain is far from its maximum value under these conditions. In voltage amplifiers, it is more common to attempt to maximise input resistance and minimise output resistance to maximise voltage gain.

File 06B

Computer Simulation Exercise 6.2

Use the circuit of Computer Simulation Exercise 6.1 to investigate the way in which the power output of the circuit is affected by the value of the load resistance R_L.

Use the simulator's sweep facility to determine the value of R_L that results in the maximum power output for a given value of R_o. Repeat this for a range of values of R_o.

6.6   Power gain

The power gain of an amplifier is the ratio of the power supplied by the amplifier to a load, to the power absorbed by the amplifier from its source. The input power can be calculated from the input voltage and the input current or, by applying Ohm's law, from a knowledge of the input resistance and either the input voltage or current. Similarly, the output power can be determined from the output voltage and output current, or from one of these and a knowledge of the load resistance.

Example 6.5

Calculate the power gain of the circuit of Example 6.1.

To determine the power gain, first we need to calculate the input power P_i and the output power P_o.

From the example, we have that $V_i = 1.82$ V and $R_i = 1$ kΩ. Therefore

$$P_i = \frac{V_i^2}{R_i} = \frac{(1.82)^2}{1000} = 3.3 \text{ mW}$$

We also know that $V_o = 15.2$ V and $R_L = 50$ Ω. Therefore

$$P_o = \frac{V_o^2}{R_L} = \frac{(15.2)^2}{50} = 4.62 \text{ W}$$

The power gain is then given by

$$\text{power gain } (A_p) = \frac{P_o}{P_i} = \frac{4.62}{0.0033} = 1400$$

Note that when calculating the input power we use R_i but when calculating the output power we use R_L (*not* R_o). We are calculating the power delivered to the load, not the power dissipated in the output resistance.

You will note from Example 6.5 that even a circuit with a relatively low voltage gain (in this case it is $15.2/1.82 = 8.35$) can have a relatively high power gain (in this case, over a thousand). The power gain of a modern electronic amplifier may be very high, gains of 10^6 or 10^7 being common. With these large numbers, it is often convenient to use a logarithmic expression of gain rather than a simple ratio. This can be done using **decibels**.

The decibel (dB) is a dimensionless figure for power gain and is defined by

$$\text{power gain (dB)} = 10 \log_{10} \frac{P_2}{P_1} \tag{6.5}$$

where P_2 is the output power and P_1 is the input power of the amplifier or other circuit.

Example 6.6

Express a power gain of 1400 in decibels.

$$\text{power gain (dB)} = 10 \log_{10} \frac{P_2}{P_1}$$

$$= 10 \log_{10}(1400)$$

$$= 10 \times 3.15$$

$$= 31.5 \text{ dB}$$

Decibels may be used to represent both amplification and attenuation, and, in addition to making large numbers more manageable, the use of decibels has several other advantages. For example, when several stages of amplification or attenuation are connected in series (this is often referred to as **cascading** circuits), the overall gain of the combination can be found simply by adding the individual gains of each stage when these are expressed in decibels. This is illustrated in Figure 6.10. The use of decibels also simplifies the description of the frequency response of circuits, and we will return to look at this further in Chapter 17 when we will study the frequency characteristics of circuits in some detail.

For certain values of gain, the decibel equivalents are easy to remember or to calculate using mental arithmetic. Since $\log_{10} n$ is simply the power to which 10 must be raised to equal n, for powers of ten it is easy to calculate. For example, $\log_{10} 10 = 1$, $\log_{10} 100 = 2$, $\log_{10} 1000 = 3$, and so on. Similarly, $\log_{10} 1/10 = -1$, $\log_{10} 1/100 = -2$ and $\log_{10} 1/1000 = -3$. Therefore, gains of 10, 100 and 1000 are simply 10 dB, 20 dB and 30 dB respectively, and attenuations of 1/10, 1/100 and 1/1000 are simply -10 dB, -20 dB and -30 dB. A circuit that doubles the power has a gain of $+3$ dB, while a circuit that halves the power has a gain of -3 dB. A circuit that leaves the power unchanged (a power gain of 1) has a gain of 0 dB. These results are summarised in Table 6.2.

Figure 6.10 Calculating the gain of several stages in series

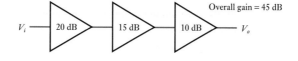

Table 6.2 Expressing power amplification and attenuation in decibels

Power gain (ratio)	Decibels (dB)
1000	30
100	20
10	10
2	3
1	0
0.5	−3
0.1	−10
0.01	−20
0.001	−30

6.7

Frequency response and bandwidth

We noted in Chapter 5 that all real systems have limitations to the range of frequencies over which they will operate, and consequently amplifiers are subject to such restrictions. In some cases, it is also *desirable* to limit the range of frequencies that are amplified by a circuit. In general, an amplifier will be required to deliver a particular amount of gain over a particular range of frequencies. The gain of the circuit within this normal operating range is termed its **mid-band gain**.

The range of frequencies over which an amplifier operates is determined by its **frequency response**, which describes how the gain of the amplifier changes with frequency. The gain of all amplifiers falls at high frequencies, although the frequency at which this effect becomes apparent will vary depending on the circuits and components used. This fall in gain at high frequencies can be caused by a number of effects, including the presence of **stray capacitance** within the circuit. We will look at the nature of these effects in Chapter 13 when we look at capacitance in detail. In some amplifiers the gain also falls at low frequencies.

In order to quantify these high- and low-frequency effects, we can define the frequencies at which the gain begins to fall. However, since this fall occurs progressively we need to define a point at which we will consider the gain to have changed appreciably. The point chosen is what is called the **half-power point**, where the power gain of the circuit has fallen to one-half of its mid-band value. From our earlier discussions, we know that this corresponds to a fall in gain of 3 dB relative to its mid-band value. Since power is proportional to the square of the voltage, a fall in power gain of one-half corresponds to a fall in voltage gain to $1/\sqrt{2}$, or about 0.707 of its mid-band value.

The frequency corresponding to the half-power point is called the **cut-off frequency** of the circuit. All circuits will have an **upper cut-off frequency**, while some will also have a **lower cut-off frequency**. This is illustrated in Figure 6.11. The diagram of Figure 6.11(a) shows the frequency response of an amplifier that has both upper and lower cut-off frequencies. In this graph the gain of the amplifier is plotted against frequency. Figure 6.11(b) shows the response of a similar amplifier, but this time the gain is plotted in decibels. Figure 6.11(c) shows the response of an amplifier that has no lower cut-off. In this last circuit, the gain of the circuit is constant at low frequencies.

The width of the frequency range of a circuit is termed its bandwidth. In circuits with both an upper and a lower cut-off frequency, this is simply the difference between these two frequencies. In circuits that have only an upper cut-off frequency, the bandwidth is the difference between this and 0 Hz and is therefore numerically equal to the upper cut-off frequency. This is shown in Figure 6.12.

As was noted in the last chapter, limitations on the frequency range of an amplifier can cause distortion of the signal being amplified. For this reason, it is important that the bandwidth and the frequency range of the amplifier are appropriate for the signals being used.

Figure 6.11 Examples of amplifier frequency response

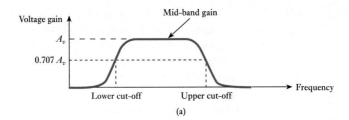

(a)

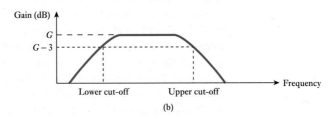

(b)

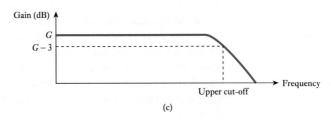

(c)

Figure 6.12 The bandwidth of an amplifier

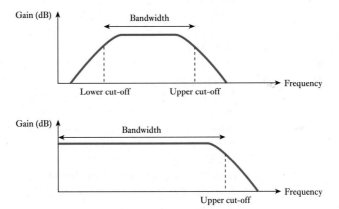

6.8 Differential amplifiers

The amplifiers considered so far take as their input a single voltage that is measured with respect to some reference voltage, which is usually the 'ground' or 'earth' reference point of the circuit (0 V). An amplifier of this form is shown in Figure 6.3. Some amplifiers have not one but two inputs and produce an output proportional to the difference between the voltages on these inputs. Such amplifiers are called **differential amplifiers**, an example of which is shown in Figure 6.13.

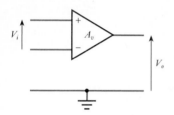

Figure 6.13 A differential amplifier

Since a differential amplifier takes as its input the difference between two input voltages, it effectively subtracts the voltage on one input from that on the other to form its input signal. Consequently, we need to differentiate between the two input terminals, and for this reason the two inputs are labelled '+' and '−'. The former is called the **non-inverting input** because a positive voltage on this input with respect to the other input will cause the output to become positive. The latter is called the **inverting input** because a positive voltage on this input with respect to the other input will cause the output to become negative.

Since a differential amplifier produces an output that is proportional to the difference between two input signals, it is clear that, if the same voltage is applied to both inputs, no output will be produced. Voltages that are common to both inputs are called **common-mode signals**, while voltage differences between the two inputs are termed **differential-mode signals**. A differential amplifier is designed to amplify differential-mode signals while ignoring (or rejecting) common-mode signals.

The ability to reject common-mode signals while amplifying differential-mode signals is often extremely useful. An example that illustrates this is the transmission of signals over great distances. Consider the situations illustrated in Figure 6.14. Figure 6.14(a) shows a sensor connected to a single-input amplifier by a long cable. Any long cable is influenced by **electromagnetic interference** (EMI), and inevitably some noise will be added to the signal from the sensor. This noise will be amplified along with the

Figure 6.14 Comparison of single and differential input methods

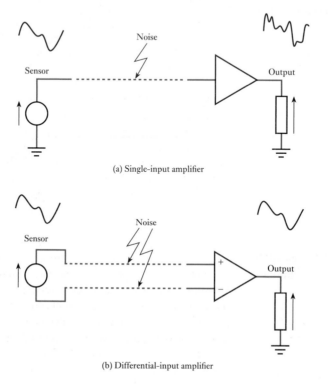

(a) Single-input amplifier

(b) Differential-input amplifier

Figure 6.15 An equivalent circuit for a differential amplifier

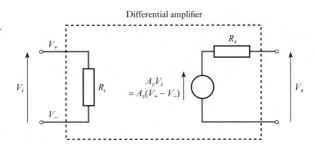

wanted signal and will therefore appear at the output along with the wanted signal. Figure 6.14(b) shows a similar sensor connected by a twin-conductor cable to the inputs of a differential amplifier. Again the cable will be affected by noise, but in this case, because of the close proximity of the two cables (which are kept as close as possible to each other), the noise picked up by each cable will be almost identical. Therefore, at the amplifier this noise appears as a common-mode signal and is ignored, while the signal from the sensor is a differential-mode signal and is amplified.

In Figure 6.9, we looked at an equivalent circuit for a single-input amplifier and we then went on to look at the uses of such an equivalent circuit in determining loading effects. We can also construct an equivalent circuit for a differential amplifier, and this is shown in Figure 6.15. You will note that it is very similar in form to that of Figure 6.9 but that the input voltage V_i is now defined as the difference between two input voltages V_+ and V_-, and that there is no connection between this second input and ground.

A common form of differential amplifier is the **operational amplifier** (or **op-amp**). These are constructed by fabricating all the necessary components on a single chip of silicon to form a monolithic **integrated circuit** or **IC**.

In Section 6.4, when looking at voltage amplifiers, we concluded that a good voltage amplifier is characterised by a high input resistance and a low output resistance. Operational amplifiers have a very high input resistance (of perhaps several megohms or even gigohms) and a low output resistance (of a few ohms or tens of ohms). They also have a very high gain, of perhaps 10^5 or 10^6. For these reasons, operational amplifiers are extremely useful 'building blocks' for constructing not only amplifiers but also a wide range of other circuits. We will therefore return to look at op-amps in more detail in Chapter 8.

Unfortunately, despite their many attractive characteristics, op-amps suffer from **variability**. That is to say, their attributes (such as gain and input resistance) tend to vary from one device to another and may also change for a particular device with variations in temperature or over time. We therefore need techniques for overcoming this variability and for tailoring the characteristics of the devices to match the requirements of particular applications. Techniques for achieving these goals are discussed in Chapter 7.

6.9   **Simple amplifiers**

Operational amplifiers are fairly complex circuits that contain a number of semiconductor devices. Amplifiers may also be formed using single transistors or other active devices. Before we look at the operation of transistors, it is perhaps worth seeing how a single control device may be used to form an amplifier.

Consider the circuit of Figure 6.16. This shows a pair of resistors arranged as a potential divider. The output voltage V_o is related to the circuit parameters by the expression

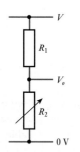

Figure 6.16   A simple potential divider

$$V_o = \frac{R_2}{R_1 + R_2} V$$

If the variable resistance R_2 is adjusted to equal R_1, the output voltage V_o will clearly be half the supply voltage V. If R_2 is reduced, V_o will also be reduced. If R_2 is increased, V_o will increase. If we replace R_2 by some, as yet undefined, control device that has an input voltage V_i that controls its resistance, varying V_i will vary the output voltage V_o. Figure 6.17 shows such an arrangement.

In fact, the control device does not have to be a voltage-controlled resistance. Any device in which the current is determined by a control input may be used in such an arrangement, and if the gain of the control device is suitable it can be used as an amplifier.

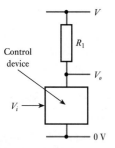

Figure 6.17   The use of a control device

Simple amplifiers of this type are used extensively in all forms of electronic circuit, and we will be returning to them when we have considered the operation of some active components that can be used as control devices in such arrangements. Unfortunately, as with operational amplifiers, these control devices suffer from variability, and again we need a method of overcoming this problem. This is the topic of the next chapter.

Key points

■ Amplification is a fundamental part of most electronic systems.

■ Amplifiers may be active or passive.

■ The power delivered at the output of a passive amplifier cannot be greater than that absorbed at its input. An example of a passive electrical amplifier is a transformer.

■ Active amplifiers take power from some external energy source and so can produce power amplification. Most electronic amplifiers are active.

■ When designing and analysing amplifiers, equivalent circuits are invaluable. They allow the interaction of the circuit with other components to be investigated without a detailed knowledge or understanding of the internal construction of the amplifier.

■ Amplifier gains are often measured in decibels (dB).

- The gain of all amplifiers falls at high frequencies. In some cases the gain also falls at low frequencies. The upper cut-off frequency and the lower cut-off frequency (if this exists) are the points at which the gain falls by 3 dB compared with its mid-band value. The difference between these two values (or between the upper cut-off frequency and zero if no lower cut-off frequency exists) defines the bandwidth of the amplifier.

- Differential amplifiers take as their input the difference between two input signals.

- Operational amplifiers (op-amps) are a common form of differential amplifier. These have many desirable attributes but suffer from variability.

- In many applications simple amplifiers, perhaps based on single transistors, may be more appropriate than more complex circuits.

Exercises

6.1 Sketch lever arrangements that represent:
 (a) a non-inverting force amplifier;
 (b) a non-inverting force attenuator;
 (c) an inverting force amplifier;
 (d) an inverting force attenuator.

6.2 Sketch a pulley arrangement that represents a non-inverting force amplifier.

6.3 Is the torque amplifier of Figure 6.2 an inverting or a non-inverting amplifier? How could this arrangement be modified to produce an amplifier of the other type?

6.4 Conventional automotive hydraulic braking systems are an example of passive amplifiers. What physical quantity is being amplified?
 Such systems may also be regarded as attenuators. What physical quantity is being attenuated?
 Power-assisted automotive brakes are active amplifiers. What is the source of power?

6.5 Identify examples (other than those given in the text and in earlier exercises) of both passive and active amplifiers for which the operation is mechanical, hydraulic, pneumatic, electrical and physiological. In each case, identify the physical quantity that is amplified and, for the active examples, the source of power.

6.6 If an amplifier has a voltage gain of 25, determine the output voltage when the input voltage is 1 V.

6.7 If an amplifier has an input voltage of 2 V and an output voltage of 0.2 V, what is its voltage gain?

6.8 An amplifier has an unloaded voltage gain of 20, an input resistance of 10 kΩ and an output resistance of 75 Ω. The amplifier is connected to a voltage source of 0.5 V, which has an output resistance of 200 Ω, and to a load resistor of 1 kΩ. What will be the value of the output voltage?

6.9 What is the voltage gain of the amplifier in the arrangement in Exercise 6.8?

6.10 Calculate the input power, the output power and the power gain of the arrangement in Exercise 6.8.

6.11 Confirm your results for Exercises 6.8 and 6.9 using computer simulation. You may wish to start with the circuit of Computer Simulation Exercise 6.1.

Exercises continued

6.12 An amplifier has an unloaded voltage gain of 500, an input resistance of 250 kΩ and an output resistance of 25 Ω. The amplifier is connected to a voltage source of 25 mV, which has an output resistance of 4 kΩ, and to a load resistor of 175 Ω. What will be the value of the output voltage?

6.13 What is the voltage gain of the amplifier in the arrangement of Exercise 6.12?

6.14 Calculate the input power, the output power and the power gain of the arrangement in Exercise 6.12.

6.15 Confirm your results for Exercises 6.12 and 6.13 using computer simulation. You may wish to start with the circuit of Computer Simulation Exercise 6.1.

6.16 A displacement sensor produces an output of 10 mV per centimetre of movement and has an output resistance of 300 Ω. It is connected to an amplifier that has an unloaded voltage gain of 15, an input resistance of 5 kΩ and an output resistance of 150 Ω. If the output of the amplifier is connected to a voltmeter with an input resistance of 2 kΩ, what voltage will be displayed on the voltmeter for a displacement of the sensor of 1 metre?

6.17 Confirm your result for Exercise 6.16 using computer simulation. You may wish to start with the circuit of Computer Simulation Exercise 6.1.

6.18 By differentiating the expression

$$P_o = A_v^2 V_i^2 \frac{R_L}{(R_L + R_o)^2}$$

with respect to R_L confirm the condition for maximum power transfer given in Section 6.5.

6.19 Under what conditions is the gain of a circuit given by the following expression?

$$\text{gain (dB)} = 20 \log_{10} \frac{V_2}{V_1}$$

6.20 Convert the following into decibels.
  (a) A power gain of 10.
  (b) A voltage gain of 1.
  (c) A power gain of 0.5.
  (d) A voltage gain of 1,000,000.

6.21 Determine the following:
  (a) The power gain of a circuit with a gain of 20 dB.
  (b) The voltage gain of a circuit with a gain of 20 dB.
  (c) The power gain of a circuit with a gain of −15 dB.
  (d) The voltage gain of a circuit with a gain of −15 dB.

6.22 An amplifier with a gain of 25 dB is connected in series with an amplifier with a gain of 15 dB and with a circuit that produces an attenuation of 10 dB (that is, a gain of −10 dB). What is the gain of the overall arrangement (in dB)?

6.23 An amplifier has a mid-band gain of 25 dB. What will be its gain at its upper cut-off frequency?

6.24 An amplifier has a mid-band voltage gain of 10. What will be its voltage gain at its upper cut-off frequency?

6.25 A circuit has a lower cut-off frequency of 1 kHz and an upper cut-off frequency of 25 kHz. What is its bandwidth?

6.26 A circuit has no lower cut-off frequency but has an upper cut-off frequency of 5 MHz. What is its bandwidth?

6.27 A differential amplifier has a voltage gain of 100. If a voltage of 18.3 V is applied to its non-inverting input, and a voltage of 18.2 V is applied to its inverting input, what will be its output voltage?

6.28 What are the minimum and maximum values of the output voltages that can be produced by the arrangement shown in Figure 6.17?

Chapter 7

Control and Feedback

Objectives

When you have studied the material in this chapter you should be able to:
- outline the basic principles of control systems;
- explain the concepts of 'open-loop' and 'closed-loop' systems and give electronic, mechanical and biological examples of each;
- discuss the role of feedback and closed-loop control in a range of automatic control systems;
- identify the major components of feedback systems;
- analyse the operation of simple feedback systems and describe the interaction between component values and system characteristics;
- describe the uses of negative feedback in overcoming problems of variability in active devices such as operational amplifiers;
- explain the importance of negative feedback in improving bandwidth, input resistance, output resistance and distortion.

7.1 Introduction

In Chapter 1 we identified **control** as one of the basic functions performed by a large number of engineering systems. In simple terms, control involves ensuring that a particular operation or task is performed correctly, and it is therefore associated with concepts such as **regulation** and **command**. While control can be associated with human activities, such as organisation and management, we are more concerned here with **control systems**, which perform tasks automatically. However, the basic principles involved are very similar.

Invariably, the goal of a control system is to determine the value, or state, of one or more physical quantities. For example, a pressure regulator might aim to control the pressure in a vessel, while a climate control system might aim to determine the temperature and humidity in a building. The control system affects the various physical quantities by using appropriate actuators, and if we choose to consider our system to include these actuators (as discussed in Chapter 1) then the output of our system can be considered to be the physical quantity, or quantities, being controlled. Thus, in the example given earlier, the output of our pressure regulation system might be considered to be the actual pressure present within the vessel. The inputs to the control system will determine the value of the output that is produced, but the form of the inputs will depend on the nature of the system.

To illustrate this latter point, let us consider two possible methods of controlling the temperature in a room. The first is to use a heater that has a control that varies the heat output. The user sets the control to give a certain heat output and hopes that this will achieve the desired temperature. If the setting is too low the room will not reach the desired value, while if the setting is too high the temperature will rise above the desired value. If an appropriate setting is chosen the room should stabilise at the right temperature, but it will become too hot or too cold if external factors, such as the outside temperature or the level of ventilation, are changed.

An alternative approach is to use a heater equipped with a temperature controller. The user then sets the required temperature and the controller increases or decreases the heat output to achieve and then maintain this value. It does this by comparing the desired and actual temperatures and using the difference between them to determine the appropriate heat output. Such a system should maintain the temperature of the room, even if external factors change.

Note that in these two methods of temperature control the inputs to the system are quite different. In the first, the input determines the heat to be produced by the heater, while, in the second, the input sets the required temperature.

7.2	Open-loop and closed-loop systems

The alternative strategies discussed above illustrate two basic forms of control, and these are illustrated in Figure 7.1. In this figure, the **user** is simply the person using the system, the **goal** is the desired result and the **output** of the system is the achieved result. In the example above, the goal is the *required* room temperature and the output is the *actual* room temperature. The **forward path** is the part of the system that affects the output in response to its input. In our examples, this is the element that produces heat. In practice, the forward path will also have inputs that provide it with power, but these are not normally shown in this form of diagram. As in the diagrams of amplifiers in the last chapter, we are interested in the functionality of the arrangement, not its implementation.

Figure 7.1(a) shows what is termed an **open-loop control system**, and this corresponds to the first of our two heating methods. Here the user of the system has a goal (in our example, this is the required temperature) and uses knowledge of the characteristics of the system to select an appropriate input to the forward path. In our example, this represents the user selecting

Figure 7.1   Open-loop and closed-loop systems

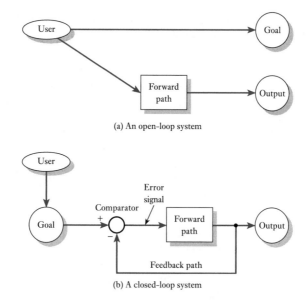

(a) An open-loop system

(b) A closed-loop system

an appropriate setting for the heat control. The forward path takes this input and generates a corresponding output (which in our example is the actual room temperature). The closeness of the output to the goal will depend on how well the input is selected. However, even if the input is chosen well, changes in the characteristics of the forward path (the heater) or in the environment (such as changes in the level of ventilation in the room) will affect the output, and perhaps make it move further from the goal.

An alternative approach is shown in Figure 7.1(b). This shows a **closed-loop control system**, which corresponds to the second of our heating methods. Again the user of the system has a goal, but in this case the user 'inputs' this goal directly into the system. Closed-loop systems make use of a **feedback path** through which information about the output is fed back for comparison with the goal. In this case, the difference between the output and the goal represents the current *error* in the operation of the system. Therefore, the output is subtracted from the goal to produce an **error signal**. If the output is *less* than the goal (in our example, the actual temperature is less than the required temperature), this will produce a positive error signal, which will cause the output to increase, reducing the error. If the output is *greater* than the goal (in our example, the actual temperature is more than the required temperature), this results in a negative error signal, which will cause the output to fall, again reducing the error. In this way, the system drives the output to match the goal. One of the many attractive characteristics of this approach is that it automatically compensates for variations in the system or in the environment. If for any reason the output deviates from the goal, the error signal will drive the system to bring the output back to the required value.

Open-loop systems rely on knowledge of the relationship between the input and the output. This relationship may be ascertained, or adjusted, by

Figure 7.3 A generalised feedback system

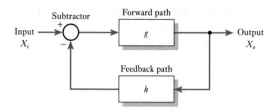

represent electrical quantities such as voltage or current. As in our earlier examples of closed-loop systems, the arrangement of Figure 7.3 consists of a forward path, a feedback path and a subtractor. The forward path will include within it the system or component that is to be controlled. This system is usually referred to as the **plant**. The forward path may also contain some additional elements that are added in order to drive the plant and to make it easier to control. These elements are referred to as the **controller**. The combined behaviour of the controller and the plant are represented in the diagram by a mathematical function g. This represents the relationship between the input and the output of the forward path and is called its **transfer function**. The feedback path represents the sensor used to detect the output and any processing that is applied to the signals it produces. The feedback path is also represented by a transfer function, which is given the symbol h.

From a knowledge of the transfer functions g and h it is possible to analyse the behaviour of the overall system of Figure 7.3 and hence predict its behaviour. The field of **control engineering** is largely concerned with the analysis of such systems, and with the design of appropriate controllers and feedback arrangements to tailor their behaviour to meet particular needs. Unfortunately, in many situations the characteristics of the physical plant are complex, perhaps including frequency dependencies, non-linearities or time delays. Consequently, the analysis of such systems is often quite involved and is beyond the scope of this book.

Fortunately, the feedback systems that we meet most often in electrical and electronic systems are much more straightforward. Here we can often assume that the transfer functions of the forward and feedback paths are simple gains, which greatly simplifies analysis. This is illustrated in Figure 7.4, where the forward path is represented by a gain of A and the feedback path by a gain of B.

Since the output of the forward path is X_o and its gain is A, its input must be given by X_o/A. Similarly, since the input to the feedback path is X_o and

Figure 7.4 A feedback system

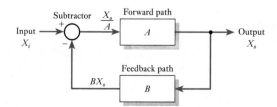

its gain is B, its output must be BX_o. From the diagram, we can see that the input to the forward path (which we have just determined to be X_o/A) is actually produced by subtracting the feedback signal (which we have just found to be BX_o) from the input X_i. Therefore

$$\frac{X_o}{A} = X_i - BX_o$$

or, by rearranging

$$\frac{X_o}{X_i} = \frac{A}{1 + AB}$$

The ratio of the output to the input is the gain of the feedback arrangement, which is usually given the symbol G. Therefore

$$G = \frac{A}{1 + AB} \tag{7.1}$$

This expression for the gain is also referred to as the **transfer function** of the feedback system.

It is common to refer to the forward gain A as the **open-loop gain** (as it is the gain that the circuit would have with the feedback disconnected) and to the overall gain G as the **closed-loop gain** (since this is the gain of the circuit with the feedback present). The overall characteristics of the system depend on the values of A and B, or more directly on the product AB.

7.4.1   If AB is negative

If either A or B are negative (but not both), the product AB will be negative. If now the term $(1 + AB)$ is less than 1, G is greater than A. In other words, the gain of the circuit will be increased by the feedback. This is termed **positive feedback**.

A special case of positive feedback occurs when $AB = -1$. Under these circumstances

$$G = \frac{A}{1 + AB} = \frac{A}{1 - 1} = \infty \text{ (infinity!)}$$

Since the gain of the circuit is infinite it has a limited range of applications, but it is useful in the production of **oscillators**.

7.4.2   If AB is positive

If A and B are either both positive or both negative, the term AB will be positive. Thus the term $(1 + AB)$ must be positive and greater than 1, and G must be less than A. In other words, the gain of the circuit with feedback is less than it would be without feedback. This is **negative feedback**.

If the product AB is not only positive but is also large compared with 1, the term $(1 + AB)$ is approximately equal to AB, and the expression

$$G = \frac{A}{1 + AB}$$

may be simplified to

$$G \approx \frac{A}{AB} = \frac{1}{B} \tag{7.2}$$

This special case of negative feedback is of great importance as we now have a system in which the overall gain is independent of the gain of the forward path, being determined solely by the characteristics of the feedback path.

The ability to produce a system where the overall gain is independent of the gain of the forward path is of great significance. In the previous chapter, we noted that devices such as operational amplifiers suffer from great variability in their gain. Negative feedback would seem to offer a way of tackling this problem. Therefore, for the remainder of this chapter we will concentrate on the uses and characteristics of negative feedback and will leave further consideration of positive feedback and oscillators until Chapter 24.

7.4.3   Notation

It should be noted that, in some textbooks, the subtractor in Figure 7.4 is replaced by an adder. This is an equally valid representation of a feedback arrangement, and a similar analysis to that given above produces an expression for the overall gain of the form

$$G = \frac{A}{1 - AB}$$

This equation clearly places different requirements on A and B to achieve positive or negative feedback than the analysis given earlier.

In this text we assume use of a subtractor in the feedback block diagram since this produces arrangements that more closely correspond to the real circuits that we will consider later. However, you should be aware that other representations exist.

7.5   Negative feedback

So far we have considered control and feedback in a generic manner. For example, in Figures 7.3 and 7.4 the input and output are given the symbols X_i and X_o to indicate that they may represent any physical quantities. However, it is now appropriate to turn our attention more specifically to the use of feedback in electrical and electronic applications.

We saw in the last chapter that one characteristic of operational amplifiers is that their gain, while being large, is also variable between devices. We also noted that their gain varies with temperature. These characteristics are common to almost all **active devices**. In contrast, passive components, such as resistors and capacitors, can be made to very high precision and can be very stable as their temperature varies.

In the last section, when looking at negative feedback, we noted that a feedback circuit in which the product AB is positive and much greater than 1 has an overall gain that is independent of the forward gain A, being determined entirely by the feedback gain B. If we construct a negative feedback system using an active amplifier as the forward path A and a passive network as the feedback path B, we can produce an amplifier with a stable overall gain independent of the actual value of A. This is illustrated in Example 7.1. Here an amplifier with a high but variable voltage gain is combined with a stable feedback network to form an amplifier with a gain of 100.

Example 7.1	Design an arrangement with a stable voltage gain of 100 using a high-gain active amplifier. Determine the effect on the overall gain of the circuit if the voltage gain of the active amplifier varies from 100,000 to 200,000.

We will base our circuit on our standard block diagram.

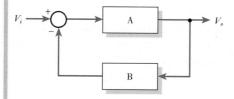

From Equation 7.2 we know that the overall gain is given by

$$G = \frac{1}{B}$$

and therefore for a gain of 100 we choose $B = 1/100$ or 0.01.

When the gain of the amplifier (A) is 100,000, the overall gain will be

$$G = \frac{A}{1 + AB} = \frac{100,000}{1 + (100,000 \times 0.01)}$$

$$= \frac{100,000}{1 + 1000}$$

$$= 99.90$$

$$\approx \frac{1}{B}$$

When the gain of the amplifier (A) is 200,000, the overall gain will be

$$G = \frac{A}{1 + AB} = \frac{200,000}{1 + (200,000 \times 0.01)}$$

$$= \frac{200,000}{1 + 2000}$$

$$= 99.95$$

$$\approx \frac{1}{B}$$

Notice that a change of 100 percent in the value of the gain of the active amplifier (A) produces a change of only 0.05 percent in the overall gain G.

Example 7.1 shows that the large variation in gain associated with active circuits can be overcome by the use of negative feedback, provided that a stable feedback arrangement can be produced. In order to make the feedback path stable, it must be constructed using only passive components. Fortunately, this is a simple task.

We have seen that the overall gain of the feedback circuit is $1/B$. Therefore, to have an overall gain of greater than 1, we require B to be less than 1. In other words, our feedback path may be a **passive attenuator**.

Construction of such a feedback arrangement using passive components is simple. If we take as an example the value used in Example 7.1, we require a passive attenuator with a voltage gain of 1/100. This can be achieved as shown in Figure 7.5. The circuit is a simple potential divider with a ratio of 99:1. The output voltage V_o is related to the input voltage V_i by the expression

$$V_o = V_i \frac{R}{R + 99R}$$

$$\frac{V_o}{V_i} = \frac{1}{100}$$

The resistor values of R and $99R$ are shown simply to indicate their relative magnitudes. In practice, R might be 1 kΩ and $99R$ would then be 99 kΩ. The actual values used would depend on the circuit configuration.

Having decided that the forward path of our feedback circuit will be a high-gain active amplifier, and that the feedback path will be a resistive attenuator, we are now in a position to complete the circuit. Continuing with the values given in Example 7.1, we can now draw our circuit diagram as shown in Figure 7.6. This shows an arrangement based on an active amplifier with a gain of A and a feedback network with a gain B of 1/100. This produces an amplifier with an overall gain G of 100 (that is, $1/B$).

Our last remaining problem is to implement the subtractor in Figure 7.6, since this does not seem to be a standard component. Fortunately, we have

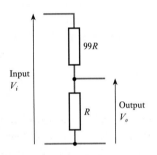

Figure 7.5 A passive attenuator with a gain of 1/100

Figure 7.6 An amplifier with a gain of 100

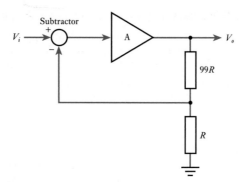

Figure 7.7 An amplifier with a gain of 100 based on an operational amplifier

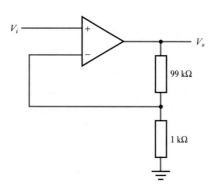

already discussed a means of providing this function, when we considered operational amplifiers in Chapter 6. Op-amps are differential amplifiers. That is, they amplify the difference between two input signals. We could visualise such an amplifier as a single-input amplifier with a subtractor connected to its input. Therefore, an operational amplifier may be used to replace both the amplifier and the subtractor in the circuit of Figure 7.6. Such an arrangement is shown in Figure 7.7.

File 07A

Computer Simulation Exercise 7.1

Simulate the circuit of Figure 7.7, basing your design on one of the operational amplifiers supported by your simulation package (and remembering to include connections to appropriate power supplies). Apply a 50 mV DC input to the circuit and measure the output voltage. Hence deduce the voltage gain of the circuit and confirm that this is as expected. Experiment with different values of the input voltage and investigate how this affects the voltage gain.

Inherent in the design of our simple amplifier is the assumption that the overall gain is equal to $1/B$. From our earlier discussions, we know that this assumption is only valid provided that the product AB is much greater than 1.

In our circuit, the forward path is implemented using an operational amplifier, and from the last chapter we know that this is likely to have a voltage gain (A) of perhaps 10^5 or 10^6. Since B is 1/100 in our example, this means that AB will have a value of about 10^3 to 10^4. Since this is much greater than 1, it would seem that our assumption that the gain is equal to $1/B$ is valid. However, let us consider another example.

Example 7.2

Design an arrangement with a stable voltage gain of 10,000 using a high-gain active amplifier. Determine the effect on the overall gain of the circuit if the voltage gain of the active amplifier varies from 100,000 to 200,000.

As before, we will base our circuit on our standard block diagram.

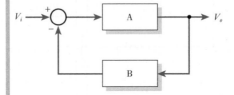

From Equation 7.2 we know that the overall gain is given by

$$G = \frac{1}{B}$$

and therefore for a gain of 10,000 we choose $B = 1/10,000$ or 0.0001.

When the gain of the amplifier (A) is 100,000, the overall gain will be

$$G = \frac{A}{1 + AB} = \frac{100,000}{1 + (100,000 \times 0.0001)}$$

$$= \frac{100,000}{1 + 10}$$

$$= 9091$$

When the gain of the amplifier (A) is 200,000, the overall gain will be

$$G = \frac{A}{1 + AB} = \frac{200,000}{1 + (200,000 \times 0.0001)}$$

$$= \frac{200,000}{1 + 20}$$

$$= 9524$$

It can be seen that the resultant gain is *not* very close to $1/B$ (10,000) and that variations in the gain of the forward path (A) have significant effects on the overall gain.

The above example shows that, in order for the gain to be stabilised by the effects of negative feedback, the product AB must be much greater than 1. In other words

$$AB \gg 1$$

or

$$A \gg \frac{1}{B} \tag{7.3}$$

Since A is the open-loop gain of the active amplifier and $1/B$ is the closed-loop gain of the complete circuit, this implies that the condition for the stabilising effects of negative feedback to work is that

$$open\text{-}loop\ gain \gg closed\text{-}loop\ gain \tag{7.4}$$

File 07B

Computer Simulation Exercise 7.2

Simulate the circuit of Example 7.2 using an idealised gain block and a subtractor (these elements are available in most simulation packages). Apply a 1 V DC input to the circuit and investigate the output voltage for different values of forward gain (A) and feedback gain (B). Hence investigate the conditions required for the overall gain to be approximately equal to $1/B$.

We have seen that negative feedback allows us to generate amplifiers with overall characteristics that are constant despite variations in the gain of the active components. In fact, negative feedback can also produce a range of other desirable effects, and it is widely used in many forms of electronic circuitry.

7.6 The effects of negative feedback

It is clear from the above that negative feedback has a profound effect on the gain of an amplifier and on the consistency of that gain. However, feedback also affects other circuit characteristics. In this section, we will look at a range of circuit parameters.

7.6.1 Gain

We have seen that negative feedback reduces the gain of an amplifier. In the absence of feedback, the gain of an amplifier (G) is simply its open-loop gain A. We know from Equation 7.1 that with feedback the gain becomes

$$G = \frac{A}{1 + AB}$$

and thus the effect of the feedback is to reduce the gain by a factor of $1 + AB$.

7.6.2 Frequency response

In Chapter 6, we noted that the gain of all amplifiers falls at high frequencies and that, in many cases, it also drops at low frequencies.

From the previous discussion of gain, we know that the closed-loop gain of a feedback amplifier is largely independent of the open-loop gain of the amplifier, provided that the latter is considerably greater than the former. Since the open-loop gain of all amplifiers falls at high frequencies (and often at low frequencies), it is clear that the closed-loop gain will also fall in these regions. However, if the open-loop gain is considerably greater than the closed-loop gain, the former will be able to fall by a considerable amount before this has an appreciable effect on the latter. Thus the closed-loop gain will be stable over a wider frequency range than that of the amplifier without feedback. This is illustrated in Figure 7.8.

The solid line in Figure 7.8 shows the variation of gain with frequency of an amplifier without feedback: that is, its open-loop frequency response. The addition of negative feedback (shown by the broken line) reduces the gain of the arrangement. The resultant closed-loop gain is constant over the range of frequencies where it is considerably less than the amplifier's open-loop gain. The addition of negative feedback thus results in an increase in the bandwidth of the amplifier.

Thus the use of negative feedback reduces gain but increases bandwidth, allowing designers to 'trade off' one characteristic against the other. Since gain is a relatively inexpensive commodity in modern electronic circuits,

Figure 7.8 The effects of negative feedback on frequency response

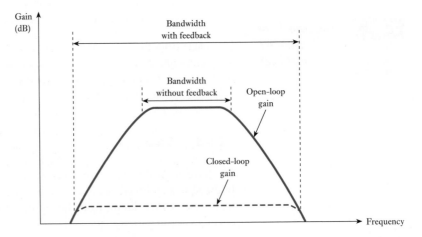

this is often a useful mechanism. We will see in the next chapter that in some cases the increase in bandwidth is directly proportional to the fall in gain. In other words, while feedback decreases the gain of an amplifier by $1 + AB$, it increases the bandwidth by $1 + AB$. In this case, the product of the gain and the bandwidth remains constant. Hence

$$\text{gain} \times \text{bandwidth} = \text{constant} \tag{7.5}$$

The value produced by multiplying these two terms is given the obvious name of the **gain–bandwidth product**.

7.6.3　Input and output resistance

One of the important characteristics of negative feedback is that it tends to keep the output constant despite changes in its environment. One way in which this manifests itself in *voltage* amplifiers is that it tends to keep the output voltage constant despite changes in the load applied to the amplifier. This is analogous to the way in which a cruise control in a car maintains a constant speed whether going up- or downhill. From our discussions of loading in the last chapter, we know that loading effects are minimised in voltage amplifiers when the output resistance is reduced or when the input resistance is increased. It is therefore no surprise to discover that negative feedback can achieve both these desirable effects. However, the situation is slightly more complex than this.

We noted earlier that the general representation of a feedback system shown in Figure 7.4 is applicable to a wide range of systems, and that the input and output can represent not only voltages but also currents or other physical quantities. Clearly, if our input and output quantities are temperatures, we would expect the feedback arrangement to maintain the output temperature despite changes in the environment. Similarly, if the input and output are currents (rather than voltages), we would expect the system to minimise the effects of loading on these currents. In order to do this, the circuit will *increase* the output resistance and *decrease* the input resistance – exactly the opposite action to that described above for a voltage amplifier.

In fact, negative feedback can either increase or decrease input resistance and either increase or decrease output resistance, depending on how it is applied. The factors that determine the effects of the feedback are the way in which the output is sensed and way in which the feedback signal is applied. If the feedback senses the output *voltage*, it will tend to make the output voltage more constant by *decreasing* the output resistance. In contrast, if the feedback senses the output *current*, it will tend to make the output current more constant by *increasing* the output resistance. Similarly, if the feedback is applied by subtracting a *voltage* related to the output from the input voltage, it will tend to make the circuit a better voltage amplifier by *increasing* the input resistance. In contrast, if the feedback is applied by subtracting a *current* from the input current, it will tend to make the circuit a better current amplifier by *decreasing* the input resistance.

To illustrate the above, let us look back at the circuit of Figure 7.7. Here the signal fed back is related to the output *voltage* (the output voltage is applied across the potential divider to produce the feedback signal), so the feedback *reduces* the output resistance of the circuit. In this case, the *voltage* from the potential divider is applied to the inverting input of the amplifier, where it is effectively subtracted from the input voltage (as in Figure 7.6). Therefore, the feedback *increases* the input resistance of the circuit. Consequently, the circuit of Figure 7.7 has a much higher input resistance and a much lower output resistance than the operational amplifier would have if used without feedback. We will return to look at the input and output resistance of this and other related circuits in the next chapter when we look at op-amps in more detail.

So far, we have noted that negative feedback can either increase or decrease input and output resistance, but we have not quantified this effect. Having noted that negative feedback decreases gain by a factor of $1 + AB$, and that it can increase bandwidth by a similar factor, it is perhaps not totally surprising to discover that the factor by which input and output resistance are increased or decreased is also $1 + AB$. The proof of this relationship is fairly straightforward, but it does not seem appropriate to repeat it here.

7.6.4   Distortion and noise

In Chapter 5, we looked at examples of the effects of distortion on a sinusoidal signal. Many forms of distortion are caused by a **non-linear amplitude response**, which means that the gain of the circuit varies with the amplitude of the input signal. Since negative feedback tends to make the gain more stable, it also tends to reduce distortion. Perhaps unsurprisingly, it can be shown that the factor by which distortion is reduced is again $1 + AB$.

The noise produced by an amplifier is also reduced by negative feedback, and again this is by a factor of $1 + AB$. However, this effect applies only to noise produced *within* the amplifier itself, not to any noise that already corrupts the input signal. The latter will be indistinguishable from the input signal and will therefore be amplified along with this signal.

7.7   Negative feedback – a summary

All negative feedback systems share a number of properties:

1. They tend to maintain their output despite variations in the forward path or in the environment.
2. They require a forward path gain that is greater than that which would be necessary to achieve the required output in the absence of feedback.
3. The overall behaviour of the system is determined by the nature of the feedback path.

When applied to electronic amplifiers, negative feedback has many beneficial effects. It stabilises the gain against variations in the open-loop gain of the amplifying device. It increases the bandwidth of the amplifier. It can be used to increase or decrease input and output resistance as required. It reduces distortion caused by non-linearities in the amplifier. It reduces the effects of noise produced within the amplifier.

In exchange for these benefits, negative feedback reduces the gain of the amplifier. In most cases this is a small price to pay, since the majority of modern amplifying devices have a high gain and are inexpensive, allowing many stages to be used if necessary. However, the use of negative feedback can have implications for the **stability** of the circuit, and we will return to look at this issue in Chapter 24.

Negative feedback plays a vital role in electronic circuits for a wide range of applications. We will look at a few examples of such circuits in the next chapter when we look in more detail at operational amplifiers.

Key points

- Feedback systems form an essential part of almost all automatic control systems, be they electronic, mechanical or biological.

- Feedback systems can be divided into two types. In negative feedback systems, the feedback tends to *reduce* the input to the forward path. In positive feedback systems, the feedback tends to *increase* the input to the forward path.

- If the gain of the forward path is A, the gain of the feedback path is B and the feedback signal is subtracted from the input, then the overall gain G of the system is given by

$$G = \frac{A}{1 + AB}$$

- If AB is positive we have negative feedback. If AB is also large compared with 1, the expression for the gain simplifies to $1/B$. In these circumstances, the overall gain is independent of the gain of the forward path.

- If AB is negative and less than 1 we have positive feedback. For the special case where AB is equal to 1, the gain is infinite. This condition is used in the production of oscillators.

- Negative feedback tends to increase the bandwidth of an amplifier at the expense of a loss of gain. In many cases, the factor by which the bandwidth is increased and the gain is reduced is $(1 + AB)$. Thus the gain–bandwidth product remains constant.

- Negative feedback also tends to improve the input resistance, output resistance, distortion and noise of an amplifier. In each case, the improvement is generally by a factor of $(1 + AB)$.

- While negative feedback brings many benefits, it can also bring with it problems of instability.

Exercises

7.1 What is meant, in engineering, by the term 'control'?

7.2 Give three examples of common control systems.

7.3 In each of the control systems identified in your answer to the last exercise, what constitutes the input to and what the output of the system?

7.4 In a car cruise control, what is the input and what is the output?

7.5 In a stock control system in a warehouse, what is the input and what is the output?

7.6 Explain the meanings of the terms 'user', 'goal', 'output', 'forward path', 'feedback path' and 'error signal' as they relate to an open- or closed-loop system.

7.7 From where does the forward path obtain its power?

7.8 Sketch a block diagram of a generalised feedback system and derive an expression for the output in terms of the input and the gains of the forward path (A) and feedback path (B).

7.9 In the expression you derived in Exercise 7.8, what range of values for AB corresponds to positive feedback?

7.10 In the expression you derived in Exercise 7.8, what range of values for AB corresponds to negative feedback?

7.11 In the expression you derived in Exercise 7.8, what range of values for AB would be used to produce an oscillator?

7.12 Explain why the characteristics of active devices encourage the use of negative feedback.

7.13 Design an arrangement with a stable voltage gain of 10 using a high-gain active amplifier. Determine the effect on the overall gain of the circuit if the voltage gain of the active amplifier varies from 100,000 to 200,000.

7.14 What is the voltage gain of the following arrangement?

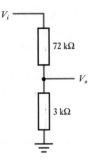

7.15 Design a passive attenuator with a gain of 1/10.

7.16 Determine the voltage gain of the following amplifier.

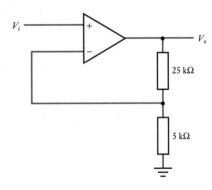

7.17 Confirm your results for Exercise 7.16 using computer simulation. You may wish to start with the circuit of Computer Simulation Exercise 7.1.

7.18 Design an amplifier with a gain of 10 based on an operational amplifier.

7.19 Confirm your results for Exercise 7.18 using computer simulation. You may wish to start with the circuit of Computer Simulation Exercise 7.1.

7.20 Design an arrangement with a stable voltage gain of 20,000 using a high-gain active amplifier. Determine the effect on the overall gain of the circuit if the voltage gain of the active amplifier varies from 100,000 to 200,000.

7.21 Confirm your results for Exercise 7.20 using computer simulation. You may wish to start with the circuit of Computer Simulation Exercise 7.2.

7.22 By what factor is the gain of an amplifier changed by the use of negative feedback? Is this an increase or a decrease in gain?

7.23 How is the bandwidth of an amplifier changed by the use of negative feedback?

7.24 Under what circumstances might it be advantageous to increase the input resistance of an amplifier? How might this be achieved?

7.25 Under what circumstances might it be advantageous to decrease the input resistance of an amplifier? How might this be achieved?

7.26 Under what circumstances might it be advantageous to decrease the output resistance of an amplifier? How might this be achieved?

7.27 Under what circumstances might it be advantageous to increase the output resistance of an amplifier? How might this be achieved?

7.28 How does negative feedback affect the distortion produced by an amplifier?

7.29 How does negative feedback affect noise present in an input signal to an amplifier?

Chapter 8

Operational Amplifiers

Objectives

When you have studied the material in this chapter you should be able to:

- outline the uses of operational amplifiers in a range of engineering applications;
- describe the physical form of a typical op-amp and its external connections;
- explain the concept of an ideal operational amplifier and describe the characteristics of such a device;
- draw and analyse a range of standard circuits for performing functions such as the amplification, buffering, addition and subtraction of signals;
- design op-amp circuits to perform simple tasks, including the specification of appropriate passive components;
- describe the ways in which real operational amplifiers differ from ideal devices;
- explain the importance of negative feedback in tailoring the characteristics of operational amplifiers to suit a particular application.

8.1 Introduction

Operational amplifiers (or **op-amps**) are among the most widely used building blocks for the construction of electronic circuits. One of the reasons for this is that they are nearly ideal voltage amplifiers, which greatly simplifies design. As a result, op-amps tend to be used not only by specialist electronic engineers but also by other engineers who want a simple solution to an instrumentation or control problem. Thus, a mechanical engineer who wishes to display the speed of rotation of an engine, or a civil engineer who needs to monitor the stress on a bridge, would be very likely to use an operational amplifier to construct the instrumentation required.

Figure 8.1 Typical operational
amplifier packages

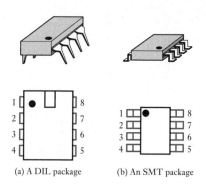

(a) A DIL package   (b) An SMT package

Operational amplifiers are a form of integrated circuit (IC). That is, they are constructed by integrating a large number of electronic devices into a single semiconductor component. In later chapters we will look at the operation of such components, but for the moment we are concerned simply with their characteristics and how they are used.

A typical op-amp takes the form of a small plastic package with an appropriate number of pins for carrying signals and power in and out of the device. Figure 8.1 shows two common forms. Figure 8.1(a) shows an eight-pin 'dual in-line' or DIL package. This is perhaps the most common form of package for a single op-amp, particularly where circuits are to be produced in relatively small numbers. Larger versions of this package, with a greater number of pins, can house two or four op-amps in a single component. Figure 8.1(b) shows an SMT (surface-mounted technology) component. These components have the advantage of a much smaller physical size but are consequently difficult to assemble manually. Such components are often used in high-volume products, where automatic assembly techniques are used. In both packages, the pins are numbered anticlockwise when viewed from the top. Pin number 1 is usually marked by a dot, or a notch, or both. The way in which the pins are connected internally is termed the 'pin-out' of the device, and Figure 8.2 illustrates typical pin-outs for a range of components. In this figure, the connections labelled V_{pos} and V_{neg} represent the positive and negative power supply voltages, respectively. You will remember that these connections are normally omitted from circuit diagrams to aid clarity, but they must be connected in a physical circuit to provide a power source for the circuit. Typical values for these quantities might be +15 V and −15 V, but these values will be discussed later when we look at real devices in Section 8.5.

One of the many attractive features of operational amplifiers is that they can be easily configured to produce a wide range of electronic circuits. These include not only amplifiers of various forms but also circuits with more specialised functions, such as adding, subtracting or modifying signals. We will look at a few basic circuits in this chapter, but we will meet a range of other op-amp circuits later in the text.

Figure 8.2 Typical operational amplifier pin-outs

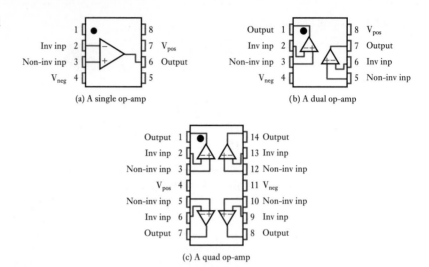

(a) A single op-amp

(b) A dual op-amp

(c) A quad op-amp

8.2

An ideal operational amplifier

We noted in the last section that operational amplifiers are nearly ideal voltage amplifiers. Design is often much simpler when using idealised components, so it is common initially to assume that our components are perfect and then to investigate the effects of any non-ideal characteristics. In order to do this in the case of operational amplifiers, first we need to have an idea of how an ideal component would behave.

In Chapter 6, we looked briefly at ideal voltage amplifiers and deduced that these would have an infinite input resistance and a zero output resistance. Under these circumstances, the amplifier would draw no current from the source, and its output voltage would be unaffected by the value of the load. Thus there would be no **loading effects** when using an ideal amplifier.

While it is relatively easy to deduce the input and output resistance of an ideal amplifier, it is less obvious what the gain of such a device would be. Clearly, the gain required of a circuit will differ with the application, and it is perhaps not clear that one particular gain is 'ideal' for all situations. However, we saw in the last chapter that negative feedback can be used to tailor the gain of an amplifier to any particular value (which is determined by the feedback gain) provided that the open-loop gain is sufficiently high. Therefore, when using negative feedback, it is advantageous to have as high an open-loop gain as possible. Thus, an ideal operational amplifier would have an infinite open-loop gain.

Therefore, an ideal op-amp would have an infinite input resistance, a zero output resistance and an infinite voltage gain. We are now in a position to draw an equivalent circuit for such a device, starting with the diagram given in Figure 6.15 and modifying the parameters accordingly. This is shown in Figure 8.3. Note that the infinite input resistance means

Figure 8.3 Equivalent circuit of an ideal operational amplifier

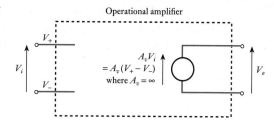

Operational amplifier

that no current flows into the device and the input terminals appear to be unconnected. Similarly, the zero output resistance means that there is no output resistor. The output voltage is equal to the voltage produced by the controlled voltage source, which is A_v times the differential input voltage V_i. In this case, the voltage gain A_v is infinite.

8.3 Some basic operational amplifier circuits

Before we look at some simple amplifier circuits, we should perhaps make sure that we are clear on some of the terminology. We noted in Chapter 6 that electronic amplifiers can be either non-inverting or inverting. If the input to a *non-inverting* amplifier is a positive voltage, then the output will also be positive. If a similar input signal is applied to an *inverting* amplifier, the output will be negative. When the input is not a fixed voltage but an alternating waveform, then the output voltage of either amplifier will also alternate. The effects of these two forms of amplification on an alternating waveform are illustrated in Figure 8.4. In this figure, the non-inverting amplifier has a gain of +2, while the inverting amplifier has a gain of −2.

In the last chapter, we derived the circuit of a non-inverting amplifier from 'first principles'. That is, we started with a generalised block diagram

Figure 8.4 Non-inverting and inverting amplifiers

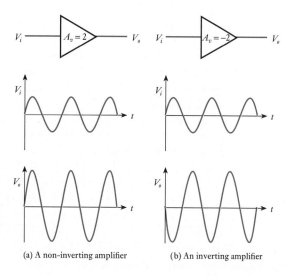

(a) A non–inverting amplifier (b) An inverting amplifier

of a feedback system and devised elements to implement the forward and feedback paths. While that circuit was based on an operational amplifier, this is not the process we normally adopt when using op-amps. More commonly, we start with a standard or 'cookbook' circuit and adapt it to suit our needs. Often, this adaptation requires no more than selecting appropriate component values. A wide range of these cookbook circuits is available to perform a wide range of tasks. We will begin by looking at a few well-known examples.

8.3.1   A non-inverting amplifier

The first of our standard circuits is that derived in the last chapter for a non-inverting amplifier. This is shown in Figure 8.5(a). Also shown, in Figure 8.5(b), is the same circuit redrawn in a different orientation. This latter form is electrically identical to the earlier circuit and has the same characteristics. It is important that readers can recognise and use this circuit in either form.

Rather than analyse the circuit from first principles, as in Chapter 7, we will look at the operation of the circuit assuming that it contains an ideal op-amp. You will see that this makes the analysis very straightforward.

First, since the gain of the op-amp is infinite, if the output voltage is finite, the input voltage to the op-amp ($V_+ - V_-$) must be zero. Therefore

$$V_- = V_+ = V_i$$

Since the op-amp has an infinite input resistance, its input current must be zero. Therefore, V_- is determined simply by the output voltage and the potential divider formed by R_1 and R_2. Thus

$$V_- = V_o \frac{R_2}{R_1 + R_2}$$

Therefore, since $V_- = V_i$

Figure 8.5   A non-inverting amplifier

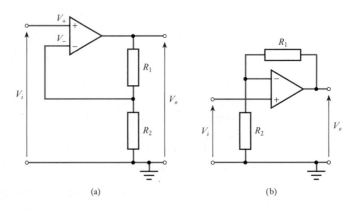

(a)   (b)

$$V_i = V_o \frac{R_2}{R_1 + R_2}$$

and the overall gain of the circuit is given by

$$G = \frac{V_o}{V_i} = \frac{R_1 + R_2}{R_2} \tag{8.1}$$

which is consistent with the circuit of Figure 7.7.

Example 8.1

Design a non-inverting amplifier with a gain of 25, based on an operational amplifier.

We start with our standard circuit.

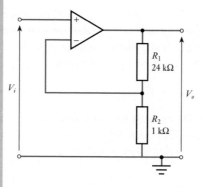

From Equation 8.1 we have

$$G = \frac{V_o}{V_i} = \frac{R_1 + R_2}{R_2}$$

Therefore, if $G = 25$,

$$\frac{R_1 + R_2}{R_2} = 25$$

$$R_1 + R_2 = 25R_2$$

$$R_1 = 24R_2$$

Since it is the ratio of the resistor values that determines the gain, we are free to choose the actual values. Here we will choose $R_2 = 1\ k\Omega$, which means that R_1 must be 24 kΩ. When using *ideal* op-amps the actual values of the resistors are unimportant – it is only the ratio of the values that is significant. However, when we use *real* components there are factors that affect our choice of component values. These will be discussed in Section 8.6.

Computer Simulation Exercise 8.1

Simulate the circuit of Example 8.1, using one of the operational amplifiers supported by your simulation package. Apply a 100 mV DC input to the circuit and measure the output voltage. Hence deduce the voltage gain of the circuit and confirm that this is as expected. Experiment with different values for the two resistors and see how this affects the voltage gain. Experiment with different values for the input voltage (including both positive and negative values) and confirm that the circuit behaves as you expect.

8.3.2   An inverting amplifier

The second of our standard circuits is that of an inverting amplifier. This is shown in Figure 8.6. As in the previous circuit, since the gain of the op-amp is infinite, if the output voltage is finite, the input voltage to the op-amp $(V_+ - V_-)$ must be zero. Therefore

$$V_- = V_+ = 0$$

Since the op-amp has an infinite input resistance, its input current must be zero. Therefore, the currents I_1 and I_2 must be equal and opposite. By applying Ohm's law to the two resistors, we see that

$$I_1 = \frac{V_o - V_-}{R_1} = \frac{V_o - 0}{R_1} = \frac{V_o}{R_1}$$

and

$$I_2 = \frac{V_i - V_-}{R_2} = \frac{V_i - 0}{R_2} = \frac{V_i}{R_2}$$

Therefore, since

$$I_1 = -I_2$$

Figure 8.6   An inverting amplifier

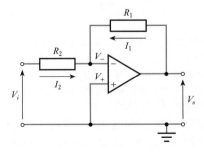

then

$$\frac{V_o}{R_1} = -\frac{V_i}{R_2}$$

and the gain G is given by the very simple result

$$G = \frac{V_o}{V_i} = -\frac{R_1}{R_2} \qquad (8.2)$$

Note the minus sign in the expression for the gain, showing that this is indeed an inverting amplifier.

In this circuit, the negative feedback maintains the voltage on the inverting input (V_-) at zero volts. This may be understood by noting that, if V_- becomes more positive than the voltage on the non-inverting input (in this case zero volts), this will cause the output of the op-amp to become negative, which will drive V_- negative through R_1. If, on the other hand, V_- becomes negative with respect to zero volts, the output will become positive, which will tend to make V_- more positive. Thus the circuit will act to keep V_- at zero, even though this terminal is not physically connected to earth. Such a point in a circuit is referred to as a **virtual earth**, and this kind of amplifier is called a **virtual earth amplifier**.

Example 8.2

Design an inverting amplifier with a gain of −25, based on an operational amplifier.

We start with our standard circuit.

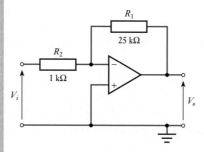

From Equation 8.2 we have

$$G = \frac{V_o}{V_i} = -\frac{R_1}{R_2}$$

Therefore, if $G = -25$,

$$-\frac{R_1}{R_2} = -25$$

$$R_1 = 25R_2$$

Since it is the ratio of the resistor values that determines the gain, we are free to choose the actual values. Here we will choose $R_2 = 1$ kΩ, which means that R_1 must be 25 kΩ. As discussed above, we will leave consideration of the choice of component values until Section 8.6.

File 08B

Computer Simulation Exercise 8.2

Simulate the circuit of Example 8.2, using one of the operational amplifiers supported by your simulation package. Apply a 100 mV DC input to the circuit and measure the output voltage. Hence deduce the voltage gain of the circuit and confirm that this is as expected. Experiment with different values for the two resistors and see how this affects the voltage gain. Experiment with different values for the input voltage (including both positive and negative values) and confirm that the circuit behaves as you expect.

You will note that the assumption that we are using an ideal operational amplifier greatly simplifies the analysis of these circuits.

8.4 Some other useful circuits

Having seen how we can use operational amplifiers to produce simple non-inverting and inverting amplifiers, we will now look at a few other standard circuits.

8.4.1 A unity-gain buffer amplifier

This is a special case of the non-inverting amplifier discussed in Section 8.3.1 with R_1 equal to zero and R_2 equal to infinity. The resulting circuit is shown in Figure 8.7. From Equation 8.1, we know that the gain of a non-inverting amplifier circuit is given by

$$G = \frac{R_1 + R_2}{R_2}$$

Figure 8.7   A unity-gain buffer amplifier

File 08C

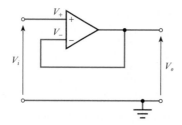

This may be rearranged to give

$$G = \frac{R_1}{R_2} + 1$$

If we substitute appropriate values for R_1 and R_2 we get

$$G = \frac{0}{\infty} + 1 = 1 \tag{8.3}$$

and we therefore have an amplifier with a gain of 1 (unity).

At first sight, this may not seem a very useful circuit, since the voltage at the output is the same as that at the input. However, one must remember that voltage is not the only important attribute of a signal. The importance of this circuit is that it has a very high input resistance and a very low output resistance, making it very useful as a **buffer**. We will look at input and output resistance later in this chapter.

8.4.2 A current-to-voltage converter

We noted in Chapter 3 that some sensors operate such that the physical quantity being measured is represented by the magnitude of the *current* produced at its output, rather than by the magnitude of a voltage. This illustrates one of many situations where we may wish to convert a varying current into a corresponding varying voltage. A circuit to perform this transformation is shown in Figure 8.8.

The analysis of this circuit is similar to that of the inverting amplifier of Section 8.3.2. Again the inverting input to the op-amp is a **virtual earth point**, and the voltage at this point (V_-) is zero. Since the currents into the virtual earth point must sum to zero, and the input current to the op-amp is zero, it follows that

$$I_i + I_R = 0$$

and

$$I_i = -I_R$$

Now, since V_- is zero, I_R is given by

$$I_R = \frac{V_o}{R}$$

Figure 8.8   A current-to-voltage converter

File 08D

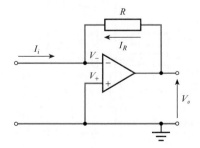

and therefore

$$I_i = -I_R = -\frac{V_o}{R}$$

or, rearranging

$$V_o = -I_i R \qquad (8.4)$$

Thus the output voltage is directly proportional to the input current. The minus sign indicates that an input current that flows in the direction of the arrow in Figure 8.8 will produce a negative output voltage.

8.4.3   A differential amplifier (subtractor)

A common requirement within signal processing is the need to subtract one signal from another. A simple circuit for performing this task is shown in Figure 8.9. Since no current flows into the inputs of the op-amp, the voltages on the two inputs are determined simply by the potential dividers formed by the external resistors. Thus

$$V_+ = V_1 \frac{R_1}{R_1 + R_2}$$

$$V_- = V_2 + (V_o - V_2)\frac{R_2}{R_1 + R_2}$$

If you have difficulty understanding the second of these expressions, you may find it useful to review the treatment of resistive potential dividers in Chapter 2.

As in earlier circuits, the negative feedback forces V_- to equal V_+, and therefore

$$V_+ = V_-$$

and

$$V_1 \frac{R_1}{R_1 + R_2} = V_2 + (V_o - V_2)\frac{R_2}{R_1 + R_2}$$

Figure 8.9   A differential amplifier or subtractor

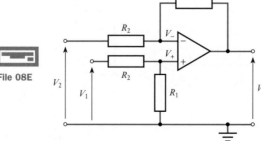

File 08E

Multiplying through by $(R_1 + R_2)$ gives

$$V_1 R_1 = V_2 R_1 + V_2 R_2 + V_o R_2 - V_2 R_2$$

which may be arranged to give

$$V_o = \frac{V_1 R_1 - V_2 R_1}{R_2}$$

and hence the ouput voltage V_o is given by

$$V_o = (V_1 - V_2)\frac{R_1}{R_2} \qquad\qquad (8.5)$$

Thus the output voltage is simply the differential input voltage $(V_1 - V_2)$ times the ratio of R_1 to R_2. Note that if $R_1 = R_2$, the output is simply $V_1 - V_2$.

8.4.4   An inverting summing amplifier (adder)

As well as subtracting one signal from another, we often need to add them together. Figure 8.10 shows a simple circuit for adding together two input signals, V_1 and V_2. This circuit can be easily expanded to sum any number of signals, simply by adding further input resistors.

The circuit is similar in form to the inverting amplifier of Section 8.3.2 with the addition of an extra input resistor. As for the earlier circuit, the inverting input to the op-amp forms a virtual earth and therefore V_- is zero. This makes the various currents in the circuit easy to calculate.

$$I_1 = \frac{V_1}{R_2}$$

$$I_2 = \frac{V_2}{R_2}$$

$$I_3 = \frac{V_0}{R_1}$$

Since no current flows into the op-amp, the external currents flowing into the virtual earth must sum to zero. Therefore

Figure 8.10 An inverting summing amplifier or adder

File 08F

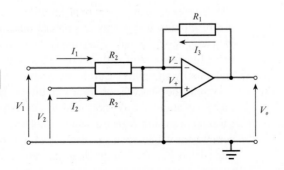

$$I_1 + I_2 + I_3 = 0$$

or, rearranging

$$I_3 = -(I_1 + I_2)$$

Substituting for the various currents then gives

$$\frac{V_0}{R_1} = -\left(\frac{V_1}{R_2} + \frac{V_2}{R_2}\right)$$

and the output voltage V_o is given by

$$V_o = -(V_1 + V_2)\frac{R_1}{R_2} \tag{8.6}$$

The output voltage is determined by the sum of the input voltages ($V_1 + V_2$) and the ratio of the resistors R_1 and R_2. The minus sign in the expression for the gain indicates that this is an inverting adder. Note that if $R_1 = R_2$ the output is simply $-(V_1 + V_2)$.

This circuit can be easily modified to add more than two input signals. Any number of input resistors may be used, and provided that they are all of value R_1 the output will become

$$V_o = -(V_1 + V_2 + V_3 + ...)\frac{R_1}{R_2} \tag{8.7}$$

Files 08C
08D
08E
08F

Computer Simulation Exercise 8.3

Simulate the various circuits described in Section 8.4, using one of the operational amplifiers supported by your simulation package. Apply appropriate input signals and confirm that the circuits operate as expected. Experiment with different values for the various resistors in your circuits and note any effects on the circuit's operation.

8.4.5 Further circuits

We have seen that operational amplifiers can be used to produce many useful circuits using only a small number of additional components. In later chapters, we will meet many other op-amp circuits for performing other functions. You will find that these are often equally simple in design and equally straightforward to analyse.

While the analysis of these circuits is generally relatively simple, in many situations we do not need to analyse these circuits at all. In many cases, we can simply take a standard cookbook circuit and select appropriate component values to customise the circuit to our needs. In such cases, we often need just the circuit diagram and an equation relating the circuit's function to the component values (as in Equations 8.1 to 8.7 above). Appendix C gives some examples of typical cookbook circuits for use in a range of situations. Some of these circuits are discussed and analysed in the text, and others are not.

We have seen in the various arrangements discussed above that the functionality of the circuit usually depends on the *relative* values of the various components rather than their absolute values. For example, in the case of the inverting amplifier circuit, the gain is given by the ratio of R_1 to R_2. This would seem to suggest that we are free to choose any values for the various resistors, provided that the ratio of their magnitudes is appropriate. This assumption would be correct if we were able to use ideal operational amplifiers in our circuits. Unfortunately, when we use real components these impose restrictions on how we must choose component values. In order to understand these restrictions, we need to know something about the nature of real devices.

8.5 Real operational amplifiers

In Section 8.2, we looked at the characteristics that we would require of an ideal operational amplifier. In that section, we deduced that an ideal device would be characterised by an infinite voltage gain, an infinite input resistance and zero output resistance.

No real op-amp can satisfy these requirements, and it is important to recognise the limitations of physical components and how these influence the design and the performance of physical circuits. In this section, we will look at various characteristics of operational amplifiers and see how these compare with those of an ideal component.

One of the problems in comparing real and ideal components is that a great many operational amplifiers are available, and the characteristics of these devices vary considerably. One of the best-known general-purpose op-amps is the **741**. This device is far from 'state of the art', but it is widely used in a range of undemanding applications. In more challenging applications it is likely that high-performance components would be used. These are often tailored to a specific class of application. For example, some components are optimised for use in situations requiring low power consumption, while others are designed to produce low levels of noise. In this section, we will look at the characteristics of general-purpose devices such as the 741, but we will also consider the range of performance achieved by other devices.

8.5.1 Voltage gain

Most operational amplifiers have a gain of between 100 and 140 dB (a voltage gain of between 10^5 and 10^7). The 741 has a gain of about 106 dB (a voltage gain of about 2×10^5), while some components have gains of

160 dB (a voltage gain of about 10^8) or more. While these gains are clearly not infinite, in many situations they are 'high enough' and gain limitations will not affect circuit operation. Unfortunately, the gain, though high, is normally subject to great variability. The gain will often vary tremendously between devices and with temperature.

8.5.2   Input resistance

The typical input resistance of a 741 is 2 MΩ, but again this quantity varies considerably between devices and may be as low as 300 kΩ. This value is low for modern op-amps, and it is not uncommon for devices that use bipolar transistors (like the 741) to have input resistances of 80 MΩ or more. In many applications, this value will be very large compared with the source resistance and may be considered high enough for loading effects to be ignored. In applications where higher input resistances are required, it is common to use devices that use field-effect transistors (FETs) in their input stages. These have a typical input resistance of about 10^{12} Ω. When using these devices, loading effects can almost always be ignored. Field-effect and bipolar transistors will be discussed in Chapters 20 and 21.

8.5.3   Output resistance

The 741 has a typical output resistance of 75 Ω, this being a typical figure for bipolar transistor op-amps. Some low-power components have a much higher output resistance, perhaps up to several thousand ohms. Often of more importance than the output resistance of a device is the maximum current that it will supply. The 741 will supply 20 mA, with values in the range 10 to 20 mA being typical for general-purpose op-amps. Special high-power devices may supply output currents of an amp or more.

8.5.4   Output voltage range

With voltage gains of several hundred thousand times, it would seem that if 1 V were to be applied to the input of an operational amplifier one would have to keep well clear of the output! However, in practice the output voltage is limited by the supply voltage. Most op-amps based on bipolar transistors (like the 741) produce a maximum output voltage swing that is slightly less than the difference between the two supply voltages. An amplifier connected to a positive supply of +15 V and a negative supply of −15 V, for example (a typical arrangement), might produce an output voltage range of about ±13 V. Op-amps based on field-effect transistors can often produce output voltage swings that go very close to both supply voltages. These are often referred to as 'rail-to-rail' devices.

8.5.5 Supply voltage range

A typical arrangement for an operational amplifier is to use supply voltages of +15 V and −15 V, although a wide range of supply voltages is usually possible. The 741, for example, can be used with supply voltages in the range ±5 V to ±18 V, this being fairly typical. Some devices allow higher voltages to be used, perhaps up to ±30 V, while others are designed for low-voltage operation, perhaps down to ±1.5 V.

Many amplifiers allow operation from a single voltage supply, which may be more convenient in some applications. Typical voltage ranges for a single supply might be 4 to 30 V, although devices are available that will operate down to 1 V or less.

8.5.6 Common-mode rejection ratio

An ideal operational amplifier would not respond to common-mode signals. In practice, all amplifiers are slightly affected by common-mode voltages, although in good amplifiers the effects are very small. A measure of the ability of a device to ignore common-mode signals is its **common-mode rejection ratio** or **CMRR**. This is the ratio of the response produced by a differential-mode signal to the response produced by a common-mode signal of the same size. The ratio is normally expressed in decibels.

Typical values for the CMRR for general-purpose operational amplifiers are between 80 and 120 dB. High-performance devices may have ratios of up to 160 dB or more. The 741 has a typical CMRR of 90 dB.

8.5.7 Input currents

For an operational amplifier to work correctly, a small input current is required into each input terminal. This current is termed the **input bias current** and must be provided by external circuitry. The polarity of this current will depend on the input circuitry used in the amplifier, and in most situations it is so small that it can be safely ignored.

Typical values for this current in bipolar op-amps range from a few microamps down to a few nanoamps or less. For the 741, this value is typically 80 nA. Operational amplifiers based on FETs have much smaller input bias currents, with values of a few picoamps being common and with values down to less than a femtoamp (10^{-15} A) being possible.

8.5.8 Input offset voltage

One would expect that, if the input voltage of the amplifier was zero, the output would also be zero. In general this is not the case. The transistors and other components in the circuit are not precisely matched, and a slight error is usually present that acts like a voltage source added to the input. This is the **input offset voltage** V_{ios}. The input offset voltage is defined as the small voltage required at the input to make the output zero.

The input offset voltage of most op-amps is generally in the range of a few hundred microvolts up to a few millivolts. For the 741, a typical value is 2 mV. This may not seem very significant, but remember that this is a voltage added to the input, and it is therefore multiplied by the gain of the amplifier. Fortunately, the offset voltage is approximately constant, so its effects can be reduced by subtracting an appropriate voltage from the input. The 741, in common with many operational amplifiers, provides connections to allow an external potentiometer to 'trim' the offset to zero. Some op-amps are **laser trimmed** during manufacture to produce a very low offset voltage without the need for manual adjustment. Unfortunately, the input offset voltage varies with temperature by a few µV/°C, making it generally impossible to remove the effects of the offset voltage completely by trimming alone.

8.5.9   Frequency response

Operational amplifiers have no lower cut-off frequency, and the gain mentioned earlier is therefore the gain of the amplifier at DC. We noted in Section 6.7 that all amplifiers have an upper cut-off frequency, and one would perhaps imagine that, to be generally useful, operational amplifiers would require very high upper cut-off frequencies. In fact this is not the case, and in many devices the gain begins to roll off above only a few hertz. Figure 8.11 shows a typical frequency response for the 741 op-amp.

The magnitude of the gain of the amplifier is constant from DC up to only a few hertz. Above this frequency, it falls steadily until it reaches unity

Figure 8.11   Typical gain vs frequency characteristic for a 741

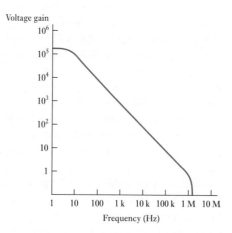

at about 1 MHz; above this frequency, the gain falls more rapidly. The upper cut-off frequency is introduced intentionally by the designer to ensure the stability of the system. We will return to the question of stability in Chapter 24.

The frequency range of an operational amplifier is usually described by the frequency at which the gain drops to unity (this is called the **transition frequency** f_T), or by its **unity-gain bandwidth**. The latter is the bandwidth over which the gain is greater than unity, and it is clear that for an operational amplifier these two measures are equal. From Figure 8.11, it can be seen that the 741 has an f_T of about 1 MHz. Typical values for f_T for other general-purpose operational amplifiers vary from a few hundred kilohertz up to a few tens of megahertz. However, a high-speed device may have an f_T of several gigahertz.

8.5.10 Slew rate

While the bandwidth determines the ability of an operational amplifier to respond to rapidly changing small signals, when large signals are used it is often the **slew rate** that is the limiting factor. This is the maximum rate at which the output voltage can change and is typically a few volts per microsecond. The effects of the slew rate are most obvious when an amplifier is required to output a large amplitude square or pulsed wave. Rather than a rapid transition from one level to another, the signal 'ramps' between the two values at a rate determined by the slew rate. The limitations of slew rate may also affect sinusoidal or other analogue signals of a large amplitude and high frequency.

8.5.11 Noise

All operational amplifiers add noise to the signals that pass through them. Noise is generated by a number of mechanisms, and these have different frequency characteristics. Some produce essentially **white noise**, meaning that it has equal power density at all frequencies (that is, the noise power within a given bandwidth is equal at all frequencies). Others produce more power in some parts of the frequency spectrum than others. For this reason, it is difficult to describe the noise performance of a given device accurately without being specific about the frequency range over which it is being used. Clearly, since noise is present at all frequencies, the amount of noise detected will depend on the bandwidth over which measurements are made. Manufacturers normally give a figure indicating the noise voltage divided by the square root of the bandwidth of measurement.

Low-noise op-amps are likely to have noise voltages of about 3 nV/√Hz. General-purpose devices may have noise voltages several orders of magnitude greater.

8.6 Selecting component values for op-amp circuits

Earlier in this chapter, we derived expressions for the gain of a range of op-amp circuits. The analysis assumed the use of an ideal amplifier and resulted in simple expressions, usually involving ratios of the values of circuit components. This implies that the absolute values of the components are unimportant. This would suggest that an inverting amplifier with a gain of 10 could be formed using resistors of 1 Ω and 10 Ω, 1 kΩ and 10 kΩ or 1 GΩ and 10 GΩ. While this would be true if we were using an ideal op-amp, it is certainly *not* true when we use real components.

In our analysis, we assumed that our operational amplifier had an infinite gain, an infinite input resistance and a zero output resistance. However, from the discussion in the last section we know that this is not true for real op-amps. Therefore, in order for our analysis to represent a reasonable model of the operation of a real circuit, we need to select external components such that the assumptions made during the calculations are reasonable. We will therefore look at each of our assumptions in turn to see what restrictions they impose on the circuit's design.

Our first assumption was that the gain of the op-amp was infinite. We used this assumption when we assumed that the input voltage to the op-amp was zero. From Chapter 7, we know that one of the requirements of effective negative feedback is that the closed-loop gain must be much less than the open-loop gain. In other words, the gain of the complete circuit with feedback must be much less than the gain of the operational amplifier without feedback.

Our second assumption was that the input resistance of the operational amplifier was infinite. We used this assumption when we assumed that the input current to the op-amp was zero. This will be a reasonable approximation provided that the currents flowing in the external components are large compared with the current into the op-amp. This will be true provided that the resistors that form the external circuitry are much smaller than the input resistance of the op-amp.

Our final assumption was that the output resistance of our operational amplifier was zero. This we used when we assumed that there would be no loading effects. This will be a reasonable assumption provided that the external resistors are much larger than the output resistance of the op-amp.

Therefore, the three assumptions will be reasonable provided that:

- We limit the gain of our circuits to a value much less than the open-loop gain of the op-amp.
- The external resistors are small compared with the input resistance of the op-amp.
- The external resistors are large compared with the output resistance of the op-amp.

From the last section, we know that the gain of our op-amp is likely to be greater than 10^5. Therefore, the assumption that the gain of the op-amp is infinite will be a reasonable approximation provided that the gain of our complete circuit is *much* less than this. Therefore, we should limit the gain of any individual circuit to 10^3 or less.

Typical values for the input resistances of bipolar operational amplifiers are in the 1 MΩ to 100 MΩ range, and a typical value for output resistance might be 10 Ω to 100 Ω. Therefore, for circuits using such devices, resistors in the 1 kΩ to 100 kΩ range would be appropriate.

Operational amplifiers based on FETs have a much higher input resistance, of the order of 10^{12} Ω or more. Circuits using these devices may therefore use higher-value resistors, of the order of 1 MΩ or more, if desired. However, resistors in the 1 kΩ to 100 kΩ range will generally produce satisfactory results with all forms of op-amp.

File 08G

Computer Simulation Exercise 8.4

Simulate the non-inverting amplifier of Example 8.1 using a 741 operational amplifier and measure its gain. Modify your circuit by replacing R_1 with a resistor of 24 Ω and R_2 with a resistor of 1 Ω, and again measure its gain. Repeat this exercise replacing the two resistors with values of 24 MΩ and 1 MΩ, and hence confirm the design rules given above. Repeat this process for the inverting amplifier of Example 8.2.

8.7 The effects of feedback on op-amp circuits

From the discussion in Chapter 7, we know that the use of negative feedback has a dramatic effect on almost all the characteristics of an amplifier. All the circuits discussed in this chapter make use of negative feedback, so it is appropriate to look briefly at its effects on various aspects of the circuits' operation.

8.7.1 Gain

Negative feedback reduces the gain of an amplifier from A to $A/(1 + AB)$. It therefore reduces the gain by a factor of $(1 + AB)$. In return for this loss of gain, feedback gives consistency since, provided that the open-loop gain is much greater than the closed-loop gain, the latter is approximately equal to $1/B$.

We have seen that an additional benefit of the use of negative feedback is that it simplifies the design process. Standard cookbook circuits can be used, and these can be analysed without needing to consider the detailed operation of the operational amplifier itself.

Figure 8.12 Gain vs frequency characteristics for a 741 with feedback

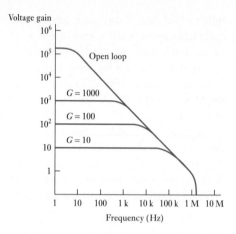

8.7.2 Frequency response

In Chapter 7, we looked at the effects of negative feedback on the frequency response and bandwidth of an amplifier. At that time, we noted that negative feedback tends to increase the bandwidth of an amplifier by maintaining its closed-loop gain constant, despite a fall in its open-loop gain.

The effects of negative feedback on a typical operational amplifier, a 741, are shown in Figure 8.12. The figure shows the frequency response of the circuit without feedback (its open-loop response) and also the response of amplifiers with different amounts of feedback. Without feedback, the amplifier has a gain of about 2×10^5 and a bandwidth of about 5 Hz. However, if feedback is used to reduce the gain to 1000, then the bandwidth increases to about 1 kHz. Decreasing the gain to 100 increases the bandwidth to about 10 kHz, while decreasing the gain to 10 increases the bandwidth to about 100 kHz. It can be seen that this behaviour illustrates the relationship discussed in Chapter 7, where we noted that in many cases

$$\text{gain} \times \text{bandwidth} = \text{constant} \tag{8.8}$$

In this case, the gain–bandwidth product is about 10^6 Hz. Note that the gain of the op-amp falls to unity at about 10^6 Hz, so in this case the gain–bandwidth product is equal to the unity-gain bandwidth.

Example 8.3

An audio amplifier is to be produced using a 741 op-amp. What is the maximum gain that can be achieved using this arrangement if the amplifier must have a bandwidth of 20 kHz?

For a 741

$$\text{gain} \times \text{bandwidth} = 10^6$$

Therefore, if the bandwidth required is 2×10^4, then the maximum gain is given by

$$\text{gain} = \frac{10^6}{\text{bandwidth}} = \frac{10^6}{2 \times 10^4} = 50$$

High-speed op-amps may have unity-gain bandwidths of a gigahertz or more, allowing the production of wide-bandwidth amplifiers that also have high gain. However, not all op-amps have a frequency response of the form shown in Figure 8.12, and in this case the relationship between gain and bandwidth is not so straightforward.

File 08H

Computer Simulation Exercise 8.5

Simulate a non-inverting amplifier with a gain of 10 based on a 741 operational amplifier, plot its frequency response and measure its bandwidth. Repeat this process with your circuit modified to produce gains of 1 and 100. In each case, calculate the product of the gain and the bandwidth, and hence investigate the relationship of Equation 8.8. Also compare the gain–bandwidth product with the unity-gain bandwidth.

8.7.3   Input and output resistance

We noted in Section 7.6 that negative feedback can be used to either increase or decrease both the input and output resistance of a circuit. We also noted that the amount by which these resistances are changed is given by the expression $(1 + AB)$. Since this is also the factor by which the gain is reduced, we can determine the value of this expression simply by dividing the open-loop gain by the closed-loop gain. For example, if an op-amp with an open-loop gain of 2×10^5 is used to produce an amplifier with a gain of 100, then $(1 + AB)$ must be equal to $2 \times 10^5/100 = 2 \times 10^3$. Note that when using negative feedback, the factor $(1 + AB)$ will always be positive. If we use an op-amp to produce an inverting amplifier with a gain of -100, we are effectively using the op-amp in a configuration where its gain is -2×10^5, so $(1 + AB)$ is equal to $-2 \times 10^5/-100 = 2 \times 10^3$ as before.

In order to determine whether the feedback increases or decreases the output resistance, we need to see whether it is the output voltage or the output current that is being used to determine the feedback quantity. In all the circuits discussed in this chapter, it is the output voltage that is being used to determine the feedback, and therefore in each case the feedback *reduces* the output resistance.

In order to determine whether the input resistance is increased or decreased, we need to determine whether it is a voltage or a current that is being subtracted at the input. In the case of the non-inverting amplifier of Section 8.3.1, it is a voltage that is subtracted from the input voltage to form the input to the op-amp. Thus in this circuit the feedback *increases* the input resistance by a factor of $(1 + AB)$. In the case of the inverting amplifier of Section 8.3.2, it is a current that is subtracted from the input current to form the input to the op-amp and therefore the feedback *decreases* the input resistance. In this particular circuit, the resistor R_2 goes from the input to the virtual earth point. Therefore, the input resistance is simply equal to R_2.

When considering other circuits, one needs to look at the quantity being fed back and the quantity being subtracted from the input to determine the effects of the feedback on the input and output resistance.

Example 8.4   Determine the input and output resistance of the following circuit, assuming that the operational amplifier is a 741.

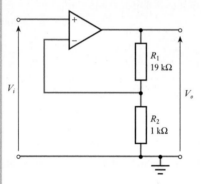

The open-loop gain of a 741 is typically 2×10^5 and the closed-loop gain of this circuit is 20. Therefore, $(1 + AB) = (2 \times 10^5)/20 = 10^4$.

The output resistance of a 741 is typically about 75 Ω, and in this circuit the output *voltage* is fed back. Thus the feedback *reduces* the output resistance by a factor of $(1 + AB)$, which becomes $75/10^4 = 7.5$ mΩ.

The input resistance of a 741 is typically about 2 MΩ, and in this circuit a feedback *voltage* is subtracted from the input voltage. Thus the feedback *increases* the input resistance by a factor of $(1 + AB)$, which becomes $2 \times 10^6 \times 10^4 = 2 \times 10^{10} = 20$ GΩ.

Example 8.5   Determine the input and output resistance of the following circuit, assuming that the operational amplifier is a 741.

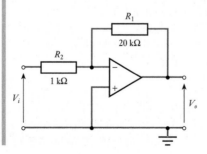

The open-loop gain of a 741 is typically 2×10^5 and the magnitude of the closed-loop gain of this circuit is 20. Therefore, $(1 + AB) = (2 \times 10^5)/20 = 10^4$.

The output resistance of a 741 is typically about 75 Ω, and in this circuit the output *voltage* is fed back. Thus the feedback *reduces* the output resistance by a factor of $(1 + AB)$, which becomes $75/10^4 = 7.5$ mΩ.

The input resistance of a 741 is typically about 2 MΩ, and in this circuit a feedback *current* is subtracted from the input current. Thus the feedback *decreases* the input resistance. In this case, the input is connected to a virtual earth point by the resistance R_2, so the input resistance is equal to R_2, which is 1 kΩ.

Examples 8.4 and 8.5 illustrate the very dramatic effects that feedback can have on the characteristics of a circuit. While the input and output resistances of an operational amplifier make it a good voltage amplifier, the use of feedback can turn it into an excellent one. This is most striking in the case of the buffer amplifier of Section 8.4.1, where feedback produces such a high input resistance and such a low output resistance that loading effects can almost always be ignored. This is shown in Example 8.6.

| **Example 8.6** | Determine the input and output resistance of the following circuit, assuming that the operational amplifier is a 741. |

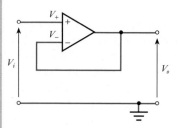

The open-loop gain of a 741 is typically 2×10^5 and the closed-loop gain of this circuit is 1. Therefore, $(1 + AB) = (2 \times 10^5)/1 = 2 \times 10^5$.

The output resistance of a 741 is typically about 75 Ω, and in this circuit the output *voltage* is fed back. Thus the feedback *reduces* the output resistance by a factor of $(1 + AB)$, which becomes $75/(2 \times 10^5) \approx 400$ $\mu\Omega$.

The input resistance of a 741 is typically about 2 MΩ, and in this circuit a feedback *voltage* is subtracted from the input voltage. Thus the feedback *increases* the input resistance by a factor of $(1 + AB)$, which becomes $(2 \times 10^6) \times (2 \times 10^5) = 4 \times 10^{11} = 400$ GΩ.

While it is clear that negative feedback can dramatically improve the input and output resistance of a circuit, it should be remembered that this improvement is brought about at the expense of a loss in gain. Since the open-loop gain of the operational amplifier changes with frequency (as shown in Figure 8.11), so will the input and output resistance. The various calculations and examples above use the low-frequency open-loop gain of the op-amp, and therefore the values obtained represent the resistances at very low frequencies. As the frequency increases, the gain of the op-amp falls, and the improvement brought about by feedback will be reduced.

8.7.4   Stability

While negative feedback can be used to tailor the characteristics of an operational amplifier for a given application, its use does have implications for the stability of the circuit. We will return to look at considerations of stability in Chapter 24.

Key points

- Operational amplifiers are among the most widely used building blocks for the construction of electronic circuits.

- Op-amps are small integrated circuits that typically take the form of a plastic package containing one or more amplifiers.

- Although they are often omitted from circuit diagrams, op-amps require connections to power supplies (typically +15 V and −15 V) in order to function.

- An ideal operational amplifier would have an infinite voltage gain, an infinite input resistance and zero output resistance.

- Designers often base their designs on a number of standard cookbook circuits. Analysis of these circuits is greatly simplified if we assume the use of an ideal op-amp.

- Standard circuits are available for various forms of amplifier, buffer, adder, subtractor and many other functions.

- Real operational amplifiers have several non-ideal characteristics. However, if we choose component values appropriately, these factors should not affect the operation of our cookbook circuits.

- When designing op-amp circuits, we normally use resistors in the range 1 kΩ to 100 kΩ.

- Feedback allows us to increase the bandwidth of the circuit dramatically by trading off gain against bandwidth.

- Feedback allows us to tailor the characteristics of an op-amp to suit a particular application. We can use feedback to overcome problems associated with the variability of the gain of the op-amp, and we can also either increase or decrease the input and output resistance depending on our requirements.

Exercises

8.1 What is meant by the term 'integrated circuit'?

8.2 Explain the abbreviations DIL and SMT as applied to IC packages.

8.3 What are typical values for the positive and negative supply voltages of an operational amplifier?

8.4 Outline the characteristics of an 'ideal' op-amp.

8.5 Sketch an equivalent circuit of an ideal operational amplifier.

8.6 Determine the gain of the following circuit.

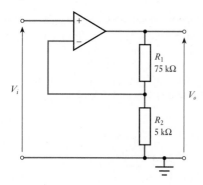

8.7 Sketch the circuit diagram of a non-inverting amplifier with a gain of 30.

8.8 Use circuit simulation to investigate your solution to the last exercise. Use one of the operational amplifiers supported by your simulation package and apply a DC input voltage of 100 mV. Hence confirm that the circuit works as expected.

8.9 Determine the gain of the following circuit.

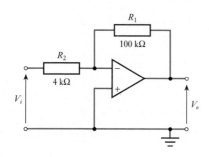

8.10 Sketch the circuit diagram of an inverting amplifier with a gain of −30.

8.11 Use circuit simulation to investigate your solution to the last exercise. Use one of the operational amplifiers supported by your simulation package and apply a DC input voltage of 100 mV. Hence confirm that the circuit works as expected.

8.12 Sketch a circuit that takes two input signals V_A and V_B and produces an output equal to $10(V_B - V_A)$.

8.13 Sketch a circuit that takes four input signals V_1 to V_4 and produces an output equal to $5(V_1 + V_2 + V_3 + V_4)$.

8.14 Derive an expression for the output V_o of the following circuit in terms of the input voltages V_1 and V_2 and hence determine the output voltage if $V_1 = 1$ V and $V_2 = 0.5$ V.

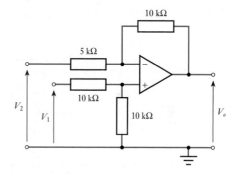

8.15 Derive an expression for the output V_o of the following circuit in terms of the input voltages V_1 and V_2 and hence determine the output voltage if $V_1 = 1$ V and $V_2 = 0.5$ V.

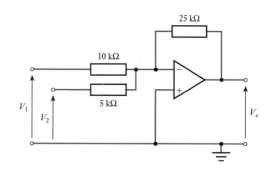

Exercises continued

8.16 What are typical ranges for the open-circuit voltage gain, input resistance and output resistance of general-purpose operational amplifiers?

8.17 What are typical ranges for the supply voltages of general-purpose operational amplifiers?

8.18 What is meant by the term 'common-mode rejection ratio'? What would be a typical CMRR for a general-purpose op-amp?

8.19 Explain the term 'input bias current'.

8.20 Define the term 'input offset voltage' and give a typical figure for this quantity. How may the effects of the input offset voltage be reduced?

8.21 Sketch a typical frequency response for a 741 op-amp. What is its upper cut-off frequency? What is its lower cut-off frequency?

8.22 Give a typical value for the gain–bandwidth product of a 741. How does this relate to the unity-gain bandwidth?

8.23 If an amplifier with a gain of 25 is constructed using a 741, what would be a typical value for the bandwidth of this circuit?

8.24 What is meant by the slew rate of an op-amp? What would be a typical value for this parameter?

8.25 What range of resistor values would normally be used for circuits based on a bipolar operational amplifier?

8.26 Estimate the gain, input resistance and output resistance of the following four circuits at low frequencies, assuming that each is constructed using an operational amplifier that has an open-loop gain of 10^6, an input resistance of $10^6\ \Omega$ and an output resistance of $100\ \Omega$.

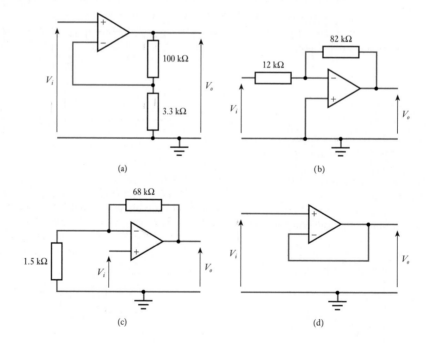

Chapter 9

Digital Systems

Objectives

When you have studied the material in this chapter you should be able to:

- define terms such as binary variable, logic state, logic gates, and logic operators such as AND, OR, NOT and Exclusive OR;
- use truth tables and Boolean algebra to represent the operation of simple logic functions;
- design arrangements of standard logic gates to perform particular functions that are specified in words or symbolically;
- perform addition and subtraction using binary arithmetic and convert numbers between a range of number bases;
- describe codes used for the representation of numeric and non-numeric quantities such as alphabetic characters.

9.1 Introduction

In Chapters 3 and 4 we considered a range of digital sensors and actuators, and in Chapter 5 we looked at the signals associated with such devices. It is now time to look at the techniques used to process digital signals and at the design of digital systems.

Although digital signals can take many forms, in this chapter we are primarily concerned with **binary** signals, since these are the most common form of digital information. Binary signals may be used individually, perhaps to represent the state of a single switch, or in combination to represent more complex quantities. Here we will start by looking at the processing of individual binary quantities and then move on to more complex arrangements.

9.2 Binary quantities and variables

A **binary quantity** is one that can take only two states. Examples include a switch that can be only ON or OFF, a hydraulic valve that can be only OPEN or CLOSED, and an electric heater that can be only ON or OFF. It is common to represent such quantities by **binary variables**, which are simply symbolic names for the quantities.

Figure 9.1   A simple binary arrangement

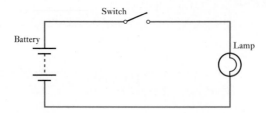

Figure 9.1 illustrates a simple binary arrangement involving a battery, a switch and a lamp. If the state of the switch is represented by the binary variable S and the state of the lamp by the binary variable L, we can represent the relationship between these two variables symbolically using a table:

S	L
OPEN	OFF
CLOSED	ON

We can also use a symbolic name for the state of each variable, so that rather than using terms such as OPEN and CLOSED, or ON and OFF, we can use symbols for the states such as '0' and '1'. If we use the symbol '0' to represent the switch being OPEN and the lamp being OFF, our table becomes

S	L
0	0
1	1

The mapping between ON and OFF and '0' and '1' is arbitrary, but the user must know what the relationship is. It is common to use '1' to represent the ON state, a switch being CLOSED or a statement being TRUE. It is common to use '0' for the OFF state, a switch being OPEN or a statement being FALSE. The table lists on the left all the possible states of the switch and indicates, on the right, the corresponding states of the lamp. Such a table is called a **truth table**, and it defines the relationship between the two variables. The order in which the possible states are listed is normally *ascending binary order*. If you are not aware of the meaning of this phrase, it will become clear when we look at number systems and binary arithmetic later in this chapter.

Figure 9.2(a) shows an arrangement incorporating two switches in series. Here it is necessary for both switches to be closed in order for the lamp to light. The relationship between the positions of the switches and

Figure 9.2 Two switches in series

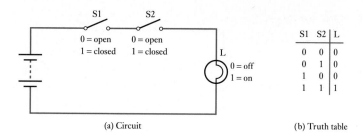

(a) Circuit

S1	S2	L
0	0	0
0	1	0
1	0	0
1	1	1

(b) Truth table

Figure 9.3 Two switches in parallel

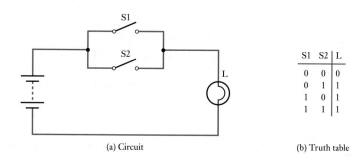

(a) Circuit

S1	S2	L
0	0	0
0	1	1
1	0	1
1	1	1

(b) Truth table

the state of the lamp is given in the truth table of Figure 9.2(b). Notice that the table now has four rows to represent all the possible combinations of the two switches. Alternatively, we could express this relationship in words as 'the lamp will be illuminated if, and only if, switch S1 is closed AND switch S2 is closed'. We can abbreviate this statement as

$$L = \text{S1 AND S2}$$

This AND relationship is very common in electronics systems and is found in a variety of everyday applications. For example, automotive brake lamps are often only illuminated if the foot brake is depressed, closing a switch, AND the ignition switch is ON.

Figure 9.3(a) shows an arrangement that has two switches in parallel. In this configuration, the lamp will light if either of the switches is closed. This function is described in the truth table of Figure 9.3(b), where the meanings of '0' and '1' are as for the previous example. As before, we can express this relationship in words as 'the lamp will be illuminated if, and only if, switch S1 is closed OR switch S2 is closed (or if both are closed)' or, in the abbreviated form

$$L = \text{S1 OR S2}$$

An example of the OR function, again an automotive application, might be the courtesy light, which is illuminated if the driver's door is open (closing a switch) OR if the passenger's door is open. This function is sometimes called the Inclusive OR function, since it includes the case where both inputs are true (that is, in this case where both switches are closed).

Our examples of the AND and OR functions can be extended to the use of three or more switches, as illustrated in Figures 9.4 and 9.5. These

Figure 9.4 Three switches in series

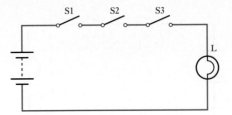

S1	S2	S3	L
0	0	0	0
0	0	1	0
0	1	0	0
0	1	1	0
1	0	0	0
1	0	1	0
1	1	0	0
1	1	1	1

Figure 9.5 Three switches in parallel

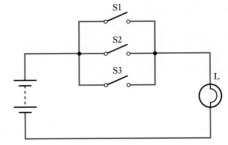

S1	S2	S3	L
0	0	0	0
0	0	1	1
0	1	0	1
0	1	1	1
1	0	0	1
1	0	1	1
1	1	0	1
1	1	1	1

Figure 9.6 A series/parallel configuration

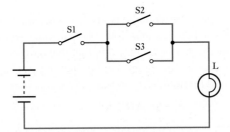

S1	S2	S3	L
0	0	0	0
0	0	1	0
0	1	0	0
0	1	1	0
1	0	0	0
1	0	1	1
1	1	0	1
1	1	1	1

figures show three switches, but the process can be expanded, allowing any number of switches to be connected in series or in parallel. Note that there are eight possible combinations of the positions of three switches, leading to eight rows in the truth tables.

Consider now the circuit of Figure 9.6. Here two switches are connected in parallel, and this combination is in series with a third switch. This produces an arrangement that can be described by the truth table as shown, or by the statement 'the lamp will be illuminated if, and only if, S1 is closed AND either S2 OR S3 is closed'. This can again be given in an abbreviated form as

$$L = \text{S1 AND (S2 OR S3)}$$

Notice the use of parentheses to make the meaning of the expression clear and to avoid ambiguity.

In the examples so far considered, we have started with a combination of switches and represented them by a truth table and a verbal description. In practice, we will generally need to perform the process in reverse, being given a function and being required to devise an arrangement to produce

Figure 9.7 Representation of an unknown network

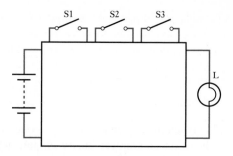

this effect. In such cases, we might be given either a truth table or a verbal description of the required system. Consider the following truth table:

S1	S2	S3	L
0	0	0	0
0	0	1	0
0	1	0	0
0	1	1	1
1	0	0	0
1	0	1	1
1	1	0	1
1	1	1	0

This represents an arrangement with three switches, S1, S2 and S3, and a lamp L. We would consider the switches to be the three *inputs* to the network and the lamp to be the *output*. It is not immediately obvious what arrangement of switches would correspond to this truth table, and perhaps it is not clear whether *any* combination of the three switches can produce the desired results. In such cases, it is often useful to consider the desired arrangement as a 'black box' with the various switches as inputs and the lamp as an output. Such an arrangement is shown in Figure 9.7.

The diagram of Figure 9.7 makes no assumptions concerning the method of interconnection of the switches and the lamp. It may be that in order to produce the desired function we will need some form of electronic circuitry in our 'black box'. Since the three switches and the lamp represent simple binary devices, we could produce a more general arrangement by showing these as simple binary variables without defining their type. You will remember that, when representing 'black box' amplifiers, such as operational amplifiers, we often omit the connections to the power supply. If we adopt a similar scheme here, we arrive at a diagram of the form shown in Figure 9.8.

We now have a symbolic representation of a network with three inputs and one output that makes no assumptions as to the form of the inputs or the output. This can clearly be extended to represent systems with any number of inputs and outputs. The inputs could represent switches, as in the earlier examples, but they could equally well be signals from binary sensors such as thermostats, level sensors or proximity switches. Similarly, the output devices could be lamps but could equally well be heaters or

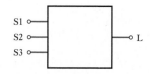

Figure 9.8 Symbolic representation of an unknown network

solenoids. To implement our digital system, we need to take the various input signals and use them to produce appropriate output signals. The normal building blocks used to achieve this are **logic gates**.

9.3   Logic gates

A logic gate is an element that takes one or more binary input signals and produces an appropriate binary output, depending on the state(s) of the input(s). There are three elementary gate types, two of which, the AND and the OR functions, we have already met. These elementary gates can be combined to form more complicated gates, which in turn may be connected to produce any required function. Each type of gate has its own **logic symbol**, which allows complex functions to be represented by a **logic diagram**. The function of each gate can also be represented by a mathematical notation known as **Boolean notation**. This allows complex functions to be manipulated, and perhaps simplified, through the use of **Boolean algebra**.

9.3.1   Elementary logic gates

The AND gate

The output of an AND gate is true (1) if, and only if, all of the inputs are true. The gate can have any number of inputs. The logic symbol and truth table for a two-input AND gate are given in Figure 9.9. The labelling of the inputs and outputs is arbitrary. The Boolean notation for the AND function is a dot. For example, the gate in Figure 9.9 could be described by the expression $C = A \cdot B$. In practice, it is common to omit the dot, and the expression is often written as $C = AB$.

The OR gate

The output of an OR gate is true (1) if, and only if, at least one of its inputs is true. It is also called the Inclusive OR gate for the reasons discussed earlier. The gate can have any number of inputs. The logic symbol and truth table for a two-input OR gate are given in Figure 9.10. The Boolean notation for the OR function is '+'. For example, the gate in Figure 9.10 could be described by the expression $C = A + B$.

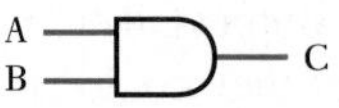

Figure 9.9   A two-input AND gate

A	B	C
0	0	0
0	1	0
1	0	0
1	1	1

$C = A \cdot B$

(a) Circuit symbol   (b) Truth table   (c) Boolean expression

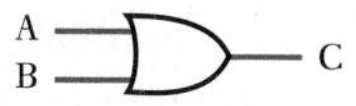

Figure 9.10 A two-input OR gate

A
B — C

A	B	C
0	0	0
0	1	1
1	0	1
1	1	1

$C = A + B$

(a) Circuit symbol   (b) Truth table   (c) Boolean expression

The NOT gate

The output of a NOT gate is true (1) if, and only if, its single input is false. This gate has the function of a **logical inverter**, since the output is the **complement** of the input. The gate is sometimes referred to as an **invert gate** or simply as an **inverter**. The circuit symbol and truth table for a NOT gate are shown in Figure 9.11. In Boolean notation, inversion is represented by putting a line (a bar) above the expression for the signal. The operation of the gate in Figure 9.11 can be written as $B = \overline{A}$, which is read as 'B equals NOT A' or as 'B equals A bar'.

The circle in the symbol for an inverter represents the process of inversion. The triangular symbol without the circle would represent a function in which the output state was identical to the input. This function is called a **buffer**. The presence of a buffer does not affect the state of a logic signal. However, when we come to consider the implementation of gates using electronic circuits, we shall see that a buffer can be used to change the electrical properties of a logic signal. It is interesting to note that the symbol for a buffer is similar to that used for a single-input analogue amplifier. Since the buffer does not produce any logical function, it is usually not considered as an elementary gate. However, for completeness its logic symbol and truth table are given in Figure 9.12. Clearly, in this case $B = A$.

Figure 9.11 A NOT gate (inverter)

A — B

A	B
0	1
1	0

$B = \overline{A}$

(a) Circuit symbol   (b) Truth table   (c) Boolean expression

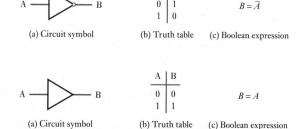

Figure 9.12 A logical buffer

A — B

A	B
0	0
1	1

$B = A$

(a) Circuit symbol   (b) Truth table   (c) Boolean expression

9.3.2 Compound gates

The elementary gates described above can be combined to form any desired logic function. However, it is often more convenient to work with slightly larger building blocks. Several compound gates are used that are simple arrangements of these elementary gates.

The NAND gate

The NAND gate is functionally equivalent to an AND gate followed by an inverter, the name being an abbreviation of Not AND. Following the example set with the symbol for an inverter, the logic symbol for a NAND gate is simply that for an AND gate with a circle at the output. The truth table for the NAND gate is similar to that for an AND gate with the output state inverted. A NAND gate can have any number of inputs. The logic symbol and truth table for a two-input NAND gate are shown in Figure 9.13. This function would be written as $C = \overline{A \cdot B}$, or simply $C = \overline{AB}$.

Figure 9.13   A two-input NAND gate

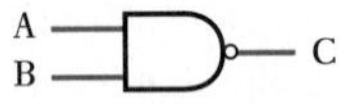

A	B	C
0	0	1
0	1	1
1	0	1
1	1	0

$$C = \overline{A \cdot B}$$

(a) Circuit symbol   (b) Truth table   (c) Boolean expression

The NOR gate

The NOR gate is functionally equivalent to an OR gate followed by an inverter, the name being an abbreviation of Not OR. Again the logic symbol is that of an OR gate with a circle at the output to indicate an inversion. A NOR gate can have any number of inputs. Figure 9.14 shows the logic symbol and truth table of a two-input NOR gate. This function would be written as $C = \overline{A + B}$.

Figure 9.14   A two-input NOR gate

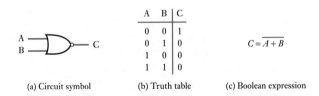

A	B	C
0	0	1
0	1	0
1	0	0
1	1	0

$$C = \overline{A + B}$$

(a) Circuit symbol   (b) Truth table   (c) Boolean expression

The Exclusive OR gate

The output of an Exclusive OR gate is true (1) if, and only if, one or other of its two inputs is true, but not if both are true. The gate gets its name from the fact that it resembles the Inclusive OR gate, except that it *excludes* the case where both inputs are true. An Exclusive OR gate always has only two inputs. The logic symbol and truth table for an Exclusive OR gate are given in Figure 9.15. The Exclusive OR function has its own Boolean symbol, which is '⊕'. The arrangement of Figure 9.15 would therefore be written as $C = A \oplus B$.

Figure 9.15 An Exclusive OR gate

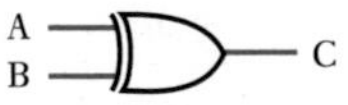

A	B	C
0	0	0
0	1	1
1	0	1
1	1	0

$C = A \oplus B$

(a) Circuit symbol   (b) Truth table   (c) Boolean expression

The Exclusive NOR gate

The last member of our group of compound gates is the Exclusive NOR gate, which, as its name suggests, is the inverse of the Exclusive OR gate. This may be considered to be an Exclusive OR gate followed by an inverter. This gate gives a true output when both inputs are 0 or when both are 1. It therefore gives a true output when the inputs are equal. For this reason, this gate is also known as an **equivalence** or an **equality gate**. The logic symbol and a truth table for the Exclusive NOR gate are shown in Figure 9.16. This would be written as $C = \overline{A \oplus B}$.

Figure 9.16 An Exclusive NOR gate

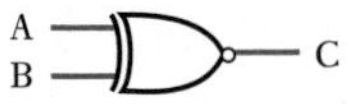

A	B	C
0	0	1
0	1	0
1	0	0
1	1	1

$C = \overline{A \oplus B}$

(a) Circuit symbol   (b) Truth table   (c) Boolean expression

9.3.3 Using logic gates

The various logic gates are summarised in Table 9.1, which shows their circuit symbols, Boolean expressions and truth tables. The table shows two circuit symbols for each gate. The first is the 'distinctive-shape' symbol given earlier, and the second is an alternative symbol defined in international standard IEC 617. Both forms are widely used, but in this text we will adopt the distinctive-shape symbols since these are probably more widely used in engineering courses at this level.

Using appropriate combinations of these gates, it is possible to implement any required relationship between a set of binary inputs and outputs. Applications might range from the use of a handful of gates to produce a simple control mechanism to the use of perhaps millions of gates to produce a complete microcomputer.

Table 9.1   Logic gates

Function	Symbol	Alternative symbol	Boolean expression	Truth Table
Buffer	A ▷ B	A [1] B	$B = A$	A | B 0 | 0 1 | 1
NOT	A ▷o B	A [1] B	$B = \overline{A}$	A | B 0 | 1 1 | 0
AND	A,B ⟹ C	A,B [&] C	$C = A \cdot B$	A B | C 0 0 | 0 0 1 | 0 1 0 | 0 1 1 | 1
OR	A,B ⟹ C	A,B [≥1] C	$C = A + B$	A B | C 0 0 | 0 0 1 | 1 1 0 | 1 1 1 | 1
NAND	A,B ⟹o C	A,B [&] C	$C = \overline{A \cdot B}$	A B | C 0 0 | 1 0 1 | 1 1 0 | 1 1 1 | 0
NOR	A,B ⟹o C	A,B [≥1] C	$C = \overline{A + B}$	A B | C 0 0 | 1 0 1 | 0 1 0 | 0 1 1 | 0
Exclusive OR	A,B ⟹ C	A,B [=1] C	$C = A \oplus B$	A B | C 0 0 | 0 0 1 | 1 1 0 | 1 1 1 | 0
Exclusive NOR	A,B ⟹o C	A,B [=1] C	$C = \overline{A \oplus B}$	A B | C 0 0 | 1 0 1 | 0 1 0 | 0 1 1 | 1

9.4 Boolean algebra

Boolean algebra defines constants, variables and functions to describe binary systems. It also defines a number of theorems that can be used to manipulate, and perhaps simplify, logic expressions.

9.4.1 Boolean constants

Boolean constants consist of '0' and '1'. The former represents the false state and the latter the true state.

9.4.2 Boolean variables

Boolean variables are quantities that can take different values at different times. They may represent the input, output or intermediate signals and are given names usually consisting of alphabetic characters, such as 'A', 'B', 'X' or 'Y'. Variables may only take the values '0' or '1'.

9.4.3 Boolean functions

Each of the elementary logic functions (such as AND, OR and NOT) are represented by unique symbols (such as '+', '·' and '‾'). These various symbols were introduced in the previous section.

9.4.4 Boolean theorems

Boolean algebra has a set of rules that define how it can be used. These consist of a set of **identities** and a set of **laws**, which are summarised in Table 9.2. Many of these rules are self-evident (given a little thought about the meaning of the relevant expression), while others are less obvious. These various rules may be used to simplify algebraic expressions, or simply to change their form to aid implementation. We will look at algebraic simplification in the next section.

9.5 Combinational logic

Digital systems can be divided into two broad categories. In the first, the outputs are determined solely by the current states of the inputs to the circuit, and such arrangements are described as **combinational logic**. In the second form of system, the outputs are determined not only by the current inputs but also by the sequence of inputs that has led to the current state. Such systems are known as **sequential logic**. In this section, we shall look

Table 9.2 Summary of Boolean algebra identities and laws

Boolean identities

AND function	OR function	NOT function
$0 \cdot 0 = 0$	$0 + 0 = 0$	$\overline{0} = 1$
$0 \cdot 1 = 0$	$0 + 1 = 1$	$\overline{1} = 0$
$1 \cdot 0 = 0$	$1 + 0 = 1$	$\overline{\overline{A}} = A$
$1 \cdot 1 = 1$	$1 + 1 = 1$	
$A \cdot 0 = 0$	$A + 0 = A$	
$0 \cdot A = 0$	$0 + A = A$	
$A \cdot 1 = A$	$A + 1 = 1$	
$1 \cdot A = A$	$1 + A = 1$	
$A \cdot A = A$	$A + A = A$	
$A \cdot \overline{A} = 0$	$A + \overline{A} = 1$	

Boolean laws

Commutative law	Absorption law
$AB = BA$	$A + AB = A$
$A + B = B + A$	$A(A + B) = A$
Distributive law	**De Morgan's law**
$A(B + C) = AB + BC$	$\overline{A + B} = \overline{A} \cdot \overline{B}$
$A + BC = (A + B)(A + C)$	$\overline{A \cdot B} = \overline{A} + \overline{B}$
Associative law	**Note also**
$A(BC) = (AB)C$	$A + \overline{A}B = A + B$
$A + (B + C) = (A + B) + C$	$A(\overline{A} + B) = AB$

at the design of combinational logic circuits and will leave sequential logic until Chapter 10.

We have seen that logic functions can be described in a number of ways. For example, we can describe the required operation of a system by a Boolean expression, in words, or by a truth table. Therefore, in order to be able to design and use logic gates effectively, we need to be able to take descriptions in any of these forms and from them generate a circuit diagram of an arrangement to perform that function. It is also useful to be able to perform the reverse operation of taking a circuit diagram and generating from it a description of its functionality.

9.5.1 Implementing a logic function from a Boolean expression

One of the many advantages of the use of Boolean algebra is that it produces an unambiguous description of a system that can be easily converted into a circuit diagram. Since a Boolean expression combines its various terms using AND, OR or NOT operations, these can be implemented directly as a logic circuit using the corresponding logic gates. This process is illustrated in Example 9.1.

Example 9.1   Implement the function $X = A + B\overline{C}$.

This expression has one output (X) and three inputs (A, B and C). X is formed by ORing together two components, A and $B\overline{C}$. The first of these is one of the inputs while the second is formed by ANDing together B and the inverse of C (remember that $B\overline{C}$ is a shorthand notation for $B \cdot \overline{C}$). Therefore, the circuit diagram is

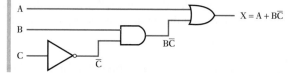

This process can also be used for more complex expressions, as illustrated in Example 9.2.

Example 9.2   Implement the function $Y = \overline{\overline{A}B + C\overline{D}}$.

This expression has one output (Y) and four inputs (A, B, C and D). In this example, two terms are ORed together and the result is inverted. These two operations may be combined by the use of a NOR gate. Y is therefore formed by NORing together two components, $\overline{A}B$ and $C\overline{D}$. These components in turn are formed by ANDing together signals derived from the inputs.

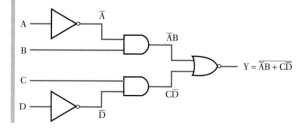

It can be seen that a simple way of implementing a Boolean expression is to identify the major elements in the expression and to note how these are combined. This defines the logic gate that will generate the output signal. We then work backwards to identify the nature of the inputs to this gate, to see how these are formed from other components. This process is then repeated until the required signals can be generated directly from the input signals. Where systems have more than one output, each output will be represented by a separate Boolean expression that can be implemented separately.

**Files 09A
    09B**

Computer Simulation Exercise 9.1

Simulate the logic circuits of Examples 9.1 and 9.2 and confirm that they produce the required outputs for all possible combinations of the inputs.

Computer Simulation Exercise 9.2

Simulate the Exclusive OR gates of Examples 9.4 and 9.5 and confirm that they each produce the required output for all possible combinations of the inputs. Hence prove the equivalence of these two implementations.

9.5.4   Implementing a logic function from a truth table

If our required system is defined by a truth table, we again produce a Boolean expression from the table and implement it as before. The task of producing an expression from a truth table can be easily understood if we bear in mind what the table actually represents, and here an example might be useful. Consider the following truth table, which is for an Exclusive NOR gate.

A	B	C
0	0	1
0	1	0
1	0	0
1	1	1

The table lists on the right-hand side the output for each possible combination of the values of the inputs. Where the output is a '1', this corresponds to a set of inputs for which the output is true. Therefore, if we list these combinations, we have a list of all the conditions for which the output is true. The function can then be described by saying that the output will be true if, and only if, the inputs correspond to one or other of the combinations in this list. In the example above, the only combinations for which the output is true are when both A and B are '0' and when they are both '1'. If A and B are both zero, then $\overline{A}$ and $\overline{B}$ must be equal to '1'. Therefore, the first condition corresponds to $\overline{A}$ AND $\overline{B}$ being equal to '1', and the second to A AND B being equal to '1'. The function can therefore be described as

$$C = \overline{A}\,\overline{B} + AB$$

Looking back at the truth table, it can be seen that there is a very simple relationship between the combinations of inputs for which the output is '1' and the resulting Boolean expression. This makes it very easy to write down the Boolean expression directly from the truth table. Once we have this expression, we can implement the function as before.

Example 9.6

Implement the function of the following truth table.

A	B	C	X
0	0	0	0
0	0	1	1
0	1	0	0
0	1	1	0
1	0	0	0
1	0	1	1
1	1	0	1
1	1	1	0

There are three combinations of inputs for which the output is true, therefore the expression will have three terms ORed together. By inspection, the expression is

$$X = \overline{A}\overline{B}C + A\overline{B}C + AB\overline{C}$$

This can be implemented as

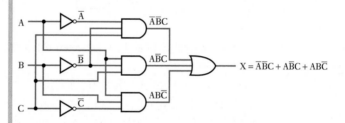

It can be seen that as circuit diagrams grow in complexity it becomes increasingly difficult to follow the interconnections. This problem can be reduced by using labels to indicate connections rather than drawing interconnecting lines. Using this approach, the diagram above becomes that shown below, which is much easier to understand.

9.5.5   Algebraic simplification

We have looked at several methods of implementing combinational logic circuits. In each case, the final stage of the implementation has been to take a Boolean expression for the required function and to represent it by a combination of logic gates in a circuit diagram. Having noted earlier that

Boolean expressions are not unique, this leads us to wonder whether a particular expression is the simplest way of describing a particular function. In some cases, it may be possible to simplify a Boolean expression, using the Boolean identities and laws listed in Table 9.2. This is illustrated in the following example.

Example 9.7

Implement the Boolean expression

$$X = ABC + \overline{A}BC + AC + A\overline{C}$$

This can be implemented directly as

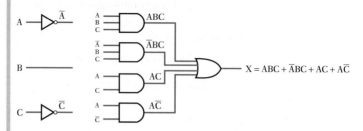

Alternatively, it can be rearranged using the commutative and distributive laws to give

$$X = BC(A + \overline{A}) + A(C + \overline{C})$$

$$= BC + A$$

which can be implemented as

File 09D

Computer Simulation Exercise 9.3

Simulate the two implementations of Example 9.7 and confirm that they each produce the required output for all possible combinations of the inputs. Hence prove the equivalence of these two implementations.

Another approach to the simplification of Boolean expressions is the use of a graphical method based on **Karnaugh maps**. However, despite this technique being extensively taught in electronics textbooks, very few engineers ever use it. This is because Karnaugh maps can only be used in applications with up to about six input variables, whereas real engineering applications invariably have many more than this. In such situations, it is possible to use computer-based methods such as **Quine–McCluskey minimisation**, but few engineers will need to come to grips with this. We will see in

Chapter 27, when looking at the implementation of digital systems, that real systems are rarely produced by assembling large numbers of individual logic gates. Where combination logic circuitry is used, it will often be constructed using array logic, which combines large numbers of gates in a single integrated circuit. When using such components, the software tools used to design the ICs perform **logic minimisation** automatically. For these reasons, we will not pursue the minimisation of logic functions further.

9.5.6 Propagation delay

So far we have considered logic gates purely from a functional viewpoint and have ignored any issues related to their implementation. In practice, physical logic gates take a finite time to respond to input signals, and there is a delay between the time when the input signals change and when the output responds. This delay is termed the **propagation delay time**. When using modern electronic components this time is very short (often less than a nanosecond) and in many cases will be unimportant. However, in some situations this delay can affect the operation of the circuit and must be taken into account in the design. We will look at propagation delay when we consider the implementation of digital components in Chapter 25.

9.6 Number systems and binary arithmetic

So far we have dealt with simple binary signals, such as those produced by switches and those required to turn lamps ON or OFF. Sometimes groups of binary signals are combined to form binary words. These words can be used to represent various forms of information, the most common being *numeric* and *alphabetic* data. When numerical information is represented, this permits arithmetic operations to be performed on the data.

9.6.1 Number systems

The decimal number system

In everyday arithmetic, we use numbers with a base of 10, this choice being almost certainly related to the fact that we have ten fingers and thumbs. This system requires ten symbols to represent the values that each digit may take, for which we use the symbols 0, 1, 2, . . . , 9. Our numbering system is 'order-dependent' in that the significance of a digit in a number depends on its position. For example, the number

   1234

means 1 *thousand*, plus 2 *hundreds*, plus 3 *tens* plus 4 *units*. Each column of the number represents a power of 10, starting with units on the right-hand side ($10^0 = 1$), and moving to increasing powers of 10 as we move to the left. Thus our number is

$$1234 = (1 \times 10^3) + (2 \times 10^2) + (3 \times 10^1) + (4 \times 10^0)$$

Digits at the left-hand side of the number are of much greater significance than those on the right-hand side. For this reason, the left-hand digit is termed the **most significant digit (MSD)**, while that on the right is termed the **least significant digit (LSD)**.

The numbering system can be extended to represent magnitudes that are not integer quantities by extending the sequence below the units column, a decimal point being placed to the right of the units column to indicate its position. Thus

$$1234.56 = (1 \times 10^3) + (2 \times 10^2) + (3 \times 10^1) + (4 \times 10^0) + (5 \times 10^{-1}) + (6 \times 10^{-2})$$

Numbers of any size can be represented by using a sufficiently large number of digits; leading zeros have no effect on the magnitude of the number, provided that they are to the left of the decimal point. Similarly, trailing zeros have no effect if they are to the right of the decimal point.

The binary number system

Binary numbers have similar characteristics to decimal numbers, except that they have a base of 2. Since each digit may now take only two values, only two symbols are required. These are usually 0 and 1. One advantage of this system is that digits can be represented by any binary quantity, such as a switch position or a lamp being ON or OFF.

Since binary quantities use symbols that are also used in decimal numbers, it is common to identify the notation being used by adding a subscript indicating the base. Thus

$$1101_2$$

is a binary number, and

$$1101_{10}$$

is a decimal number. In many cases the base is known, or is obvious, in which case the subscript is usually omitted.

Like their decimal counterparts, the digits in a binary word are also position-dependent. As before, the digits represent ascending powers of the base, such that

$$1101_2 = (1 \times 2^3) + (1 \times 2^2) + (0 \times 2^1) + (1 \times 2^0)$$

Therefore, rather than having units, tens, hundreds and thousands columns as in decimal numbers, we have 1s, 2s, 4s, 8s, 16s . . . columns.

Fractional parts may also be represented, as

$$1101.01_2 = (1 \times 2^3) + (1 \times 2^2) + (0 \times 2^1) + (1 \times 2^0) + (0 \times 2^{-1}) + (1 \times 2^{-2})$$

The position of the units column is now indicated by a **binary point** (rather than a decimal point), and columns to the right of this point represent magnitudes of $1/2, 1/4, \ldots, 1/2^n$.

The term *bi*nary digi*t* is often abbreviated to **bit**. Thus a binary number consisting of eight digits would be referred to as an 8-bit number.

Other number systems

Although there are clear reasons for using both decimal and binary numbers, any integer may be used as the base of a number system. For reasons that are unimportant at this stage, common numbering systems include those using bases of 8 (octal) and 16 (hexadecimal or simply hex). Octal numbers require eight symbols and use $0, 1, \ldots, 7$. Hexadecimal numbers require sixteen symbols and use $0, 1, \ldots, 9$, A, B, C, D, E and F. From the above discussion of decimal and binary numbers, it is clear that

$$123_8 = (1 \times 8^2) + (2 \times 8^1) + (3 \times 8^0)$$

and that

$$123_{16} = (1 \times 16^2) + (2 \times 16^1) + (3 \times 16^0)$$

Table 9.3 gives the numbers 0 to 20_{10} in decimal, binary, octal and hexadecimal.

Table 9.3 Number representations

Decimal	Binary	Octal	Hexadecimal
0	0	0	0
1	1	1	1
2	10	2	2
3	11	3	3
4	100	4	4
5	101	5	5
6	110	6	6
7	111	7	7
8	1000	10	8
9	1001	11	9
10	1010	12	A
11	1011	13	B
12	1100	14	C
13	1101	15	D
14	1110	16	E
15	1111	17	F
16	10000	20	10
17	10001	21	11
18	10010	22	12
19	10011	23	13
20	10100	24	14

9.6.2 Number conversion

Most scientific calculators can perform conversions between number bases automatically, but it is perhaps useful to see how this process can be performed 'manually'. This is instructive, if only because it gives an insight into the relationships between the various number systems.

Conversion from binary to decimal

Converting binary numbers into decimal is straightforward. It is achieved simply by adding up the decimal values of each '1' in the number.

For small numbers this conversion can be performed quite simply using mental arithmetic. Larger numbers take a little longer. Numbers with fractional parts can be converted in the same manner by adding the decimal equivalent of each term.

Example 9.8 | Convert 11010_2 to decimal.

$$11010_2 = (1 \times 2^4) + (1 \times 2^3) + (0 \times 2^2) + (1 \times 2^1) + (0 \times 2^0)$$

$$= 16 + 8 + 0 + 2 + 0$$

$$= 26_{10}$$

Conversion from decimal to binary

Conversion from decimal to binary is effectively the reverse of the above process, although the similarity is not at first apparent. It is achieved by repeatedly dividing the number by 2 and noting any remainder. This procedure is repeated until the number vanishes.

Example 9.9 | Convert 26_{10} to binary.

	Number	Remainder
Starting point	26	
÷ 2	13	0
÷ 2	6	1
÷ 2	3	0
÷ 2	1	1
÷ 2	0	1

read number from this end
= 11010

Thus

$$26_{10} = 11010_2$$

Numbers with fractional parts are converted in parts, the integer part being converted as above and the fractional part being converted by repeated multiplication by 2, noting, and then discarding, the overflow beyond the binary point after each multiplication. As with fractional parts in decimal numbers, the number of places used to the right of the binary point depends on the accuracy required.

Example 9.10

Convert 34.6875_{10} to binary.

First the whole number part (34) is converted as before

	Number	Remainder
Starting point	34	
÷ 2 =	17	0
÷ 2 =	8	1
÷ 2 =	4	0
÷ 2 =	2	0
÷ 2 =	1	0
÷ 2 =	0	1

read number from this end
= 100010

then the fraction part (0.6875) is converted

	Overflow	Number	
read number from top		.6875	× 2 =
= 0.1011	1	.375	× 2 =
	0	.75	× 2 =
	1	.5	× 2 =
	1	.0	

Thus

$$34.6875_{10} = 100010.1011_2$$

Conversion from hexadecimal to decimal

Conversion from hexadecimal to decimal is similar to the conversion from binary to decimal, except that powers of 16 are used in place of powers of 2.

Example 9.11

Convert $A013_{16}$ to decimal.

$$A013_{16} = (A \times 16^3) + (0 \times 16^2) + (1 \times 16^1) + (3 \times 16^0)$$

$$= (10 \times 4096) + (0 \times 256) + (1 \times 16) + (3 \times 1)$$

$$= 40960 + 0 + 16 + 3$$

$$= 40979_{10}$$

Conversion from decimal to hexadecimal

Conversion from decimal to hexadecimal is likewise similar to the conversion from decimal to binary, except that divisions and multiplications are by 16 rather than by 2.

Example 9.12

Convert 7046_{10} to hexadecimal.

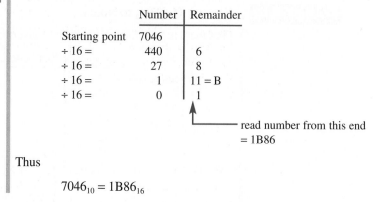

	Number	Remainder
Starting point	7046	
÷ 16 =	440	6
÷ 16 =	27	8
÷ 16 =	1	11 = B
÷ 16 =	0	1

read number from this end
= 1B86

Thus

$$7046_{10} = 1B86_{16}$$

Conversions between other bases

Conversions between other number bases and decimal are similar to those described for binary and hexadecimal numbers using the appropriate power or multiplication factor. Conversion between two non-decimal number bases can be achieved using decimal as an intermediate stage. This involves converting a number in one number system into decimal form and then converting it from decimal into the target number system.

It is also possible to convert directly between number systems, although this is sometimes tricky since most of us are strongly bound to thinking in decimal numbers. Examples of direct conversions that are easy to achieve include translations from binary to hexadecimal and vice versa. These conversions are straightforward because each hexadecimal digit corresponds to exactly four binary digits (4 bits). This allows each hexadecimal digit to be converted directly. All that is required is a knowledge of the binary equivalent of each of the sixteen hexadecimal digits (as given in Table 9.3) and the translation is trivial.

Example 9.13

Convert $F851_{16}$ to binary.

$$F851_{16} = (1111)(1000)(0101)(0001)$$

$$= 1111100001010001_2$$

| Example 9.14 | Convert 1111100001010001_2 to hexadecimal. |

$$111011011000100_2 = (0111)(0110)(1100)(0100)$$

$$= 76C4_{16}$$

Note that, when arranging binary numbers into groups of four for conversion into hexadecimal, the grouping begins with the right-most digit (the LSD) and extra leading zeros are added at the left-hand side as necessary.

From the above example, it is clear that large binary numbers are unwieldy and difficult to remember. Because it is so easy to convert between binary and hexadecimal, it is very common to use the latter in preference to the former for large numbers. 76C4 is much easier to write and remember than 111011011000100.

9.6.3 Binary arithmetic

One of the many advantages of using binary rather than decimal representations of numbers is that arithmetic is much simpler. To see why this is true, one only has to consider that in order to perform decimal long multiplication one needs to know all the products of all possible pairs of the ten decimal digits. To perform binary long multiplication one only needs to know that $0 \times 0 = 0$, $0 \times 1 = 1 \times 0 = 0$ and that $1 \times 1 = 1$. This simplicity is a characteristic of all forms of binary arithmetic but for the moment let us consider just addition.

Binary addition

The addition of two single-digit binary quantities is a very simple task, the rules of which may be summarised as

$$0 + 0 = 0$$
$$0 + 1 = 1$$
$$1 + 0 = 1$$
$$1 + 1 = 10$$

The simplicity of this arrangement suggests that it should be relatively easy to construct an electronic circuit that could perform this operation for us – an **adder circuit**. This would take two inputs representing the two binary digits to be added and produce two outputs representing their sum. Note that two outputs are required since it can be seen that the addition of two single-digit numbers can give rise to a two-digit number. The least significant digit of the answer (the right-hand digit) is termed the *sum* output, and the most significant digit is termed the *carry* output (since this digit would carry over to the next column in a multi-digit sum). This arrangement is shown in Figure 9.17(a). As discussed in Section 9.5, we will start our design by constructing the truth table of our required function; this is

Figure 9.17 A binary half adder

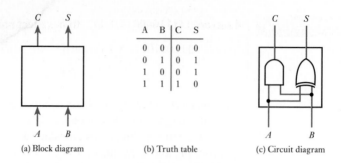

A	B	C	S
0	0	0	0
0	1	0	1
1	0	0	1
1	1	1	0

(a) Block diagram   (b) Truth table   (c) Circuit diagram

shown in Figure 9.17(b). This shows that both the outputs are very easy to produce. The C output is simply the AND function of the two inputs, while the S output is the Exclusive OR of the inputs. A circuit to implement this arrangement is shown in Figure 9.17(c). For reasons that will become clear shortly, such a circuit is referred to as a **half adder**.

While it may be useful to be able to add together single digits, in many situations we wish to add together multiple-bit quantities. For example, we might wish to add together two 8-bit words. One approach to this problem would be to design a circuit that has sixteen inputs to represent the two 8-bit words. Unfortunately, the design of such a circuit would prove difficult using the techniques we adopted above, since our truth table would have over 65,000 rows! When *we* perform addition manually we add the digits separately, and this seems a sensible approach in this situation.

In order to add together two multiple-digit words we need individual circuits to add each corresponding bit. Such an arrangement is shown in Figure 9.18, which shows a circuit for adding two 4-bit words. The least significant bits of each word (the right-hand bits) are added together by a half adder circuit similar to that shown in Figure 9.17. The sum output S

Figure 9.18 An arrangement to add two 4-bit binary numbers

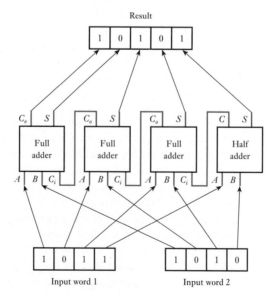

from this circuit forms the least significant bit of the result, while the carry output C is fed forward to the next 'column' in the addition. This next stage needs to add together the second digit from each input word, plus the carry digit from the previous stage. Therefore, we need a slightly more complicated circuit than the half adder discussed earlier. This circuit must add three input signals and is termed a **full adder**. Since the full adder has a carry input from the previous stage and a carry output that feeds to the next stage, these are labelled C_i and C_o to avoid confusion. By adding an appropriate number of full adders it is possible to add together words of any length. In each case, the S output produces the corresponding element in the output result. The carry output from the final stage then forms the most-significant bit of the result.

The design of a full adder can be tackled in a similar manner to that of the half adder described earlier, and this is left as an exercise for the reader.

Binary subtraction

A similar approach to that described above can be used to produce both **half subtractors** and **full subtractors**, and as before these can be combined to form a circuit that will subtract binary numbers of any desired length. Again the design of such circuits is left as an exercise for the reader.

Arithmetic in electronic systems

We have seen that it is relatively easy to construct circuits to perform binary addition and subtraction. It is also possible to create electronic circuitry that will perform multiplication and division (although these are considerably more complex). However, in modern systems it is far more common for arithmetic operations to be performed in **microprocessors** or **programmable logic devices** (**PLDs**) than it is for these operations to be performed in custom-designed circuits. Microprocessors and sophisticated PLDs contain **arithmetic logic units** (**ALUs**), which can perform addition and subtraction as well as a range of other operations. More complex procedures, such as multiplication and division, can then be performed by combinations of simple operations. Some devices also have hardware multiply circuitry, which greatly speeds up complex processes.

9.7 Numeric and alphabetic codes

9.7.1 Binary code

By far the most common method of representing numeric information in digital systems is by the use of the simple binary code described earlier. This has the advantages of simplicity of arithmetic and efficiency of storage. However, there are some applications in which other representations are used for specific purposes.

9.7.2 Binary-coded decimal code

Binary-coded decimal (BCD) code, as its name implies, is formed by converting each digit of a decimal number individually into its binary form.

Example 9.15

Convert 9450_{10} to BCD.

$$9450_{10} = (1001)(0100)(0101)(0000)_{BCD}$$

Conversion from BCD to decimal is just as simple and is achieved by dividing the number into groups of four, starting with the least significant digit and then converting each digit into decimal. Additional leading zeros can be used to complete the last group, if required.

Example 9.16

Convert 11100001110110_{BCD} to decimal.

$$11100001110110_{BCD} = (0011)(1000)(0111)(0110)_{BCD}$$

$$= 3876_{10}$$

BCD requires more digits than the straight binary form and is therefore less efficient; however, it has the advantage of very simple conversions to, and from, decimal. It is therefore widely used in situations where input and output data are in a decimal form, such as in pocket calculators.

9.7.3 ASCII code

So far we have concentrated on codes that are used to represent numeric quantities. Often it is also necessary to store and transmit alphabetic data in digital form, for example for storing text in a computer. Various standard codes are used for this purpose, but by far the most widely used is the **American Standard Code for Information Interchange**, which is normally abbreviated to **ASCII** (pronounced 'ass-key').

The full standard represents each character by a 7-bit code, allowing 128 possible values. Codes are defined for both upper and lower case alphabetic characters; the digits 0 to 9; punctuation marks such as commas, full stops and question marks; and various non-printable codes that are used as control characters. Since codes are included for both alphabetic and numeric characters, codes of this form are often referred to as **alphanumeric** codes.

9.7.4 Error detection and correction techniques

We have seen in earlier chapters that all electronic systems suffer from noise. One possible effect of noise in digital systems is the corruption of data. This is a particular problem when data must be transmitted from one place to another.

One method of tackling the problem of data corruption during transmission is to add some form of redundant information to the data at the transmitter that allows it to be checked at the receiving end. A simple example of this technique is the use of **parity checking**. Here an additional bit of information (the parity bit) is added to each word at the transmitter. The polarity of this added bit is chosen so that the number of 1's in the word (including the added parity bit) is either always even (*even parity*) or always odd (*odd parity*). On reception, the parity of each received word is checked, any change in parity indicating an error.

An alternative method of checking the correctness of data is to use a **checksum**. This provides a test of the integrity of a block of data rather than of individual words. When a group of words is to be transmitted, the words are summed at the transmitter and the sum is transmitted after the data. At the receiver, the words are again summed and the result compared with the sum produced by the transmitter. If the results agree the data is probably correct. If they do not, an error has been detected.

The parity and checksum techniques both send a small amount of redundant information to allow the integrity of the data to be tested. If one is prepared to send additional redundant information, it is possible to construct codes that not only detect the presence of errors but also indicate their location within a word, allowing them to be corrected. The performance of these codes in terms of their ability to detect and correct multiple errors depends on the amount of redundant information that can be tolerated. The more redundancy that is incorporated, the greater is the rate at which data must be sent and the more complicated the system. It should also be remembered that it is not possible to construct a code that will allow an unlimited number of errors. This would imply that the system could produce the correct output with a random input – clearly an impossibility.

Error-correcting codes are employed extensively in many forms of digital transmission and storage, and are used in compact discs (CDs), mobile phones, computer disk drives and many other applications. The use of error detection plays a large part in guaranteeing the high fidelity of the information in these systems.

Key points

- To simplify the description of binary variables it is common to represent their two states by the symbols '1' and '0'. These might represent ON and OFF, TRUE and FALSE or any other pair of binary conditions.

- In some simple cases it is possible to implement binary systems using switches. However, it is more generally useful to design such systems using logic gates.

- Our basic building blocks are a small number of simple gates. Three elementary forms, AND, OR and NOT, can be used to form any logic function, although it is sometimes more useful to work with compound gates such as NAND, NOR and Exclusive OR.

- Digital arrangements where the outputs are determined solely by the current states of the inputs are described as combinational logic circuits.

- Combinational circuits can be described by a truth table that lists all the possible combinations of the inputs and indicates the corresponding values of the outputs.

- It is also possible to define a logic function using Boolean algebra. This notation and set of rules and identities allows binary relationships to be described and simplified.

- The implementation of logic circuits normally starts with a description of the required function in words, a truth table or a Boolean expression. In each case the implementation is straightforward.

- In addition to binary variables, digital systems often use many-valued quantities, which are represented by binary words of an appropriate length.

- Several number systems are used to represent numerical quantities, the most common being decimal, binary, octal and hexadecimal.

- Since binary numbers use only two digits, 0 and 1, arithmetic is simpler than in decimal.

- Codes are also used for non-numeric information, such as the ASCII code, which is used for alphanumeric data.

- Some coding techniques allow error detection and possibly correction.

Exercises

9.1 Show how a power source, a lamp and a number of switches can be used to represent the following logical functions

$$L = A \cdot B \cdot C$$

$$L = A + B + C$$

$$L = (A \cdot B) + (C \cdot D)$$

$$L = A \oplus B$$

9.2 Derive expressions for the following arrangements using AND, OR and NOT operations.

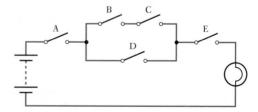

Exercises continued

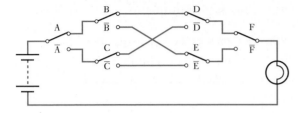

9.3 If the two circuits given in the previous exercise were described by truth tables, how many rows would each table require?

9.4 Sketch the truth table of a three-input NAND gate.

9.5 Sketch the truth table of a three-input NOR gate.

9.6 Show that the two circuits (a) and (b) below are equivalent by drawing truth tables for each circuit.

(a)   (b)

9.7 Repeat the operations of Exercise 9.6 for the following circuits

(a)   (b)

9.8 Simulate the pairs of circuits in Exercises 9.6 and 9.7 and confirm that each pair produces the same output for every possible combination of the inputs.

9.9 List all the possible values of a Boolean constant.

9.10 List all the possible values of a Boolean variable.

9.11 What symbols are used in Boolean algebra to represent the functions AND, OR, NOT and Exclusive OR.

9.12 Write the function of a three-input NOR gate as a Boolean expression.

9.13 Given that A is a Boolean variable, evaluate and hence simplify the following expressions: $A \cdot 1$; $A \cdot \overline{A}$; $1 + A$; $A + \overline{A}$; $1 \cdot 0$; $1 + 0$.

9.14 Exercises 9.6 and 9.7 illustrate fundamental laws of Boolean algebra. What is the name given to these laws?

9.15 What is the difference between combination and sequential logic?

9.16 Implement the following expressions using standard logic gates.

$$X = (\overline{A + B}) \cdot C$$

$$Y = A\overline{B}C + \overline{A}D + C\overline{D}$$

$$Z = \overline{(A \cdot B) + (\overline{C + D})}$$

9.17 Derive a Boolean expression for the following circuit.

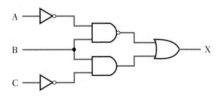

9.18 Design a logic circuit to take three inputs, A, B and C, and produce a single output X, such that X is true if, and only if, precisely two of its inputs are true. *Hint:* you might find it helpful to produce a truth table of this arrangement before implementing it.

9.19 Use circuit simulation to investigate your solution to Exercise 9.18 and hence demonstrate that it behaves as required.

9.20 Convert the following binary numbers into decimal: 1100, 110001, 10111, 1.011.

9.21 Convert the following decimal numbers into binary: 56, 132, 67, 5.625.

Exercises continued

9.22 Convert the following hexadecimal numbers into decimal: A4C3, CB45, 87, 3FF.

9.23 Convert the following decimal numbers into hexadecimal: 52708, 726, 8900.

9.24 Convert $A4C7_{16}$ into binary.

9.25 Convert 10110010100101_2 into hexadecimal.

9.26 Perform the following binary arithmetic:

$$
\begin{array}{cccc}
1\,0\,1\,1\,1 & 1\,1\,0\,1\,0\,1 & 1\,0\,1\,1 & 1\,0\,1\,0\,1\,0 \\
+1\,0\,0\,1 & -1\,1\,0\,1\,0 & \times 1\,1\,1 & \div 1\,1\,0 \\
\hline
\end{array}
$$

Chapter 10

Sequential logic

Objectives

When you have studied the material in this chapter you should be able to:

- describe the characteristics of a wide range of sequential logic circuits, including bistables, monostables and astables;
- explain the differences between various forms of bistable, including latches, flip-flops and pulse-triggered versions;
- discuss the use of bistables in the construction of memory registers and shift registers;
- design simple binary ripple counters and modulo-*N* counters of any length;
- appreciate the role of specialised sequential integrated circuits such as monostables, astables and timers.

10.1 Introduction

We have seen that in combination logic the outputs are determined only by the current state of the inputs. In sequential logic, however, the outputs are determined not only by the current inputs but also by the sequence of inputs that led to the current state. In other words, the circuit has the characteristic of **memory**.

When constructing combinational logic, our basic building blocks are normally the various gates described in Section 9.3. When constructing sequential logic, we generally use slightly larger building blocks, which are often some form of **multivibrator**. This term describes a range of circuits that share certain characteristics. They each have two outputs, which are the inverse of each other and which are conventionally given the names Q and $\bar{Q}$. Having only these two outputs means that the circuits have only two possible output states, namely $Q = 1$, $\bar{Q} = 0$ and $Q = 0$, $\bar{Q} = 1$. Different forms of multivibrator are defined by the behaviour of the circuits in these two states.

- **Bistable multivibrators** are stable in each of the output states.
- **Monostable multivibrators** have one stable and one metastable state.
- **Astable multivibrators** have no stable states.

Of these, bistable circuits are by far the most important and widely used. We will therefore start by looking at a few basic forms of bistable and then go on to look at some of their uses.

10.2 Bistables

10.2.1 The S–R latch

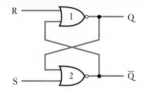

Figure 10.1 An S–R latch

An example of a simple bistable is shown in Figure 10.1. The circuit consists of two interconnected NOR gates, and it has two inputs (S and R) and two outputs (Q and $\overline{Q}$). To demonstrate that this circuit has two stable states, consider the situation when S and R are both at logical 0. If we assume that Q is 1, then the inputs to gate 2 are 1 and 0, and its output, $\overline{Q}$, will be 0. The inputs to gate 1 are therefore both 0, so its output, Q, will be 1 – which was our original assumption. Therefore, this condition is self-consistent and the arrangement is stable. If we now assume that Q is 0, then both inputs to gate 2 are 0, so its output, $\overline{Q}$, will be 1. This means that one of the inputs to gate 1 is 1, so its output will be 0. Again, this is self-consistent and the arrangement is stable. Therefore, while S and R are both 0, the circuit has two stable states and it will simply stay in whichever state it finds itself. We sometimes refer to this as the **memory state** of the circuit. Note that in each of its stable states the two outputs are the inverse of each other, justifying our labelling them as Q and $\overline{Q}$.

Now consider the effect of taking S to 1, while R remains at 0. Regardless of the previous state of the circuit, if S goes to 1, $\overline{Q}$ will be driven to 0. Both inputs to gate 1 will now be 0, forcing Q to 1. If S now returns to 0, the circuit will remain in this stable state. Thus taking S to 1 forces Q to 1 and $\overline{Q}$ to 0. If we consider the effect of taking R to 1 while S remains at zero, we will find that this has the effect of driving Q to 0 and $\overline{Q}$ to 1.

The behaviour of the circuit can be described by saying that, when S is 'activated' (or 'asserted') by taking it to 1, it *sets* the Q output to 1, and that, when R is activated by taking it to 1, it *resets* the Q output to 0. When neither input is active (that is, they are both at 0) the circuit simply stays in the state it is in. It can be seen that the circuit has the characteristic that when both inputs are 0 it remembers which of its two inputs was last taken to 1. We could therefore consider our circuit to be a **single-bit memory element**. Because of the way in which it responds to its two inputs, the circuit of Figure 10.1 is called a **SET–RESET latch** or simply an **S–R latch**.

It should be noted that if both S and R are active at the same time (that is, they are both at 1) then both outputs will be driven to 0. Under these conditions, the outputs are no longer the inverse of each other and the circuit

Figure 10.2 S–R latch logic symbols

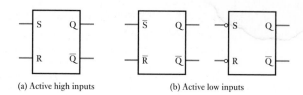

(a) Active high inputs   (b) Active low inputs

Figure 10.3 Sample input and output waveforms for an S–R latch

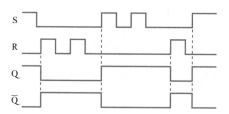

no longer functions as a multivibrator. For this reason, this combination of inputs is normally prohibited.

Circuits of the form described above are referred to as having **active high** inputs, since they produce their intended function when they are taken high (that is, to logic '1'). Other circuits have inputs that are normally kept high, and they create their intended function when they are taken low (to logic '0'). Such gates are said to have **active low** inputs. To avoid confusion, inputs are labelled to indicate whether they are active high or active low. For example, an active high reset input would be labelled R, while an active low reset input would be labelled $\bar{R}$.

Since bistables are used as basic circuit elements, we normally concern ourselves with their interconnection rather than their internal circuitry. Therefore, the circuit is given its own circuit symbol as shown in Figure 10.2. This figure shows the symbols used for circuits with both active high and active low inputs. Note that active low inputs can be represented in two ways: either by labelling the inputs as $\bar{S}$ and $\bar{R}$; or by labelling the inputs as S and R and showing inverters (represented by circles) at the inputs.

To illustrate the operation of the S–R latch, Figure 10.3 shows the relationship between signals on the inputs and the outputs of such a gate, for a set of sample input signals. A possible use of an S–R bistable is shown in the following example.

| Example 10.1 | Design of a burglar alarm. |

A burglar alarm consists of a series of switches connected to doors and windows throughout a building. Opening any door or window opens the corresponding switch, which should sound the alarm. It is essential that the alarm continues to sound if the door or window concerned is subsequently closed. Some method must be incorporated to silence the alarm when the building has been checked. The following arrangement satisfies the requirements.

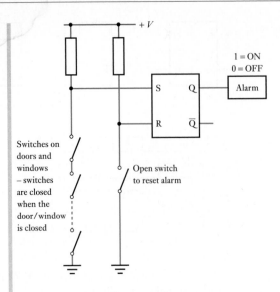

An S–R latch has two pull-up resistors connected to its S and R inputs. The various door and window switches are connected in series and wired so that, when all are closed, they short the S input to ground. The R input is similarly shorted to ground by a RESET switch, which is normally closed.

Initially, the system is reset by momentarily opening the RESET switch with all the sensor switches closed. The latch will be reset with $Q = 0$ and the alarm will be off. Once the RESET switch has been closed the system is armed. If one of the sensor switches is opened, by the opening of a door or window, the S input will go high, setting Q to 1 and sounding the alarm. If now the sensor switch is closed, the system will remain in the alarm state until it is reset by opening the RESET switch.

While S–R latches illustrate the basic operation of bistable elements, they are rarely used. Most applications use slight variations on this basic circuit that offer more attractive circuit characteristics. Many forms of bistable are available, but it would be inappropriate to look at the detailed design of all these circuits. Most engineers are concerned with the external behaviour of such circuits rather than the details of their construction. Therefore, we will simply look at the general characteristics of some of the more important types. While all bistables normally have two outputs, Q and $\overline{Q}$, they differ in the number and form of their inputs.

10.2.2   The D latch

The **D latch** or **data latch** is one of the most widely used forms of bistable and is also known as the **transparent D latch**. Its circuit symbol is given in Figure 10.4(a), which shows that it has two inputs, the data input D and

Figure 10.4   A D latch

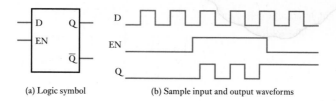

(a) Logic symbol   (b) Sample input and output waveforms

the enable input *EN*. When the signal on the enable input is high (1), the output *Q* is equal to the signal on the *D* input. In this state the gate is effectively transparent (hence its alternative name) and the output follows the input. However, when the enable input goes low (0), the output is held constant at whatever value it has at that instant. Thus the circuit stores the value of *D* at the time when *EN* goes low. Sample waveforms for a D latch are shown in Figure 10.4(b). This, and later waveform diagrams, show only the *Q* output, since the $\overline{Q}$ output is simply its inverse.

D latches are often used in groups to store words of information. It is common to combine a number of such devices in a single integrated circuit to give, perhaps, four bits (a quad latch) or eight bits (an octal latch) of storage in a single device.

10.2.3   Edge-triggered devices and the D flip-flop

In many situations, it is necessary to synchronise the operation of a number of different circuits, and it is useful to be able to control precisely when a circuit will change state. Some bistables are constructed so that they only change state upon the application of a **trigger** signal. This trigger signal is defined as the rising or falling edge of an input signal termed the **clock**. These devices are termed **edge-triggered bistables** or, more commonly, **flip-flops**. These are divided into those that are triggered by the rising edge of the clock signal (so-called positive edge-triggered devices) and those that are triggered on the falling edge of the clock (negative edge-triggered devices).

Flip-flops are available in a number of different forms, including the S–R flip-flop and the D flip-flop, which are edge-triggered versions of the latches discussed earlier. The circuit symbols used for these circuits are similar to those of the corresponding latch, except that the enable input is replaced with a clock input. The clock input is conventionally indicated by a triangle; an inverting circle is used to show a negative edge-triggered device. Figure 10.5 shows circuit symbols for both a positive edge-triggered and a

Figure 10.5   D flip-flops

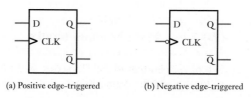

(a) Positive edge-triggered   (b) Negative edge-triggered

Figure 10.6 Sample waveforms
for a positive edge-triggered D
flip-flop

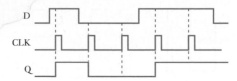

negative edge-triggered D flip-flop, and Figure 10.6 shows a set of sample
waveforms for a positive edge-triggered device.

10.2.4 J–K flip-flop

The J–K flip-flop is perhaps one of the most versatile and widely used
forms of bistable, since it can be configured to perform a range of tasks. As
its name suggests, it has two inputs, J and K, and these are similar in some
respects to the S and R inputs of an S–R flip-flop. When asserted alone, the
J input sets Q to 1, while when K is asserted it resets Q to 0. As in the S–R
device, when neither input is asserted the circuit is in its memory state, but
the operation of the arrangement is different in the case where both inputs
are asserted simultaneously. This is an ambiguous situation in the case of
an S–R bistable and is therefore avoided. In the case of the J–K device,
when both inputs are active the circuit changes state (or toggles) upon the
application of a trigger event. Figure 10.7 shows the circuit symbol for a
negative edge-triggered J–K flip-flop and a set of sample waveforms. Since
this is an edge-triggered device, the state of the inputs is only of importance
at the instant of the trigger event, which in this case is the falling edge of
the clock. Therefore, Figure 10.7(b) marks these events with dotted lines
and indicates the state of J and K at these times, together with the corresponding action.

Figure 10.7 A negative edge-
triggered J–K flip-flop

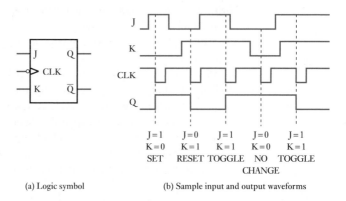

(a) Logic symbol   (b) Sample input and output waveforms

10.2.5 Asynchronous inputs

We have seen that in flip-flops the *control inputs* (for example the J and K
inputs of a J–K flip-flop) affect the operation of the circuit only at the

Figure 10.8   A J–K flip-flop
with PRESET and CLEAR

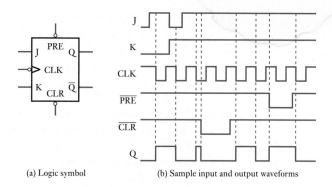

(a) Logic symbol   (b) Sample input and output waveforms

moment of an appropriate transition of the clock signal (*CLK*). We therefore refer to these inputs as **synchronous**, since their operation is synchron-
ised to the clock input.

In many applications it is advantageous to be able to set or clear the
output at other times, independent of the clock. Therefore, some devices
have additional inputs to perform these functions. These are termed
asynchronous inputs, since they are not bound by the state of the clock.
Unfortunately, IC manufacturers are unable to agree on common names for
these inputs: they may be called PRESET and CLEAR; DC SET and DC
CLEAR; SET and RESET; or DIRECT SET and DIRECT CLEAR. Here
we will use the names PRESET (*PRE*) and CLEAR (*CLR*). As with control
input, these lines can be active high or active low, although it is more common for them to be active low. Figure 10.8(a) illustrates how these inputs
are shown in the circuit symbol of a J–K flip-flop, and Figure 10.8(b) gives
sample waveforms.

10.2.6   Propagation delay and races

We noted in the last chapter that logic gates take a finite time to operate.
Since bistables are constructed from gates, these will also exhibit a delay
between their inputs and outputs. As in logic gates, this delay is termed the
propagation delay time of the device, and under certain circumstances this
can lead to problems. Consider the circuit of Figure 10.9. This shows a
situation where two edge-triggered devices are connected to the same clock

Figure 10.9   A possible race
condition

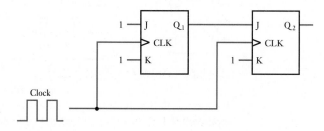

signal, while the output of one device forms an input to the other. On the rising edge of the clock signal, the first flip-flop will respond by changing its output. This process will take a finite time determined by the propagation delay time of the device. During this time the second flip-flop will also be responding to the clock signal, which it shares with the first device. If it is fast enough it may have responded before Q_1 changes but, if not, its input will change before it has had a chance to act. The final output state of the second flip-flop is far from certain and will depend on the relative speeds of the two devices. This uncertainty of operation is referred to as a **race** condition, since the outcome is determined by a race between the two components.

Race hazards may be tackled by careful design and, in fact, the design of edge-triggered devices aims to prevent such problems by arranging that the input signals must be stable for a certain time (the **hold time**) before the clock event.

10.2.7 Pulse-triggered bistables or master/slave flip-flops

Another method of overcoming race problems is the use of **master/slave flip-flops** rather than edge-triggered devices. These are also known as **pulse-triggered flip-flops**.

The construction of these devices involves using two bistables in series, a master and then a slave device. However, the resultant circuit behaves like a single bistable in which the outputs are *determined* while the clock signal is high, but where the outputs *change* only when the clock falls. Thus on the falling edge of the clock the outputs take up the value determined by the state of the inputs a short time before.

Master/slave versions are available for a range of bistables, such as S–R, D and J–K types. The logic symbol usually includes the label 'M/S', and the triangle used on the clock of edge-triggered devices is omitted. Figure 10.10 shows the circuit symbol of a J–K master/slave flip-flop together with some sample waveforms.

Figure 10.10   A J–K
master/slave flip-flop

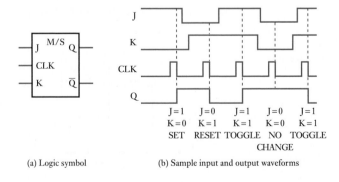

(a) Logic symbol   (b) Sample input and output waveforms

10.3   Memory registers

Having looked at a few forms of bistable, we are now in a position to look at some circuits that make use of them.

Registers of one kind or another are extensively used in almost all fields of digital electronics. One of the most widely used forms of register is that used to store words of information in computers and calculators. These registers can be used directly in calculations, as in the case of the *accumulator* in a processor or calculator, or they can be used for general memory applications, where thousands, or perhaps millions, of registers are used to store programs and data.

A simple memory register is shown in Figure 10.11. Here four D master/slave flip-flops are used to create a 4-bit register. Clearly, additional flip-flops could be used to create a register of any required length.

Memory registers can be created from a range of different types of bistable, but in practice we are less concerned with their internal circuitry than with their external behaviour. Registers are normally constructed as a single circuit and are used as building blocks. Figure 10.12 shows a typical 8-bit memory register. This could be used independently, or together with other devices to produce a longer register. Note that it is normal to number the individual bits of a digital word from 0, starting with the least significant bit (LSB), and to draw them with the most significant bit on the left, as in a conventional number.

Figure 10.11   A 4-bit memory register

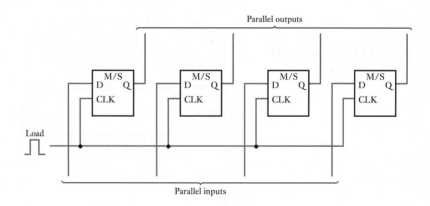

Figure 10.12   An 8-bit memory register

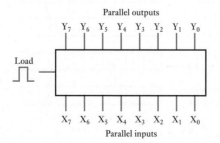

10.4   Shift registers

A slightly different configuration of bistables will produce a shift register. Like a memory register this can store a parallel word of information, but it has the additional feature that it can *shift* the data from element to element along its length. Some devices can shift in just one direction, while others are bidirectional. The process is illustrated in Figure 10.13.

Upon application of a pulse to the *load* input, the pattern of 1's and 0's on the inputs (X_7–X_0) are written into the register and appear at the outputs (X_7–X_0). If now a pulse is applied to the *shift* input, this will cause the pattern stored in the register to move one place to the right. The space produced at the left-hand end of the register is filled by the current value of the signal applied to the *serial data input* line (represented by D in the diagram). Application of further pulses will progressively move the data down the register, gradually filling the register with data from the serial input. You will notice that the rightmost bit of the output represents each bit of the stored data in turn.

In Chapter 5, we noted that digital information can be transmitted in either a parallel or a serial form. The shift register allows us to convert between these forms by performing either **parallel-to-serial** or **serial-to-parallel** conversion. It can be seen that parallel data loaded into the register is available in serial form at the rightmost end of the register as it is *shifted out* of the register. Similarly, data applied to the serial data input line is *shifted in* to the register and is then available at the parallel outputs. The use of these techniques is illustrated in the following example.

Figure 10.13 A simple shift register

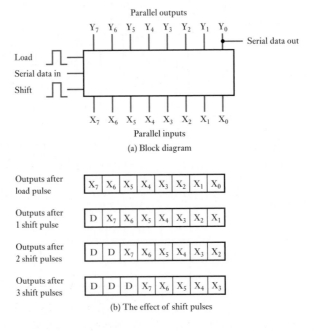

(a) Block diagram

(b) The effect of shift pulses

| Example 10.2 | **Application of shift registers.** |

One of the most common uses of shift registers is in **serial communications** systems. This involves converting parallel data into a serial form at the *transmitter*, conveying the serial data over some distance and then converting it back into a parallel form at the *receiver*. The process is illustrated below.

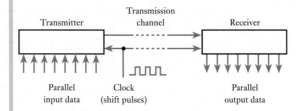

At the heart of the transmitter is a shift register, which loads the input data in parallel and then outputs it in a serial form at a rate determined by a local clock signal. The serial data stream is then transmitted over some form of *transmission channel* to the receiver. This channel may take the form of a piece of wire, a radio signal, a series of laser pulses or some other information medium. At the receiver, a second shift register loads the serial data and outputs the information in parallel form. To enable it to load the information, it must receive not only the serial data but also the clock signal to allow it to *synchronise* with the transmitter.

The main advantage of this method of transmission is that it requires fewer lines for the information to be communicated, requiring only two lines (one for data and one for the clock), rather than one line for each bit of the parallel data. Serial techniques are used extensively for long-distance communication. They are also used for short-range applications, sometimes down to a few inches. In some systems, the requirement to transmit the clock signal along with the data is removed by generating (or recovering) the clock signal at the receiver. This reduces the number of signal lines required to one.

| 10.5 | **Counters** |

Among the most important classes of sequential circuits are the various forms of counter. These can be used to count events but can also be used to count regular clock 'ticks' and hence to measure time. Counters form the basis of a wide range of timing and sequencing applications, in everything from quartz watches to digital computers.

10.5.1 Ripple counters

Consider the circuit of Figure 10.14. The circuit consists of four negative edge-triggered J–K flip-flops, with the Q output of each device forming

Figure 10.14   A simple ripple counter

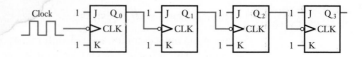

Figure 10.15   Waveform diagram for the simple ripple counter

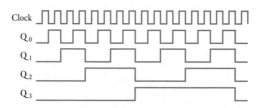

the clock input to the next. The J and K input of each device are connected to 1, so each will toggle on the negative-going edge of the signal connected to its clock input (when connected in this way the devices are acting as **T-type** or **toggle bistables**, which toggle their output in response to each clock trigger).

Figure 10.15 shows the resultant waveforms in the circuit when a square-wave clock signal is applied (a square wave is a repetitive pulse waveform with equal-length pulses and spaces). It can be seen that Q_0 toggles on each falling edge of the clock, producing a square waveform at half the clock frequency. Q_1 toggles at each falling edge of Q_0 and therefore produces a square waveform at half the frequency of Q_0, or one-quarter the frequency of the clock. Similarly, each further stage divides the signal frequency by a factor of 2, producing successively lower frequencies. Such a circuit may be thought of as a **frequency divider**, and each stage represents a **frequency halver**.

Example 10.3	Application of a frequency divider.

A common example of the use of a frequency divider is found in a digital watch. Most digital watches use a crystal oscillator, which produces a stable timing waveform of 32,768 Hz. This particular frequency is used because it is an exact power of 2 ($32,768 = 2^{15}$), which simplifies the process of frequency division. Following the oscillator, a fifteen-stage binary divider is used to produce a 1 Hz signal, which is suitable for driving a stepper motor (for watches with an analogue display using hands) or a digital display.

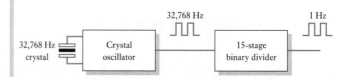

Figure 10.16 The output sequence for the ripple counter

Number of clock pulses	Q_3	Q_2	Q_1	Q_0
0	0	0	0	0
1	0	0	0	1
2	0	0	1	0
3	0	0	1	1
4	0	1	0	0
5	0	1	0	1
6	0	1	1	0
7	0	1	1	1
8	1	0	0	0
9	1	0	0	1
10	1	0	1	0
11	1	0	1	1
12	1	1	0	0
13	1	1	0	1
14	1	1	1	0
15	1	1	1	1
16	0	0	0	0
17	0	0	0	1
18	0	0	1	0
19	0	0	1	1
20	0	1	0	0

It is also interesting to look at the sequence of 1's and 0's that appear on the outputs of the flip-flops. These can be deduced from Figure 10.15, and Figure 10.16 shows the values of the outputs after each clock pulse. It can be seen that the pattern of the outputs represents the binary code for the number of pulses applied to the circuit. The arrangement therefore represents a *counter*. In this particular circuit, the effects of the input are propagated along the series of flip-flops, the outputs changing sequentially along the line. For this reason, this form of counter is called a **ripple counter**. One consequence of the operation of this form of circuit is that the various stages change at slightly different times. For this reason, they are known as **asynchronous counters**.

From Figure 10.16 it can be seen that the circuit counts up to the binary equivalent of 15 and then restarts at zero. The counter therefore takes sixteen distinct values. We refer to such a counter as a **modulo-16 counter** or, sometimes, simply as a **mod-16 counter**. This circuit is also called a **4-bit counter**, since its outputs represent a 4-bit digital number. By using more or fewer stages we can construct a range of circuits that count modulo-2^n, where n is the number of flip-flops used. These counters will count from 0 to 2^{n-1} and then repeat.

File 10A

Computer Simulation Exercise 10.1

Simulate the circuit of Figure 10.14 and apply a square-wave input. Confirm that this produces waveforms of the form shown in Figure 10.15. Now modify your circuit by adding additional stages and confirm that you can produce frequency dividers of arbitrary length.

Figure 10.17   A decade counter

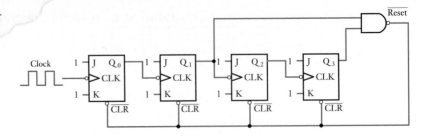

10.5.2   Modulo-*N* counters

We have seen that by varying the number of stages we can modify our simple ripple counter to produce circuits that count up to different numbers before restarting at zero. However, varying the number of stages only allows us to choose values for the modulus of the counter that are powers of 2. In many applications, we wish to count up to particular numbers that may not fulfil this requirement. We therefore need a method of constructing a generalised counter that can count up to any number. Such a circuit is usually called a **modulo-*N* counter**.

In order to produce a counter with a modulus of N, we simply need to ensure that on the clock pulse after the counter reaches $N-1$ it returns to zero. An example of such a circuit, for a value of $N = 10$, is given in Figure 10.17. Counters that count modulo-10 are referred to as **decade counters**. Decade counters that output binary numbers in the range 0000 to 1001 are often called **binary-coded decimal counters** or simply **BCD counters**.

The circuit of the decade counter is similar to that of the simple ripple counter of Figure 10.14, but it has an extra reset circuit to clear all the flip-flops when they reach a count of 10. The clearing operation is achieved by using flip-flops that have **CLEAR** inputs. The signal applied to this input comes from circuitry that detects a count of 10. Since 10 is binary 1010, a simple two-input NAND gate can be used to detect the first occasion on which bits 1 and 3 are high (remember the bits start from bit 0). As soon as this is detected the $\overline{RESET}$ line goes low, clearing the counter to zero, which also sets $\overline{RESET}$ back to 1. The counter then continues from zero as before.

A modulo-*N* counter can be constructed for any value of N by forming a counter with n stages, where $2^n > N$, and then adding a reset circuit that detects a count of N.

File 10B

Computer Simulation Exercise 10.2

Simulate the circuit of Figure 10.17 and apply a square-wave input. Confirm that this functions as a decade counter. Modify the circuit to produce a modulo-12 counter.

10.5.3 Down and up/down counters

The counters described above start at zero and count up; unsurprisingly, such circuits are called **up counters**. A slight modification to the counter can produce a circuit that will count downwards (a **down counter**). Alternatively, a slightly more complicated arrangement will produce an **up/down counter** that can count in either direction. In this latter circuit, the direction of counting is determined by an up/$\overline{\text{down}}$ control signal. When this is high (1) the circuit counts up on each active edge of the clock, and when it is low (0) it counts down. Up/down counters have a range of applications, one of which is shown in the following example.

Example 10.4 | Application of an up/down counter.

In Section 3.6, we looked at the signals produced by an incremental position encoder and noted that these take the form of two square waves that are phase shifted with respect to each other. In Example 3.1, we looked at the design of a microcomputer mouse and devised an arrangement that produced a similar pair of output signals for each axis of motion. The form of the signals is shown below.

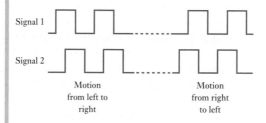

Signal 1

Signal 2

Motion	Motion
from left to	from right
right	to left

We can see that motion in one direction causes signal 1 to lead signal 2, while motion in the other direction causes signal 2 to lead signal 1. We require a mechanism to count the number of 'steps' in one direction and to subtract from it the number of steps in the other direction to give a measure of the absolute position.

To see how this objective may be met, we need to look at the timing of the two signals. In particular, look at the state of signal 2 on the negative-going edge of signal 1. You will see that for motion from left to right signal 2 is high on the falling edge of signal 1. However, for motion from right to left signal 2 is low on the falling edge of signal 1. This allows us to use an up/down counter to determine the absolute position.

Signal 1 is fed to the clock of the counter, and signal 2 is connected to the up/down control input. Motion from left to right will now cause the counter to count up and motion from right to left will cause it to count down. Depending on the application, some mechanism might be required to zero the counter to set its absolute value.

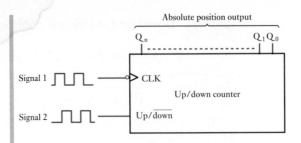

In practice, the pulses produced by a mouse are counted in the microcomputer rather than by an external hardware counter. However, the principles involved are the same.

File 10C

Computer Simulation Exercise 10.3

Modify your circuit for Computer Simulation Exercise 10.1 by connecting the clock inputs of flip-flops 1, 2 and 3 to the $\overline{Q}$ output of the previous stage rather than the Q output. Take the Q_3 to Q_0 signals as the outputs of the circuit and confirm that this arrangement now functions as a down counter.

10.5.4   Propagation delay in ripple counters

Although ripple counters are extremely simple to construct, they do have a major disadvantage, which is particularly apparent when high-speed operation is required. Since the output of one flip-flop is triggered by the change of state of the previous stage, delays produced by each flip-flop are summed along the chain. Each flip-flop takes a finite time to respond to changes at its inputs, this being its propagation delay time t_{PD}. In a ripple counter of n stages, it will take $n \times t_{PD}$ for the counter to respond. If the counter is read during this time the value will be garbled, as some stages will have changed while others will not. This produces a fundamental limit to the maximum clock frequency that can be used with the counter. If clock pulses are received by the first stage before the last has responded, at no time will the counter read the correct value.

10.5.5   Synchronous counters

One solution to the problems of propagation delay in ripple counters is to use synchronous techniques. This is achieved by connecting all the flip-flops in a counter to a common clock signal so that they all change state at the same time. This ensures that, a short time after the clock signal changes, the counter can be read with confidence that all stages will have responded.

Clearly, if all the stages of a counter are connected to the same clock, then some method must be used to determine which stages change state and which remain the same. This is achieved using a small amount of combinational logic. The result is a circuit that is slightly more complicated than a ripple counter but which can count at much higher speeds. Synchronous counters are available in many forms, including up, down, up/down and modulo-N counters.

10.5.6 Integrated circuit counters

Although it is feasible to construct counters using individual gates or combinations of flip-flops, it is more common to use specialised integrated circuits that contain all the functions of a counter in a single package. Both synchronous and asynchronous types are available in a number of sizes and with a range of features. Binary, decade and BCD (binary-coded decimal) counters are available, as are up, down and up/down versions. Typical circuits might provide 4- to 14-bit counters, or several independent counters in a single package. Some counters allow their contents to be cleared or preset (loaded) with a particular value.

Most integrated counters are designed so that they can be cascaded to form counters of greater length. This is achieved with ripple counters simply by taking the most significant bit of one stage as the clock input for the next. With synchronous counters, a common clock is used for all stages and appropriate signals are provided to allow a number of counters to be joined together.

10.6 Monostables or one-shots

Having spent some time looking at the nature and uses of bistables, we will now look briefly at another form of multivibrator, namely the **monostable**.

As with other forms of multivibrator, the monostable has two complementary outputs, Q and $\overline{Q}$, and thus has two possible output states. As its name suggests, one of these output states is stable, while the other is **metastable**. In the absence of any input signal, the circuit stays in its stable state with its Q output at 0. The application of an appropriate signal on its **trigger input** (T) will cause the circuit to enter its metastable state (with $Q = 1$), where it will stay for a fixed period of time determined by circuit components. It will then automatically revert to its stable state. The result is a circuit that responds to a trigger signal by producing a single, fixed-length pulse. This behaviour gives rise to the circuit's alternative name of **one-shot**. The circuit symbol for a monostable is shown in Figure 10.18.

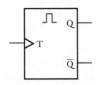

Figure 10.18 Logic symbol for a monostable

Figure 10.19 Monostable
waveforms

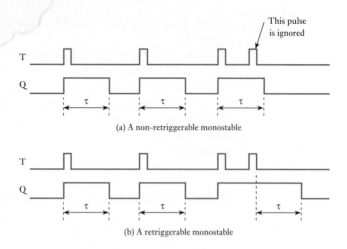

(a) A non-retriggerable monostable

(b) A retriggerable monostable

Monostables can be divided into two types in terms of the way in which they respond to trigger inputs. **Non-retriggerable monostables** ignore any trigger pulses that occur while the circuit is outputting a pulse, while **retriggerable monostables** extend the output pulse if a second trigger occurs. The length of the output pulse τ is normally set by the values of a resistor and a capacitor that form part of the circuit. Figure 10.19 shows typical monostable output waveforms.

10.7   Astables

The last member of the multivibrator family is the **astable**. This has two metastable states and therefore produces the function of a **digital oscillator**. The circuit spends a fixed time in each state determined by circuit values. With an appropriate choice of components, the circuit can be made to spend equal times in each state and thus produce a square waveform.

10.8   Timers

While monostables and astables are available in integrated circuit form, a generally more useful component is the integrated circuit timer, which can perform a range of functions. An example of such a device is the **555 timer**, which can be used as a monostable or an astable, as well as for a number of other applications. Such circuits are very versatile and can be configured to perform a range of functions using just a couple of external passive components. For most engineers the internal construction of these circuits is unimportant – all the information required to use such components is given in the data sheet.

Key points

■ Sequential logic circuits have the characteristic of *memory* in that their outputs are affected by the sequence of events that led up to the current set of inputs.

■ Among the most important groups of sequential logic elements are the various forms of multivibrator. These circuits can be divided into three types:

  1. bistables, which have two stable states;
  2. monostables, which have one stable and one metastable state;
  3. astables, which have no stable states but two metastable states.

■ The most widely used class of multivibrator is the bistable. These may be divided into:

  1. latches;
  2. edge-triggered flip-flops;
  3. pulse-triggered (master/slave) flip-flops.

■ Each class of device can then be divided into a range of devices with different operating characteristics. These are often described by symbolic names, such as S–R, J–K or D-type devices.

■ Bistables are frequently used in groups to form registers or counters.

■ Registers form the basis of computer memories, and special forms of register (shift registers) are also used for serial to parallel conversion.

■ Counters, in their various forms, are used extensively for timing and sequencing functions. They are produced using two basic circuit techniques.

  1. Asynchronous or ripple counters, in which the clock for one stage is generated from the output of the previous stage. The result is a ripple effect as each stage changes in sequence.
  2. Synchronous counters, in which all stages are clocked simultaneously so that all the outputs change at the same time.

■ Both techniques can be used to produce counters that can count up or down.

■ Modulo-N counters can also be produced.

■ Standard integrated circuit building blocks are available to simplify the construction of counters and registers. These can normally be cascaded to form units of any desired length.

■ Monostables and astables are available as standard integrated circuits. Also available are integrated circuit timers that can be configured to perform a range of functions.

Exercises

10.1   Explain the distinction between combinational and sequential circuits.

10.2   Define the terms 'bistable', 'monostable' and 'astable'.

10.3   Explain the origins of the labels S and R in an S–R bistable.

10.4   In the circuit of Figure 10.1, are the inputs active high or active low?

10.5   An S–R bistable can be produced by replacing the two NOR gates in Figure 10.1 with NAND gates. Investigate the operation of this circuit and deduce whether its inputs are active high or active low.

10.6   Deduce the waveform at the Q output of the following circuits.

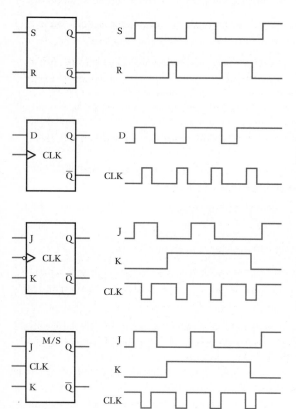

10.7   What is meant by a race in sequential logic?

10.8   Explain how master/slave bistables overcome problems associated with races.

10.9   Design an 8-bit memory register using D master/slave flip-flops, being careful to number your inputs and outputs appropriately.

10.10 Describe the operation of a simple shift register and explain how this can be used to perform serial to parallel and parallel to serial conversion.

10.11 Design a modulo-6 ripple counter using negative edge-triggered J–K flip-flops.

10.12 Simulate your circuit for Exercise 10.11 and confirm that it functions as expected.

10.13 In Figure 10.16, we looked at the output sequence for a 4-bit up counter. Sketch a corresponding output sequence for a 4-bit down counter.

10.14 Design a modulo-32 ripple down counter using negative edge-triggered J–K flip-flops. Hint: you may find it helpful to look at Computer Simulation Exercise 10.3.

10.15 Simulate your circuit for Exercise 10.14 and confirm that it functions as expected.

10.16 Describe the effects of propagation delay on the maximum operating speed of a ripple counter. How are these problems tackled in a synchronous counter?

10.17 Explain the difference between a retriggerable and a non-retriggerable monostable.

10.18 What form of multivibrator has the characteristics of a digital oscillator?

Chapter 11

Measurement of Voltages and Currents

Objectives

When you have studied the material in this chapter you should be able to:

- **describe several forms of alternating waveform, such as sine waves, square waves and triangular waves;**
- **define terms such as peak value, peak-to-peak value, average value and r.m.s. value as they apply to alternating waveforms;**
- **convert between these various values for both sine waves and square waves;**
- **write equations for sine waves to represent their amplitude, frequency and phase angle;**
- **configure moving-coil meters to measure currents or voltages within a given range;**
- **describe the problems associated with measuring non-sinusoidal alternating quantities using analogue meters and explain how to overcome these problems;**
- **explain the operation of digital multimeters and describe their basic characteristics;**
- **discuss the use of oscilloscopes in displaying waveforms and measuring parameters such as phase shift.**

11.1 Introduction

In the previous chapters we have looked at a range of electrical and electronic arrangements and noted their properties and characteristics. An understanding of the operation of these circuits will assist you in later chapters as we move on to analyse the behaviour of these arrangements in more detail. In order to do this, first we need to look at the measurement of voltages and currents in electrical circuits, and in particular at the measurement of alternating quantities.

Alternating currents and voltages vary with time and periodically change their direction. Figure 11.1 shows examples of some alternating waveforms. Of these, by far the most important is the **sinusoidal** waveform or

Figure 11.1 Examples of alternating waveforms

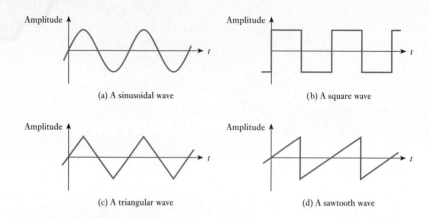

(a) A sinusoidal wave

(b) A square wave

(c) A triangular wave

(d) A sawtooth wave

sine wave. Indeed, in many cases, when engineers use the terms 'alternating current' or 'alternating voltage' they are referring to a sinusoidal quantity. Since sine waves are so widely used, it is important that we understand the nature of these waveforms and the ways in which their properties are defined.

11.2　　Sine waves

In Chapter 2, we noted that the length of time between corresponding points in successive cycles of a sinusoidal waveform is termed its **period** T and that the number of cycles of the waveform within 1 second is termed its **frequency** f. The frequency of a waveform is related to its period by the expression

$$f = \frac{1}{T}$$

The maximum amplitude of the waveform is termed its **peak** value, and the difference between the maximum positive and maximum negative values is termed its **peak-to-peak** value. Because of the waveform's symmetrical nature, the peak-to-peak value is twice the peak value.

Figure 11.2 shows an example of a sinusoidal voltage signal. This illustrates that the period T can be measured between any convenient corresponding points in successive waveforms. It also shows the peak voltage V_p and the peak-to-peak voltage V_{pk-pk}. A similar waveform could be plotted for a sinusoidal *current* waveform indicating its peak current I_p and peak-to-peak current I_{pk-pk}.

Figure 11.2 A sinusoidal voltage signal

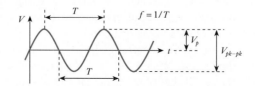

Example 11.1 Determining sine-wave parameters. Determine the period, frequency, peak voltage and peak-to-peak voltage of the following waveform.

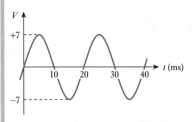

From the diagram the period is 20 ms or 0.02 s, so the frequency is $1/.02 = 50$ Hz. The peak voltage is 7 V and the peak-to-peak voltage is therefore 14 V.

11.2.1 Instantaneous value

The shape of a sine wave is defined by the sine mathematical function. Thus we can describe such a waveform by the expression

$$y = A \sin \theta$$

where y is the value of the waveform at a particular point on the curve, A is the peak value of the waveform and θ is the angle corresponding to that point. It is conventional to use lower case letters for time-varying quantities (such as y in the above equation) and upper case letters for fixed quantities (such as A).

In the voltage waveform of Figure 11.2, the peak value of the waveform is V_p, so this waveform could be represented by the expression

$$v = V_p \sin \theta$$

One complete cycle of the waveform corresponds to the angle θ going through one complete cycle. This corresponds to θ changing by 360°, or 2π radians. Figure 11.3 illustrates the relationship between angle and magnitude for a sine wave.

Figure 11.3 Relationship between instantaneous value and angle for a sine wave

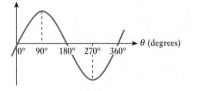
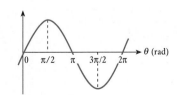

11.2.2   Angular frequency

The frequency f of a waveform (in hertz) is a measure of the number of cycles of that waveform that pass within 1 second. Each cycle corresponds to 2π radians, and it follows that there will be $2\pi f$ radians per second. The number of radians per second is termed the **angular frequency** of the waveform and is given the symbol ω. Therefore

$$\omega = 2\pi f \text{ rad/s} \tag{11.1}$$

11.2.3   Equation of a sine wave

The angular frequency can be thought of as the rate at which the angle of the sine wave changes. Therefore, the phase angle at a particular point in the waveform, θ, is given by

$$\theta = \omega t \text{ rad}$$

Thus our earlier expression for a sine wave becomes

$$y = A \sin \theta$$

$$= A \sin \omega t$$

and the equation of a sinusoidal voltage waveform becomes

$$v = V_p \sin \omega t \tag{11.2}$$

or

$$v = V_p \sin 2\pi f t \tag{11.3}$$

A sinusoidal current waveform might be described by the equation

$$i = I_p \sin \omega t \tag{11.4}$$

or

$$i = I_p \sin 2\pi f t \tag{11.5}$$

Example 11.2

Determining sine wave equations. Determine the equation of the following voltage signal.

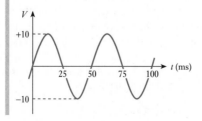

From the diagram the period is 50 ms or 0.05 s, so the frequency is $1/.05 = 20$ Hz. The peak voltage is 10 V. Therefore, from Equation 11.3

$$v = V_p \sin 2\pi ft$$
$$= 10 \sin 2\pi 20t$$
$$= 10 \sin 126t$$

11.2.4 Phase angles

The expressions of Equations 11.2 to 11.5 assume that the angle of the sine wave is zero at the origin of the time measurement ($t = 0$) as in the waveform of Figure 11.2. If this is not the case, then the equation is modified by adding the angle at $t = 0$. This gives an equation of the form

$$y = A \sin(\omega t + \phi) \tag{11.6}$$

where ϕ is the phase angle of the waveform at $t = 0$. It should be noted that at $t = 0$ the term ωt is zero, so $y = A \sin \phi$. This is illustrated in Figure 11.4.

Figure 11.4 The effects of phase angles

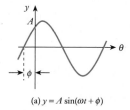

(a) $y = A \sin(\omega t + \phi)$

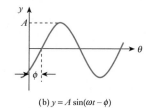

(b) $y = A \sin(\omega t - \phi)$

Example 11.3

Determining sine wave equations. Determine the equation of the following voltage signal.

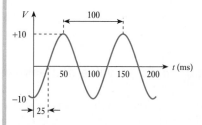

In this example, the period is 100 ms or 0.1 s, so the frequency is $1/0.1 = 10$ Hz. The peak voltage is 10 V. Here the point corresponding to zero degrees of the sine wave occurs at $t = 25$ ms, so at $t = 0$ the phase angle (ϕ) is given by $-25/100 \times 360° = -90°$ (or $\pi/2$ rad). Therefore

$$v = V_p \sin(2\pi ft + \phi)$$
$$= 10 \sin(2\pi 10t + \phi)$$
$$= 10 \sin(63t - \pi/2)$$

11.2.5   Phase differences

Two waveforms of the same frequency may have a constant **phase difference** between them, as shown in Figure 11.5. In this case, we will often say that one waveform is **phase-shifted** with respect to the other. To describe the phase relationship between the two, we often take one of the waveforms as our reference and describe the way in which the other *leads* or *lags* this waveform. In Figure 11.5, waveform A has been taken as the reference in each case. In Figure 11.5(a), waveform B reaches its maximum value some time *after* waveform A. We therefore say that B lags A. In this example B lags A by 90°. In Figure 11.5(b), waveform B reaches its maximum value *before* waveform A. Here B leads A by 90°. In the figure the phase angles are shown in degrees, but they could equally well be expressed in radians.

   It should be noted that the way in which the phase relationship is expressed is a matter of choice. For example, if A leads B by 90°, then clearly B lags A by 90°. These two statements are equivalent, and the one used will depend on the situation and personal preference.

   Figure 11.5 illustrates phase difference using two waveforms of the same magnitude, but this is not a requirement. Phase difference can be measured between any two waveforms of the same frequency, regardless of their relative size. We will consider methods of measuring phase difference later in this chapter.

Figure 11.5   Phase difference between two sine waves

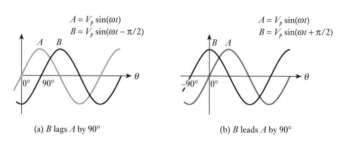

(a) B lags A by 90°   (b) B leads A by 90°

11.2.6   Average value of a sine wave

Clearly, if one measures the average value of a sine wave over one (or more) complete cycles, this average will be zero. However, in some situations we are interested in the average magnitude of the waveform independent of its polarity (we will see an example of this later in this chapter). For a symmetrical waveform such as a sine wave, we can visualise this calculation as taking the average of just the positive half-cycle of the waveform. In this case, the average is the area within this half-cycle divided by half the period. This process is illustrated in Figure 11.6(a). Alternatively, one can view the calculation as taking the average of a **rectified sine wave** (that is, a sine wave where the polarity of the negative half-cycles has been reversed). This is shown in Figure 11.6(b).

Figure 11.6 Calculation of the average value of a sine wave

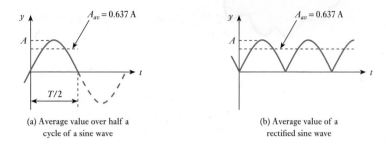

(a) Average value over half a
cycle of a sine wave

(b) Average value of a
rectified sine wave

We can calculate this average value by integrating a sinusoidal quantity over half a cycle and dividing by half the period. For example, if we consider a sinusoidal voltage $v = V_p \sin \theta$, the period is equal to 2π, so

$$V_{av} = \frac{1}{\pi} \int_0^\pi V_p \sin \theta \, d\theta$$

$$= \frac{V_p}{\pi}[-\cos \theta]_0^\pi$$

$$= \frac{2V_p}{\pi}$$

Therefore

$$V_{av} = \frac{2}{\pi} \times V_p = 0.637 \times V_p \tag{11.7}$$

and

$$I_{av} = \frac{2}{\pi} \times I_p = 0.637 \times I_p \tag{11.8}$$

11.2.7   r.m.s. value of a sine wave

Often of more interest than the average value is the **root-mean-square** or **r.m.s.** value of the waveform. This is true not only for sine waves but also for other alternating waveforms.

In Chapter 2, we noted that when a voltage V is applied across a resistor R this will produce a current I (determined by Ohm's law), and that the power dissipated in the resistor will be given by three equivalent expressions:

$$P = VI \qquad P = I^2R \qquad P = \frac{V^2}{R}$$

If the voltage has a varying magnitude, then the *instantaneous* power will be related to the instantaneous voltage and instantaneous current in a similar manner. As before, we use lower case characters to represent varying quantities, so the instantaneous power p is related to the instantaneous voltage v and instantaneous current i by the expressions:

$$p = vi \qquad p = i^2R \qquad p = \frac{v^2}{R}$$

The *average* power will be given by the average (or *mean*) values of these expressions. Since the resistance is constant, we could say that the average power is given by

$$P_{av} = \frac{[\text{average (or mean) of } v^2]}{R} = \frac{\overline{v^2}}{R}$$

or

$$P_{av} = [\text{average (or mean) of } i^2]R = \overline{i^2}R$$

Placing a line (a *bar*) above an expression is a common notation for the mean of that expression. The term $\overline{v^2}$ is referred to as the **mean-square voltage** and $\overline{i^2}$ as the **mean-square current**.

While the mean-square voltage and current are useful quantities, we more often use the square root of each quantity. These are termed the root-mean-square voltage (V_{rms}) and the root-mean-square current (I_{rms}) where

$$V_{rms} = \sqrt{\overline{v^2}}$$

and

$$I_{rms} = \sqrt{\overline{i^2}}$$

We can evaluate each of these expressions by integrating a corresponding sinusoidal quantity over a complete cycle and dividing by the period. For example, if we consider a sinusoidal voltage $v = V_p \sin \omega t$, we can see that

$$V_{rms} = \left(\frac{1}{T} \int_0^T V_p^2 \sin^2 \omega t \, dt \right)^{1/2}$$

$$= \left(\frac{V_p^2}{T} \int_0^T \frac{1}{2}(1 - \cos 2\omega t) \, dt \right)^{1/2}$$

$$= \frac{V_p}{\sqrt{2}}$$

Therefore

$$V_{rms} = \frac{1}{\sqrt{2}} \times V_p = 0.707 \times V_p \qquad\qquad (11.9)$$

and similarly

$$I_{rms} = \frac{1}{\sqrt{2}} \times I_p = 0.707 \times I_p \qquad\qquad (11.10)$$

Combining these results with the earlier expressions gives

$$P_{av} = \frac{\overline{v^2}}{R} = \frac{V_{rms}^{\;2}}{R}$$

and

$$P_{av} = \overline{i^2}R = I_{rms}^{\;2}R$$

If we compare these expressions with those for the power produced by a constant voltage or current, we can see that the r.m.s. value of an alternating quantity produces the same power as a constant quantity of the same magnitude. Thus for alternating quantities

$$P_{av} = V_{rms}I_{rms} \tag{11.11}$$

$$P_{av} = \frac{V_{rms}^{\;2}}{R} \tag{11.12}$$

$$P_{av} = I_{rms}^{\;2}R \tag{11.13}$$

This is illustrated in the following example.

Example 11.4	Calculating power. Calculate the power dissipated in a 10 Ω resistor if the applied voltage is:

(a) a constant 5 V;
(b) a sine wave of 5 V r.m.s.;
(c) a sine wave of 5 V peak.

$$\text{(a)} \qquad P = \frac{V^2}{R} = \frac{5^2}{10} = 2.5 \text{ W}$$

$$\text{(b)} \qquad P_{av} = \frac{V_{rms}^{\;2}}{R} = \frac{5^2}{10} = 2.5 \text{ W}$$

$$\text{(c)} \qquad P_{av} = \frac{V_{rms}^{\;2}}{R} = \frac{\left(V_p/\sqrt{2}\right)^2}{R} = \frac{V_p^2/2}{R} = \frac{5^2/2}{10} = 1.25 \text{ W}$$

11.2.8 Form factor and peak factor

The **form factor** of any waveform is defined as

$$\text{form factor} = \frac{\text{r.m.s. value}}{\text{average value}} \tag{11.14}$$

For a sine wave

$$\text{form factor} = \frac{0.707\,V_p}{0.637\,V_p} = 1.11 \tag{11.15}$$

The significance of the form factor will become apparent in Section 11.5.

The **peak factor** (also known as the **crest factor**) for a waveform is defined as

$$\text{peak factor} = \frac{\text{peak value}}{\text{r.m.s. value}} \tag{11.16}$$

For a sine wave

$$\text{peak factor} = \frac{V_p}{0.707 \, V_p} \tag{11.17}$$

$$= 1.414$$

Although we have introduced the concepts of average value, r.m.s. value, form factor and peak factor for sinusoidal waveforms, it is important to remember that these measures may be applied to any repetitive waveform. In each case, the meanings of the terms are unchanged, although the numerical relationships between these values will vary. To illustrate this, we will now turn our attention to square waves.

11.3   Square waves

11.3.1   Period, frequency and magnitude

Frequency and period have the same meaning for all repetitive waveforms, as do the peak and peak-to-peak values. Figure 11.7 shows an example of a square-wave voltage signal and illustrates these various parameters.

Figure 11.7 A square wave voltage signal

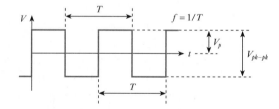

11.3.2   Phase angle

We can if we wish divide the period of a square wave into 360° or 2π radians, as in a sine wave. This might be useful if we were discussing the phase difference between two square waveforms, as shown in Figure 11.8. Here two square waves have the same frequency but have a phase difference of 90° (or $\pi/2$ radians). In this case B lags A by 90°. An alternative way of describing the relationship between the two waveforms is to give the time delay of one with respect to the other.

Figure 11.8 Phase-shifted
square waves

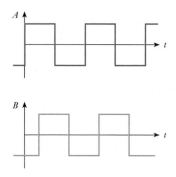

11.3.3 Average and r.m.s. values

Since the average value of a symmetrical alternating waveform is its average value over the positive half-cycle, the average value of a symmetrical square wave (as in Figure 11.1(b)) is equal to its peak value. Thus for a voltage waveform the average value is V_p and for a current waveform it is I_p.

Since the instantaneous value of a symmetrical square wave is always equal to either its positive or its negative peak value, the square of this value is constant. For example, for a voltage waveform the instantaneous value will always be either $+V_p$ or $-V_p$ and in either case the square of this value will be constant at V_p^2. Thus the mean of the voltage squared will be V_p^2, and the square root of this will be V_p. Therefore, the r.m.s. value of a square wave is simply equal to its peak value.

11.3.4 Form factor and peak factor

Using the definitions given in Section 11.2.8, we can now determine the form factor and peak factor for a square wave. Since the average and r.m.s. values of a square wave are both equal to the peak value, it follows that

$$\text{form factor} = \frac{\text{r.m.s. value}}{\text{average value}}$$

$$= 1.0$$

$$\text{peak factor} = \frac{\text{peak value}}{\text{r.m.s. value}}$$

$$= 1.0$$

The relationship between the peak, average and r.m.s. values depends on the shape of a waveform. We have seen that this relationship is very different for a square wave and a sine wave, and further analysis would show similar differences for other waveforms, such as triangular waves.

11.4   Measuring voltages and currents

A wide range of instruments is available for measuring voltages and currents in electrical circuits. These include analogue ammeters and voltmeters, digital multimeters, and oscilloscopes. While each of these devices has its own characteristics, there are some issues that are common to the use of each of these instruments.

11.4.1   Measuring voltage in a circuit

To measure the voltage between two points in a circuit, we place a voltmeter (or other measuring instrument) between the two points. For example, to measure the voltage drop across a component we connect the voltmeter *across* the part as shown in Figure 11.9(a).

Figure 11.9   Measuring voltage and current

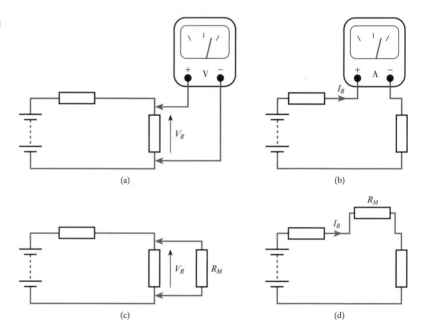

11.4.2   Measuring current in a circuit

To measure the current flowing through a conductor or a component, we connect an ammeter *in series* with the element, as shown in Figure 11.9(b). Note that the ammeter is connected so that conventional current flows from the positive to the negative terminal.

11.4.3   Loading effects

In Chapter 6, we looked at the effects of applying a load to a circuit and noted that under some circumstances this can change the behaviour of that circuit. These **loading** effects can also occur when a voltmeter or an ammeter is connected to a circuit. The result is that the process of measurement actually changes the quantity being measured.

These loading effects are illustrated in Figures 11.9(c) and 11.9(d), which show equivalent circuits for the measurement processes of Figures 11.9(a) and 11.9(b). In each case, the measuring instrument is replaced by its equivalent resistance R_M, and it is clear that the presence of these additional resistances will affect the operation of the circuits. When measuring voltages (as in Figure 11.9(c)), the presence of the voltmeter reduces the effective resistance of the circuit and therefore tends to lower the voltage between these two points in the circuit. To minimise this effect, the resistance of the voltmeter should be as *high* as possible to reduce the current that it passes. When measuring currents (as in Figure 11.9(d)), the ammeter tends to increase the resistance in the circuit and therefore tends to reduce the current flowing. To minimise this effect, the ammeter should have as *low* a resistance as possible to reduce the voltage drop across it.

When using analogue voltmeters and ammeters (as described in the next section), loading effects should always be considered. Instruments will normally indicate their effective resistance (which will usually be different for each range of the instrument), and this information can be used to quantify any loading errors. If these are appreciable it may be necessary to make corrections to the measured values. When using digital voltmeters or oscilloscopes, loading effects are usually less of a problem but should still be considered.

11.5   Analogue ammeters and voltmeters

Most modern analogue ammeters and voltmeters are based on moving-coil meters as described in Section 4.4. These produce an output in the form of movement of a pointer, where the displacement is directly proportional to the current through the meter. Meters are characterised by the current required to produce **full-scale deflection (f.s.d.)** of the meter and their effective resistance R_M. Typical meters produce a full-scale deflection for a current of between 50 μA and 1 mA and have a resistance of between a few ohms and a few kilohms.

11.5.1   Measuring direct currents

Since the deflection of the meter's pointer is directly proportional to the current through the meter, currents up to the f.s.d. value can be measured directly. For larger currents, **shunt resistors** are used to scale the meter's effective sensitivity. This is illustrated in Figure 11.10, where a meter with an f.s.d. current of 1 mA is used to measure a range of currents.

In Figure 11.10(a), the meter is being used to measure currents in the range 0–1 mA, that is, currents up to its f.s.d. value. In Figure 11.10(b), a shunt resistor R_{SH} of $R_M/9$ has been placed in parallel with the meter. Since

Figure 11.10 Use of a meter as an ammeter

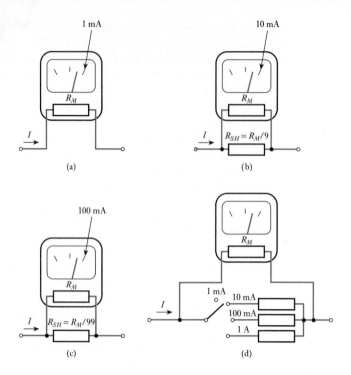

the same voltage is applied across the meter and the resistor, the current through the resistor will be nine times greater than that through the meter. To put this another way, only one-tenth of the input current I will pass through the meter. Therefore, this arrangement has one-tenth the sensitivity of the meter alone and will produce an f.s.d. for a current of 10 mA. Figure 11.10(c) shows a similar arrangement for measuring currents up to 100 mA, and clearly this technique can be extended to measure very large currents. Figure 11.10(d) shows a **switched-range ammeter** arrangement, which can be used to measure a wide range of currents. It can be seen that the effective resistance of the meter is different for each range.

Example 11.5

Selecting an ammeter shunt resistor. A moving-coil meter produces a full-scale deflection for a current 1 mA and has a resistance of 25 Ω. Select a shunt resistor to turn this device into an ammeter with an f.s.d. of 50 mA.

We need to reduce the sensitivity of the meter by a factor of

$$\frac{50 \text{ mA}}{1 \text{ mA}} = 50$$

Therefore, we want 1/50 of the current to pass through the meter. Therefore, R_{SH} must be equal to $R_M \div 49 = 510$ mΩ.

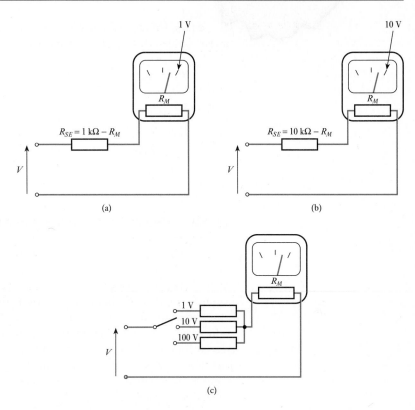

Figure 11.11   Use of a meter as a voltmeter

11.5.2   Measuring direct voltages

To measure direct voltages, we place a resistor in series with the meter and measure the resultant current, as shown in Figure 11.11. In Figure 11.11(a), the meter has an f.s.d. current of 1 mA and the series resistor R_{SE} has been chosen such that $R_{SE} + R_M = 1 \text{ k}\Omega$. The voltage V required to produce a current of 1 mA is given by Ohm's law and is $1 \text{ mA} \times 1 \text{ k}\Omega = 1 \text{ V}$. Therefore, a full-scale deflection of the meter corresponds to an input voltage of 1 V.

In Figure 11.11(b), the series resistor has been chosen such that the total resistance is 10 kΩ, and this will give a full-scale deflection for an input voltage of 10 V. In this way, we can tailor the sensitivity of the arrangement to suit our needs. Figure 11.11(c) shows a switched-range voltmeter that can be used to measure a wide range of voltages. As with the ammeter, the effective resistance of the meter changes as the ranges are switched.

| Example 11.6 | Selecting a voltmeter series resistor. A moving-coil meter produces a full-scale deflection for a current of 1 mA and has a resistance of 25 Ω. Select a series resistor to turn this device into a voltmeter with an f.s.d. of 50 V. |

The required total resistance of the arrangement is given by the f.s.d. current divided by the full-scale input voltage. Hence

$$R_{SE} + R_M = \frac{50 \text{ V}}{1 \text{ mA}} = 50 \text{ k}\Omega$$

Therefore

$$R_{SE} = 50 \text{ k}\Omega - R_M$$

$$= 49.975$$

$$\approx 50 \text{ k}\Omega$$

11.5.3   Measuring alternating quantities

Moving-coil meters respond to currents of either polarity, each producing deflections in opposite directions. Because of the mechanical inertia of the meter, it cannot respond to rapid changes in current and so will average the readings over time. Consequently, a symmetrical alternating waveform will cause the meter to display zero.

In order to measure an alternating current, we can use a **rectifier** to convert it into a unidirectional current that *can* be measured by the meter. This process was illustrated for a sine wave in Figure 11.6(b). The meter responds by producing a deflection corresponding to the average value of the rectified waveform.

We noted in Section 11.2 that when measuring sinusoidal quantities we are normally more interested in the r.m.s. value than in the average value. Therefore, it is common to calibrate AC meters so that they effectively multiply their readings by 1.11, this being the form factor of a sine wave. The result is that the meter (which responds to the average value of the waveform) gives a direct reading of the r.m.s. value of a sine wave. However, a problem with this arrangement is that it gives an incorrect reading for non-sinusoidal waveforms. For example, we noted in Section 11.3 that the form factor for a square wave is 1.0. Consequently, if we measure the r.m.s. value of a square wave using a meter designed for use with sine waves, it will produce a reading that is about 11 percent too high. This problem can be overcome by adjusting our readings to take account of the form factor of the waveform we are measuring.

Like all measuring devices, meters are only accurate over a certain range of frequencies determined by their frequency response. Most devices will work well at the frequencies used for AC power distribution (50 or 60 Hz), but all will have a maximum frequency at which they can be used.

11.5.4   Analogue multimeters

General-purpose instruments use a combination of switches and resistors to achieve a large number of voltage and current ranges within a single unit.

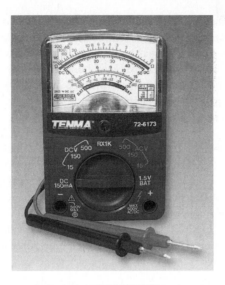

Such units are often referred to as analogue **multimeters**. A rectifier is also
used to permit both unidirectional and alternating quantities to be measured, and additional circuitry is used to allow resistance measurement.
While such devices are very versatile, they often have a relatively low input
resistance and therefore can have considerable loading effects on the circuits to which they are connected. A typical analogue multimeter is shown
in Figure 11.12.

11.6 Digital multimeters

A standard measuring instrument in any electronics laboratory is a **digital
multimeter** (DMM). This combines high accuracy and stability in a device
that is very easy to use. It also normally has a very high input resistance
when used as a voltmeter and a very low input resistance when measuring
currents, so minimising loading effects. While these instruments are capable of measuring voltage, current and resistance, they are often (inaccur-
ately) referred to as **digital voltmeters** or simply **DVMs**. At the heart of
the meter is an **analogue-to-digital converter** (ADC), which takes as its
input a voltage signal and produces as its output a digital measurement that
is used to drive a numeric display. We will look at the operation of such
ADCs in Chapter 26.

Measurements of voltage, current and resistance are achieved by using
appropriate circuits to generate a voltage proportional to the quantity to be
measured. When measuring voltages, the input signal is connected to an
attenuator, which can be switched to vary the input range. When measuring
currents, the input signal is connected across an appropriate shunt resistor,
which generates a voltage proportional to the input current. The value of
the shunt resistance is switched to select different input ranges. In order to

measure resistance the inputs are connected to an **ohms converter**, which passes a small current between the two input connections. The resultant voltage is a measure of the resistance between these terminals.

In simple DMMs, an alternating voltage is rectified, as in an analogue multimeter, to give its average value. This is then multiplied by 1.11 (the form factor of a sine wave) to display the corresponding r.m.s. value. As discussed earlier, this approach gives inaccurate readings when the alternating input signal is not sinusoidal. For this reason, more sophisticated DMMs use a **true r.m.s. converter**, which accurately produces a voltage proportional to the r.m.s. value of an input waveform. Such instruments can be used to make measurements of alternating quantities even when they are not sinusoidal. However, all DMMs are accurate over only a limited range of frequencies.

Figure 11.13(a) shows a typical hand-held digital multimeter and Figure 11.13(b) is a simplified block diagram of a such a device.

Figure 11.13   A digital multimeter (DMM)

(a) A typical digital multimeter

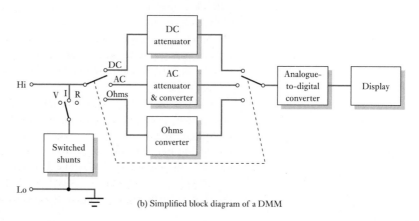

(b) Simplified block diagram of a DMM

An oscilloscope is an instrument that allows a voltage to be measured in terms of the deflection that it produces in a spot of light on a **cathode ray tube**. Usually, a **timebase** circuit is used to scan the spot repeatedly from left to right across the screen at a constant speed by applying a saw-tooth waveform to the horizontal deflection circuitry. An input signal is then used to generate a vertical deflection proportional to the magnitude of the input voltage. In this way, the oscilloscope effectively acts as an automated 'graph plotter' that plots the input voltage against time. Most oscilloscopes can display two input quantities by switching the vertical deflection circuitry between two input signals. This can be done by displaying one complete 'trace' of one waveform, then displaying one trace of the other (**ALT** mode), or by rapidly switching between the two waveforms during each trace (**CHOP** mode). The choice between these modes is governed by the timebase frequency, but in each case the goal is to switch between the two waveforms so quickly that both are displayed steadily and with no noticeable flicker or distortion. In order to produce a stable trace the timebase circuitry includes a **trigger** circuit, which attempts to synchronise the beginning of the timebase sweep so that it always starts at the same point in a repetitive waveform, thus producing a stationary trace.

Simple oscilloscopes use analogue circuitry to implement the various functions. However, more sophisticated instruments use digital techniques to store and manipulate the input data. Digital oscilloscopes are particularly useful when looking at very slow waveforms or short transients, where their ability to store information enables them to display a steady trace. Figure 11.14(a) shows a typical analogue laboratory oscilloscope, and Figure 11.14(b) shows a simplified block diagram of such an instrument.

Oscilloscopes can be used to measure direct and alternating voltages in place of an analogue or digital voltmeter; however, the accuracy of such measurements is generally relatively low. Voltages are measured by comparing the displacement of the trace with a scale printed on the face of the display, and this process generally limits the accuracy of readings to a few percent. Direct or alternating currents can also be measured by sensing the voltages on either side of a resistor and using Ohm's law to calculate the current. Again the accuracy of such measurements will be limited.

While oscilloscopes do not produce very accurate measurements, they do permit measurements to be made over a much wider frequency range than is possible with analogue or digital multimeters. This often extends to some hundreds of megahertz or perhaps several gigahertz. Another great advantage is that an oscilloscope allows the user to see the shape of a waveform. This is invaluable in determining whether circuits are functioning correctly and may also permit distortion or other problems to be detected. Oscilloscopes are also useful when signals have both DC and AC components. This last point is illustrated in Figure 11.15. The figure shows a voltage signal that has a large DC component with a small AC component superimposed on it. If this signal were applied to an analogue or digital multimeter, the reading would simply reflect the magnitude of the DC

Figure 11.14 An analogue oscilloscope

(a) A typical analogue oscilloscope

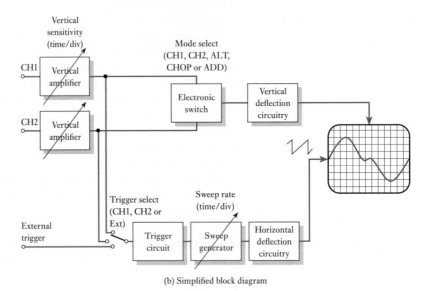

(b) Simplified block diagram

Figure 11.15 A waveform having both DC and AC components

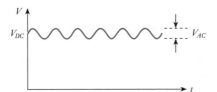

component. However, using an oscilloscope the presence of the alternating component is visible, and the true nature of the signal becomes apparent. Oscilloscopes also allow the DC component to be 'blocked' by selecting an **AC-coupled input**, permitting the AC component to be seen and measured easily. We will see how capacitors can be used to block DC signals in Chapter 17.

When used with AC signals, it is normal to measure the peak-to-peak voltage of a waveform when using an oscilloscope, since this is the quantity that is most readily observed. Care must be taken when comparing such

Figure 11.16 Measurement of phase difference using an oscilloscope

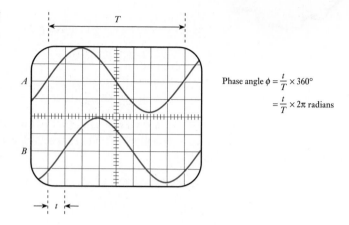

Phase angle $\phi = \dfrac{t}{T} \times 360°$

$= \dfrac{t}{T} \times 2\pi$ radians

readings with those taken using a multimeter, since the latter will normally give r.m.s. values.

Oscilloscopes also allow the direct comparison of waveforms and permit the temporal relationship between them to be investigated. For example, we might use the two traces to display the input and output signals to a module, and hence determine the phase difference between the input and the output.

The measurement of phase difference is illustrated in Figure 11.16. The horizontal scale (which corresponds to time) is used to measure the period of the waveforms (T) and also the time difference between corresponding points in the two waveforms (t). The term t/T now represents the fraction of a complete cycle by which the waveforms are phase-shifted. Since one cycle represents 360° or 2π radians, the phase difference ϕ is given by

$$\text{phase difference } \phi = \frac{t}{T} \times 360° \qquad (11.18)$$

$$= \frac{t}{T} \times 2\pi \text{ radians}$$

In Figure 11.16, waveform B lags waveform A by approximately one-eighth of a complete cycle, or about 45° ($\pi/4$ rad).

Key points

- Alternating waveforms vary with time and periodically change their direction. By far the most important form of alternating waveform is the sine wave.

- The frequency of a periodic waveform f is equal to the reciprocal of its period T.

- The magnitude of an alternating waveform can be described by its *peak* value, its *peak-to-peak* value, its *average* value or its *r.m.s.* value.

- A sinusoidal voltage signal can be described by the expressions

$$v = V_p \sin(2\pi f t + \phi)$$

or

$$v = V_p \sin(\omega t + \phi)$$

where V_p is the peak voltage, f is the frequency (in hertz), ω is the angular frequency (in radians/second) and ϕ is the angle of the waveform at $t = 0$.

- A sinusoidal current signal can be described by the expressions

$$i = I_p \sin(2\pi f t + \phi)$$

or

$$i = I_p \sin(\omega t + \phi)$$

where I_p is the peak current and the other terms are as before.

- Two waveforms of the same frequency may have a constant phase difference between them. One waveform is said to *lead* or *lag* the other.

- The average value of a repetitive alternating waveform is defined as the average over the positive half-cycle.

- The root-mean-square (r.m.s.) value of an alternating waveform is the value that will produce the same power as an equivalent direct quantity.

- For a sinusoidal signal, the *average* voltage or current is $2/\pi$ (or 0.637) times the corresponding *peak* value, and the *r.m.s.* voltage or current is $1/\sqrt{2}$ (or 0.707) times the corresponding *peak* value.

- For square waves, the average and r.m.s. values of voltage and current are equal to the corresponding peak values.

- Simple analogue ammeters and voltmeters are often based on moving-coil meters. These can be configured to measure currents or voltages over a range of magnitudes through the use of series or shunt resistors.

- Meters respond to the average value of a rectified alternating waveform and are normally calibrated to read the r.m.s. value of a sine wave. These will give inappropriate readings when used with non-sinusoidal signals.

- Digital multimeters are easy to use and offer high accuracy. Some have a true r.m.s. converter, allowing them to be used with non-sinusoidal alternating signals.

- Oscilloscopes display the form of a signal and allow distortion to be detected and measured. They also allow comparison between signals and the measurement of parameters such as phase shift.

Exercises

11.1 Sketch three common forms of alternating waveform.

11.2 A sine wave has a period of 10 s. What is its frequency (in hertz)?

11.3 A square wave has a frequency of 25 Hz. What is its period?

11.4 A triangular wave (of the form shown in Figure 11.1) has a peak amplitude of 2.5 V. What is its peak-to-peak amplitude?

11.5 What is the peak-to-peak current of the waveform described by the following equation?

$$i = 10 \sin \theta$$

11.6 A signal has a frequency of 10 Hz. What is its angular frequency?

11.7 A signal has an angular frequency of 157 rad/s. What is its frequency in hertz?

11.8 Determine the peak voltage, the peak-to-peak voltage, the frequency (in hertz) and the angular frequency (in rad/s) of the following waveform.

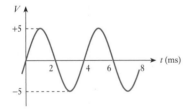

11.9 Write an equation to describe a voltage waveform with an amplitude of 5 V peak and a frequency of 50 Hz.

11.10 Write an equation to describe a current waveform with an amplitude of 16 V peak-to-peak and an angular frequency of 150 rad/s.

11.11 What are the frequency and peak amplitude of the waveform described by the equation.

$$v = 25 \sin 471t$$

11.12 Determine the equation of the following voltage signal.

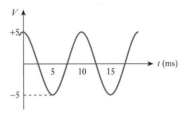

11.13 A sine wave has a peak value of 10. What is its average value?

11.14 A sinusoidal current signal has an average value of 5 A. What is its peak value?

11.15 Explain what is meant by the mean-square value of an alternating waveform. How is this related to the r.m.s. value?

11.16 Why is the r.m.s. value a more useful quantity than the average value?

11.17 A sinusoidal voltage signal of 10 V peak is applied across a resistor of 25 Ω. What power is dissipated in the resistor?

11.18 A sinusoidal voltage signal of 10 V r.m.s. is applied across a resistor of 25 Ω. What power is dissipated in the resistor?

11.19 A sinusoidal waveform with an average voltage of 6 V is measured by an analogue multimeter. What voltage will be displayed?

11.20 A square wave voltage signal has a peak amplitude of 5 V. What is its average value?

11.21 A square wave of 5 V peak is applied across a 25 Ω resistor. What will be the power dissipated in the resistor?

11.22 A moving-coil meter produces a full-scale deflection for a current of 50 μA and has a resistance of 10 Ω. Select a shunt resistor to turn this device into an ammeter with an f.s.d. of 250 mA.

Exercises continued

11.23 A moving-coil meter produces a full-scale deflection for a current of 50 μA and has a resistance of 10 Ω. Select a series resistor to turn this device into a voltmeter with an f.s.d. of 10 V.

11.24 What percentage error is produced if we measure the voltage of a square wave using an analogue multimeter that has been calibrated to display the r.m.s. value of a sine wave.

11.25 A square wave of 10 V peak is connected to an analogue multimeter that is set to measure alternating voltages. What voltage reading will this show?

11.26 Describe the basic operation of a digital multimeter.

11.27 How do some digital multimeters overcome the problem associated with different alternating waveforms having different form factors?

11.28 Explain briefly how an oscilloscope displays the amplitude of a time-varying signal.

11.29 How is an oscilloscope able to display two waveforms simultaneously?

11.30 What is the difference between the ALT and CHOP modes on an oscilloscope?

11.31 What is the function of the trigger circuitry in an oscilloscope?

11.32 A sinusoidal waveform is displayed on an oscilloscope and has a peak-to-peak amplitude of 15 V. At the same time, the signal is measured on an analogue multimeter that is set to measure alternating voltages. What value would you expect to be displayed on the multimeter?

11.33 Comment on the relative accuracies of the two measurement methods outlined in the last exercise.

11.34 What is the phase difference between waveforms A and B in the following oscilloscope display. Which waveform is leading and which lagging?

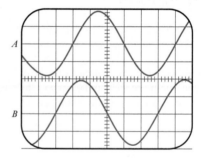

Chapter 12

Resistance and DC Circuits

Objectives

When you have studied the material in this chapter you should be able to:

- **define terms such as current, charge, electromotive force, potential difference, resistance and power, and write equations relating them;**
- **apply Ohm's law and Kirchhoff's voltage and current laws in a range of situations;**
- **derive simple equivalent circuits for electrical networks to aid in their analysis;**
- **explain the principle of superposition and its use in circuit analysis;**
- **describe the processes of nodal analysis and mesh analysis, and explain their uses and importance;**
- **use nodal analysis and mesh analysis to determine currents and voltages in electrical networks;**
- **discuss the selection of analytical techniques for use with electrical circuits.**

12.1   Introduction

We have seen in the earlier chapters that many electrical and electronic circuits can be analysed, and in some cases designed, using little more than Ohm's law. However, in some cases additional techniques are required and in this chapter we will start to look in more detail at the analysis of electrical circuits. We will begin by reviewing some of the basic elements that we have been using to describe our circuits and provide a more detailed understanding of their characteristics. We will then look at a range of techniques for modelling and analysing electrical and electronic circuits.

12.2 Current and charge

An electrical **current** represents a flow of electric **charge** and therefore

$$I = \frac{dQ}{dt} \tag{12.1}$$

where I is the current in amperes, Q is the charge in coulombs and dQ/dt represents the rate of flow of charge (with units of coulombs per second). Conventionally, current is assumed to represent a flow of positive charge.

At an atomic level, a current represents a flow of **electrons**, each of which carries a minute *negative* charge of about 1.6×10^{-19} coulombs. For this reason, the flow of a conventional current in one direction actually represents the passage of electrons in the opposite direction. However, unless we are looking at the physical operation of devices, this distinction is unimportant.

Rearranging the expression of Equation 12.1, we can obtain an expression for the charge passed as a result of the flow of a current.

$$Q = \int I \, dt \tag{12.2}$$

If the current is constant, this results in the simple relationship that charge is equal to the product of the current and time.

$$Q = I \times t$$

12.3 Voltage sources

A voltage source produces an electromotive force (e.m.f.), which causes a current to flow within a circuit. Despite its name, an e.m.f. is not a force in the conventional sense but represents the energy associated with the passage of charge through the source. The unit of e.m.f. is the **volt**, which is defined as the potential difference between two points when one joule of energy is used to move one coulomb of charge from one point to the other.

Real voltage sources, such as batteries, have resistance associated with them, which limits the current that they can supply. When analysing circuits, we often use the concept of an **ideal voltage source** that has no resistance. Such sources can represent constant or alternating voltages, and they can also represent voltages that vary in response to some other physical quantity (**controlled** or **dependent voltage sources**). We saw examples of this last group when we considered the modelling of amplifiers in Chapter 6. Figure 12.1 shows examples of the symbols used to represent various forms of voltage source.

Unfortunately, a range of notations is used to represent voltages in electrical circuits. Most textbooks published in America adopt a notation where the polarity of a voltage is indicated using a '+' symbol. In the UK, and many other countries, it is more common to use the notation shown in Figure 12.1, where an arrow is used to indicate polarity. Here the label

Figure 12.1 Voltage sources

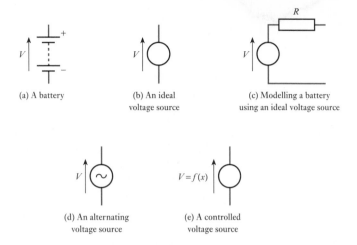

(a) A battery

(b) An ideal
voltage source

(c) Modelling a battery
using an ideal voltage source

(d) An alternating
voltage source

(e) A controlled
voltage source

associated with the arrow represents the voltage at the head of the arrow with respect to the voltage at its tail. An advantage of this notation is that the label can unambiguously represent a positive, negative or alternating quantity.

12.4 Current sources

In addition to the concept of an ideal voltage source, it is sometimes convenient to model an **ideal current source**. As with its voltage counterpart, such a component is not physically realisable, but the use of such a conceptual model can greatly simplify some forms of circuit analysis. Just as an ideal voltage source produces a certain voltage no matter what is connected to it, so an ideal current source will always pass a particular current. This current could be constant or alternating (depending on the nature of the current source), or it might be determined by some physical quantity within a circuit (a **controlled** or **dependent current source**). The circuit symbol for a current source is shown in Figure 12.2.

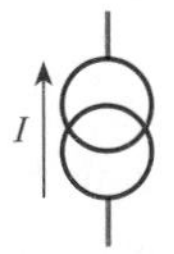

It is interesting to note that, while an ideal voltage source has *zero* output resistance, an ideal current source has *infinite* output resistance. This is evident if we consider loading effects and the situation required for the output current to remain constant regardless of variations in load resistance.

Figure 12.2 An ideal current source

12.5 Resistance and Ohm's law

Readers are already familiar with one of the best-known relationships in electrical engineering, which is that the voltage across a conductor is directly proportional to the current flowing in it (Ohm's law).

$$V \propto I$$

The constant of proportionality of this relationship is termed the **resistance** of the conductor (R), which gives rise to the well-known expressions:

$$V = IR \qquad I = \frac{V}{R} \qquad R = \frac{V}{I}$$

We noted in Chapter 2 that the unit of resistance is the ohm (Ω), which can be defined as the resistance of a circuit in which a current of 1 A produces a **potential difference** of 1 V.

When current flows through a resistance, power is dissipated in it. This power is dissipated in the form of heat. The power (P) is related to V, I and R by the expressions:

$$P = IV \qquad P = \frac{V^2}{R} \qquad P = I^2R$$

Components designed to provide resistance in electrical circuits are termed **resistors**. The resistance of a given sample of material is determined by its dimensions and by the electrical characteristics of the material used in its construction. The latter is described by the **resistivity** of the material ρ (Greek letter *rho*) or sometimes by its **conductivity** σ (Greek letter *sigma*), which is the reciprocal of the resistivity. Figure 12.3 shows a piece of resistive material with electrical contacts on each end. If the body of the component is uniform, its resistance will be directly related to its length (l) and inversely related to its cross-sectional area (A). Under these circumstances, the resistance of the device will be given by

$$R = \frac{\rho l}{A} \tag{12.3}$$

The units of resistivity are ohm-metres (Ω-m). Copper has a resistivity of about 1.6×10^{-8} Ω-m at 0 °C, while carbon has a resistivity of 6500×10^{-8} Ω-m at 0 °C.

Since the flow of current through a resistor produces heat, this will cause the temperature of the resistor to rise. Unfortunately, the resistance of most materials changes with temperature, this variation being determined by its **temperature coefficient of resistance** α. Pure metals have positive temperature coefficients, meaning that their resistance increases with temperature. Many other materials (including most insulators) have negative coefficients. The materials used in resistors are chosen to minimise these temperature-related effects. In addition to altering its resistance, an excessive increase in temperature would inevitably result in damage to a resistor. Consequently, any particular component has a maximum **power rating**,

Figure 12.3 The effects of component dimensions on resistance

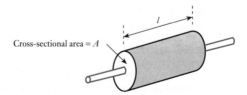

Cross-sectional area = A

which should not be exceeded. Larger components have a greater surface area and can therefore dissipate heat more effectively. Consequently, power ratings tend to increase with the physical size of resistors (although it is also affected by other factors). A small general-purpose resistor might have a power rating of an eighth or a quarter of a watt, while larger components might handle several watts.

12.6 Resistors in series and parallel

In Chapter 2, we noted the effective resistance produced by connecting a number of resistors in series or in parallel. Before moving on, it is perhaps appropriate to ensure that the reasons for these relationships are clear.

Figure 12.4(a) shows an arrangement in which a voltage V is applied across a series arrangement of resistors $R_1, R_2, \ldots, R_N$. The voltage across each individual resistor is given by the product of the current (I) and its resistance. The applied voltage V must be equal to the sum of the voltages across the resistors, and therefore

$$V = IR_1 + IR_2 + \ldots + IR_N$$

$$= I(R_1 + R_2 + \ldots + R_N)$$

$$= IR$$

where $R = (R_1 + R_2 + \ldots + R_N)$. Therefore, the circuit behaves as if the series of resistors were replaced by a single resistor with a value equal to their sum.

Figure 12.4(b) shows an arrangement where several resistors are connected in parallel. The voltage across each resistor is equal to the applied voltage V, so the current in each resistor is given by this voltage divided by its resistance. The total current I is equal to the sum of the currents in the individual resistors, and therefore

Figure 12.4 Resistors in series and parallel

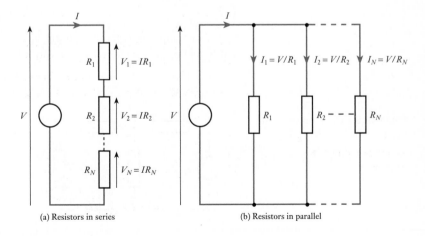

(a) Resistors in series   (b) Resistors in parallel

$$I = \frac{V}{R_1} + \frac{V}{R_2} + \ldots + \frac{V}{R_N}$$

$$= V\left(\frac{1}{R_1} + \frac{1}{R_2} + \ldots \frac{1}{R_N}\right)$$

$$= V\left(\frac{1}{R}\right)$$

where $1/R = 1/R_1 + 1/R_2 + \ldots + 1/R_N$. Therefore, the circuit behaves as if the combination of resistors were replaced by a single resistor whose value is given by the reciprocal of the sum of the reciprocals of their values.

12.6.1 Notation

Parallel combinations of resistors are very common in electrical circuits, so there is a notation to represent the effective resistance of resistors in parallel. This consists of the resistor names or values separated by '//'. Therefore, $R_1//R_2$ would be read as 'the effective resistance of R_1 in parallel with R_2'. Similarly, 10 kΩ//10 kΩ simply means the resistance of two 10 kΩ resistors connected in parallel (this is 5 kΩ).

12.7 Kirchhoff's laws

A point in a circuit where two or more circuit components are joined together is termed a **node**, while any closed path in a circuit that passes through no node more than once is termed a **loop**. A loop that contains no other loop is called a **mesh**. These definitions are illustrated in Figure 12.5. Here points A, B, C, D, E and F are nodes in the circuit, while the paths ABEFA, BCDEB and ABCDEFA represent loops. It can be seen that the first and second of these loops are also meshes, while the last is not (since it contains smaller loops).

We can apply Kirchhoff's current law to the various nodes in a circuit and Kirchhoff's voltage law to the various loops and meshes.

Figure 12.5   Circuit nodes and loops

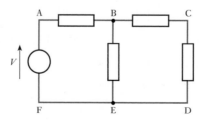

Figure 12.6 Application of Kirchhoff's current law

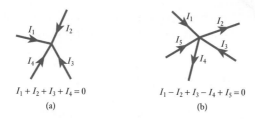

$$I_1 + I_2 + I_3 + I_4 = 0$$

(a)

$$I_1 - I_2 + I_3 - I_4 + I_5 = 0$$

(b)

12.7.1 Current law

Kirchhoff's current law says that at any instant the algebraic sum of all the currents flowing into any node in a circuit is zero. If we consider currents flowing *into* a node to be positive and currents flowing *out of* that node to be negative, then all the various currents must sum to zero. That is

$$\sum I = 0$$

This is illustrated in Figure 12.6. In Figure 12.6(a), the currents are each defined as flowing *into* the node and therefore their magnitudes simply sum to zero. It is evident that one or more of these currents must be negative in order for this to be possible (unless they are all zero). A current of $-I$ flowing into a node is clearly equivalent to a current of I flowing out from it. In Figure 12.6(b), some currents are defined as flowing *into* the node while others flow *out*. This results in the equation shown.

Example 12.1 | Using Kirchhoff's current law. Determine the magnitude of I_4 in the following circuit.

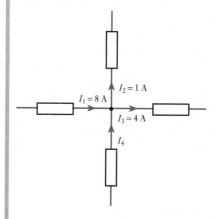

Summing the currents flowing *into* the node, we have

$$I_1 - I_2 - I_3 + I_4 = 0$$

$$8 - 1 - 4 + I_4 = 0$$

$$I_4 = -3 \text{ A}$$

Therefore, I_4 is equal to -3 A: that is, a current of 3 A flowing in the opposite direction to the arrow in the diagram.

12.7.2   Voltage law

Kirchhoff's voltage law says that at any instant the algebraic sum of all the voltages around any loop in a circuit is zero.

$$\Sigma V = 0$$

Our only difficulty in applying this law is in ensuring that we get the polarities of the voltages correct in our arithmetic. A simple way of ensuring this is to use arrows within our circuit diagrams to represent the polarity of each e.m.f. or potential difference (as in earlier circuits). We then move around the loop in a clockwise direction and any arrow in this direction represents a positive voltage while any arrow in the opposite direction represents a negative voltage. This is illustrated in Figure 12.7.

In Figure 12.7(a), all the e.m.f.s and potential differences are defined in a clockwise direction. Therefore, their magnitudes simply sum to zero. Note that the directions of the arrows show how the voltages are defined (or measured) and do *not* show the polarity of the voltages. If E has a positive value, the top of the voltage source in Figure 12.7(a) will be positive with respect to the bottom. If E has a negative value, the polarity will be reversed. Similarly, E could represent a varying or alternating voltage, but the relationship shown in the equation would still hold. In Figure 12.7(b), some of the e.m.f.s and potential differences are defined in a clockwise direction and some are defined in the opposite direction. This results in the equation shown, where clockwise terms are added and anticlockwise terms are subtracted.

Figure 12.7   Applying Kirchhoff's voltage law

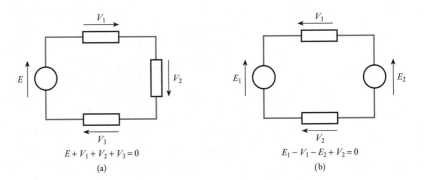

$$E + V_1 + V_2 + V_3 = 0$$

(a)

$$E_1 - V_1 - E_2 + V_2 = 0$$

(b)

| Example 12.2 | Using Kirchhoff's voltage law. Determine the magnitude of V_2 in the following circuit. |

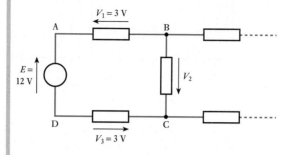

Summing the voltages clockwise around the loop ABCDA, we have

$$E - V_1 + V_2 - V_3 = 0$$
$$12 - 3 + V_2 - 3 = 0$$
$$V_2 = -6 \text{ V}$$

Therefore, V_2 has a value of -6 V: that is, a potential difference of 6 V with node B being more positive than node C. Had we chosen to define V_2 by an arrow pointing in the opposite direction, our calculations would have found it to have a value of $+6$ V, which would again have represented a potential difference of 6 V with node B being more positive than node C.

12.8 Thévenin's theorem and Norton's theorem

In Chapter 6, we saw that it is often convenient to represent electrical circuits by simpler **equivalent circuits** that model their behaviour. At that time, we noted that we could represent a real voltage source (such as the output of a sensor or an amplifier) by an ideal voltage source and a series resistor. This representation is an example of what is termed a **Thévenin equivalent circuit**. This is an arrangement based on **Thévenin's theorem**, which may be paraphrased as:

> As far as its appearance from outside is concerned, any two terminal networks of resistors and energy sources can be replaced by a series combination of an ideal voltage source V and a resistor R, where V is the open-circuit voltage of the network and R is the resistance that would be measured between the output terminals if the energy sources were removed and replaced by their internal resistance.

It can be seen that this simple equivalent circuit can be used to represent not only the outputs of amplifiers or sensors but also any arrangement of resistors, voltage sources and current sources that have two output terminals. However, it is important to note that the equivalence is valid only 'as far as its appearance from outside' is concerned. The equivalent circuit does not represent the internal characteristics of the network, such as its power consumption.

Figure 12.8 Thévenin and
Norton equivalent circuits

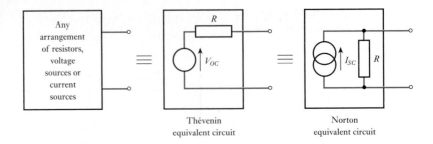

Thévenin
equivalent circuit

Norton
equivalent circuit

While Thévenin equivalent circuits are useful in a wide range of applications, there are situations where it would be more convenient to have an equivalent circuit that uses a current source rather than a voltage source. Fortunately, such circuits are described by **Norton's theorem**, which may be paraphrased as:

> *As far as its appearance from outside is concerned, any two terminal networks of resistors and energy sources can be replaced by a parallel combination of an ideal current source I and a resistor R, where I is the short-circuit current of the network and R is the resistance that would be measured between the output terminals if the energy sources were removed and replaced by their internal resistance.*

The implications of these two theorems are summarised in Figure 12.8. This shows that any such circuit can be represented by *either* form of equivalent circuit. The form that is used depends on the application. When we were modelling voltage amplifiers in Chapter 6, we found it convenient to use a Thévenin form. We shall see in later chapters that when we consider devices such as transistors it is often more convenient to use a Norton form.

Since the three arrangements of Figure 12.8 are equivalent, it follows that they should each produce the same output in all circumstances. If we connect nothing to the terminals of each circuit their outputs should be the same: this is the **open-circuit voltage** V_{OC}. Similarly, if we connect the output terminals together, each circuit should produce the same current: the **short-circuit current** I_{SC}. These equivalences allow us to deduce relationships between the various values used in the equivalent circuits.

From the theorems above, it is clear that the same resistance R is used in each equivalent circuit. If we look at the Thévenin circuit and consider the effect of joining the output terminals together it is clear, from Ohm's law, that the resultant current I_{SC} would be given by

$$I_{SC} = \frac{V_{OC}}{R}$$

Similarly, looking at the Norton circuit we can see, again from Ohm's law, that the open circuit voltage V_{OC} is given by

$$V_{OC} = I_{SC}R$$

Rearranging either of these relationships gives us the same result, which is that the resistance in the two equivalent circuits is given by

$$R = \frac{V_{OC}}{I_{SC}} \tag{12.4}$$

Therefore, the resistance can be determined from the open-circuit voltage and the short-circuit current, or the resistance seen 'looking into' the output of the circuit with the effects of the voltage and current sources removed.

Values for the components required to model a particular circuit can be found by analysis of its circuit diagram or by taking measurements on a physical circuit. These approaches are illustrated in the following examples.

Example 12.3

Drawing equivalent circuits from a circuit diagram. Determine the Thévenin and Norton equivalent circuits of the following arrangement.

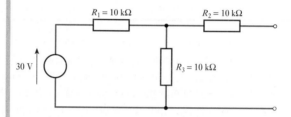

If nothing is connected across the output, no current can pass through R_2 and there will be no voltage drop across it. Therefore, the output voltage will be determined simply by the voltage source and the potential divider formed by R_1 and R_2. Since the two resistors are equal, the output voltage will be half the voltage of the source, so

$$V_{OC} = \frac{30}{2} = 15 \text{ V}$$

If the output terminals are shorted together, R_2 is effectively in parallel with R_3, so their combined resistance is $R_2//R_3 = 10 \text{ k}\Omega//10 \text{ k}\Omega = 5 \text{ k}\Omega$. The total resistance connected across the voltage source is therefore $R_1 + 5 \text{ k}\Omega = 15 \text{ k}\Omega$, and the current taken from the source will be 30 V/15 kΩ = 2 mA. Since R_2 and R_3 are the same size and are in parallel, the current through each will be the same. Therefore, the current through each resistor will be 2 mA/2 = 1 mA. Here the current through R_2 is the output current (in this case the short-circuit output current), so

$$I_{SC} = 1 \text{ mA}$$

From Equation 12.4, we know that the resistance in Thévenin and Norton equivalent circuits is given by the ratio of V_{OC} to I_{SC}, therefore

$$R = \frac{V_{OC}}{I_{SC}} = \frac{15 \text{ V}}{1 \text{ mA}} = 15 \text{ k}\Omega$$

Alternatively, R could be obtained by noting the effective resistance seen looking into the output of the circuit with the voltage source replaced by its internal resistance. The internal resistance of an ideal voltage source is zero, so this produces a circuit where R_1 is effectively in parallel with R_3. The resistance seen at the output is therefore $R_2 + (R_1//R_3) = 10\ \text{k}\Omega + (10\ \text{k}\Omega//10\ \text{k}\Omega) = 15\ \text{k}\Omega$, as before.

Therefore our equivalent circuits are

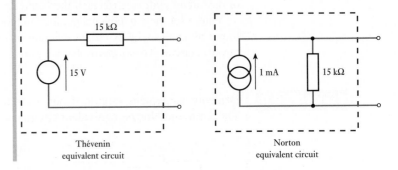

Thévenin
equivalent circuit

Norton
equivalent circuit

To obtain the equivalent for an actual circuit (rather than from its circuit diagram), we make appropriate measurements of its characteristics. Since we can obtain the necessary values from a knowledge of the open-circuit voltage V_{OC} and the short-circuit current I_{SC}, one approach is simply to measure these quantities. If a high-resistance voltmeter is placed across the output of the circuit this will give a reasonable estimate of its open-circuit output voltage, provided that the input resistance of the meter is high compared with the output resistance of the circuit. However, direct measurement of the short-circuit current is often more difficult, since shorting out the circuit may cause damage. An alternative approach is to take other measurements and to use these to deduce V_{OC} and I_{SC}.

Example 12.4 Determining equivalent circuit values from measurements. A two-terminal network, which has an unknown internal circuit, is investigated by measuring its output voltage when connected to different loads. It is found that when a resistance of 25 Ω is connected the output voltage is 2 V, and when a load of 400 Ω is connected the output voltage is 8 V. Determine the Thévenin and Norton equivalent circuits of the unknown circuit.

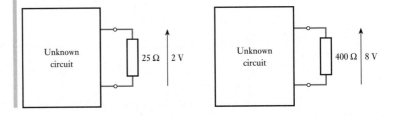

Method 1

One approach to this problem is to plot a graph of the output current against the output voltage. When the output voltage is 2 V the output current is 2 V/25 Ω = 80 mA, and when the output voltage is 8 V the output current is 8 V/400 Ω = 20 mA. This gives the following graph.

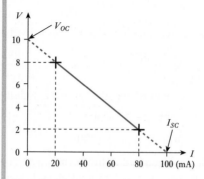

Extrapolating this line shows that when the current is zero the voltage is 10 V (the open-circuit voltage), and when the voltage is zero the current is 100 mA (the short-circuit current). From Equation 12.4

$$R = \frac{V_{OC}}{I_{SC}} = \frac{10 \text{ V}}{100 \text{ mA}}$$

$$= 100 \text{ } \Omega$$

Therefore the equivalent circuits are

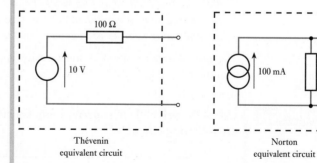

Thévenin
equivalent circuit

Norton
equivalent circuit

Method 2

Non-graphical methods are also available. For example, if we assume that the circuit is replaced by a Thévenin equivalent circuit consisting of a voltage source V_{OC} and a resistor R, we have the following arrangements.

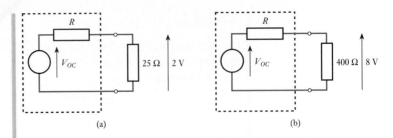

(a)   (b)

Applying the equation of a potential divider to (a) and (b) gives

$$V_{OC}\frac{25}{R+25} = 2 \quad \text{and} \quad V_{OC}\frac{400}{R+400} = 8$$

which can be rearranged to give the simultaneous equations

$$25V_{OC} = 2R + 50$$

$$400V_{OC} = 8R + 3200$$

which can be solved in the normal way to give $V_{OC} = 10$ and $R = 100$. The value of I_{SC} can now be found using Equation 12.4, and we have the same values as before.

12.9   Superposition

When a circuit contains more than one energy source, we can often simplify the analysis by applying the **principle of superposition**. This allows the effects of each voltage and current source to be calculated separately and then added to give their combined effect. More precisely, the principle states:

> *In any linear network of resistors, voltage sources and current sources, each voltage and current in the circuit is equal to the algebraic sum of the voltages or currents that would be present if each source were to be considered separately. When determining the effects of a single source, the remaining sources are replaced by their internal resistance.*

The use of superposition is most easily appreciated through the use of examples.

Example 12.5

Use of the principle of superposition. **Calculate the output voltage V of the following circuit.**

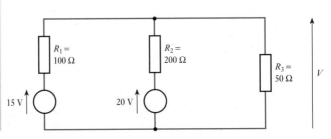

First we consider the effect of the 15 V source. When doing this we replace the other voltage source by its internal resistance, which for an ideal voltage source is zero (that is, a short circuit). This gives us the following circuit.

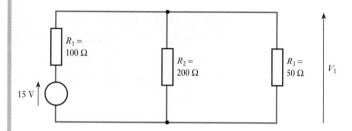

This is a potential divider circuit formed by R_1 and the parallel combination of R_2 and R_3. Using the equations for resistors in parallel and for a potential divider, this gives

$$V_1 = 15\frac{200//50}{100 + 200//50}$$

$$= 15\frac{40}{100 + 40}$$

$$= 4.29 \text{ V}$$

Next we consider the effect of the 20 V source, replacing the 15 V source with a short circuit. This gives

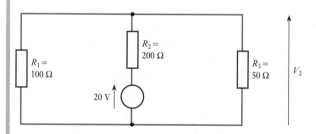

This is a potential divider circuit formed by R_2 and the parallel combination of R_1 and R_3, which gives

$$V_2 = 20\frac{100//50}{200 + 100//50}$$

$$= 20\frac{33.3}{200 + 33.3}$$

$$= 2.86 \text{ V}$$

Note that R_1 and R_3 are in parallel – they could equally well be drawn side by side.

The output of the original circuit is now found by summing these two voltages

$$V = V_1 + V_2 = 4.29 + 2.86 = 7.15 \text{ V}$$

File 12A

Computer Simulation Exercise 12.1

Use circuit simulation to investigate the circuit of Example 12.5. Determine the magnitude of the voltage V and confirm that this is as expected.

The effective internal resistance of a current source is infinite. Therefore, when removing the effects of a current generator, we replace it by an open circuit. This is illustrated in the following example.

Example 12.6

A further example of the use of superposition. Calculate the output current I in the following circuit.

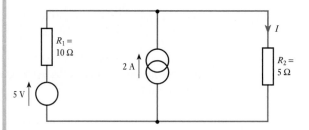

First we consider the voltage source. When doing this we replace the current source by its internal resistance, which for an ideal current source is infinite (that is, an open circuit). This gives the following circuit.

Therefore

$$I_1 = \frac{5 \text{ V}}{10 \text{ } \Omega + 5 \text{ } \Omega} = 0.33 \text{ A}$$

Next we consider the effect of the current source, replacing the voltage source with a short circuit. This gives

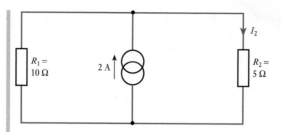

The two resistors are effectively in parallel across the current source. Since their combined resistance is $10\ \Omega//5\ \Omega = 3.33\ \Omega$, the voltage across the two resistors will be $2\ \mathrm{A} \times 3.33\ \Omega = 6.66\ \mathrm{V}$. Therefore, the current I_2 is given by

$$I_2 = \frac{6.66\ V}{5\ \Omega} = 1.33\ \mathrm{A}$$

The output current of the original circuit is now found by summing these two currents

$$I = I_1 + I_2 = 0.33 + 1.33 = 1.66\ \mathrm{A}$$

File 12B

Computer Simulation Exercise 12.2

Use circuit simulation to investigate the circuit of Example 12.6. Determine the magnitude of the current I and confirm that this is as expected.

Example 12.7

Using superposition in an op-amp circuit. Calculate the voltage gain of the following circuit.

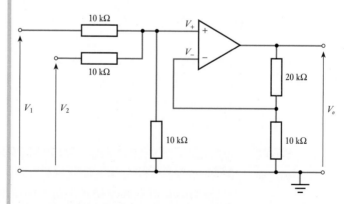

The analysis is similar to that used in the examples in Chapter 8. First we note that, as in earlier circuits, the negative feedback forces V_- to equal V_+, and therefore

$$V_- = V_+$$

Since no current flows into the inputs of the op-amp, V_- and V_+ are determined by the potential dividers formed by the resistors.

V_- is easy to calculate and is given by

$$V_- = V_o \frac{10 \text{ k}\Omega}{10 \text{ k}\Omega + 20 \text{ k}\Omega} = \frac{V_o}{3}$$

V_+ is slightly more complicated to compute, since it is determined by the two input voltages. However, applying the principle of superposition, we know that the voltage on V_+ will be equal to the sum of the voltages that would be generated if each input voltage were applied separately.

If V_1 is applied while V_2 is set to zero, then the resistor connected to V_2 effectively goes to ground and is in parallel with the existing 10 kΩ resistor that goes from V_+ to ground. Therefore,

$$V_+ = V_1 \frac{10 \text{ k}\Omega // 10 \text{ k}\Omega}{10 \text{ k}\Omega // 10 \text{ k}\Omega + 10 \text{ k}\Omega} = V_1 \frac{5 \text{ k}\Omega}{5 \text{ k}\Omega + 10 \text{ k}\Omega} = \frac{V_1}{3}$$

If now V_2 is applied while V_1 is set to zero, we have a directly equivalent situation and clearly, because of the symmetry of the circuit

$$V_+ = V_2 \frac{10 \text{ k}\Omega // 10 \text{ k}\Omega}{10 \text{ k}\Omega // 10 \text{ k}\Omega + 10 \text{ k}\Omega} = V_2 \frac{5 \text{ k}\Omega}{5 \text{ k}\Omega + 10 \text{ k}\Omega} = \frac{V_2}{3}$$

Therefore if both inputs are applied simultaneously we have

$$V_+ = \frac{V_1}{3} + \frac{V_2}{3}$$

Now since

$$V_- = V_+$$

we have

$$\frac{V_o}{3} = \frac{V_1}{3} + \frac{V_2}{3}$$

and

$$V_o = V_1 + V_2$$

Thus the circuit is a non-inverting adder. This circuit can be extended to have any number of inputs (see Exercise 12.16 and Appendix C).

File 12C

Computer Simulation Exercise 12.3

Use circuit simulation to investigate the circuit of Example 12.7 using one of the operational amplifiers supported by your simulation package. Apply appropriate input signals and confirm that the circuit operates as expected.

12.10   Nodal analysis

In Section 12.7, we saw that we can apply Kirchhoff's current law to any node in a circuit and Kirchhoff's voltage law to any loop. The analysis of real circuits often requires us to apply these laws to a group of nodes or loops, and this produces a series of simultaneous equations that must be solved to find the various voltages and currents in the circuit. Unfortunately, as the complexity of the circuit increases, the number of nodes and loops increases and the analysis becomes more involved. In order to simplify this process, we often use one of two systematic approaches to the production of these simultaneous equations, namely **nodal analysis** and **mesh analysis**. We will look at nodal analysis in this section and at mesh analysis in the next.

Nodal analysis is a systematic method of applying Kirchhoff's current law to nodes in a circuit in order to produce an appropriate set of simultaneous equations. The technique involves six distinct steps.

1. One of the nodes in the circuit is chosen as a reference node. This selection is arbitrary, but it is normal to select the ground or earth node as the reference point, and all voltages will then be measured with respect to this node.
2. The voltages on the remaining nodes in the circuit are then labelled V_1, V_2, V_3, etc. Again, the numbering of these node voltages is arbitrary.
3. If the voltages on any of the nodes are known (due to the presence of fixed-voltage sources), then these values are added to the diagram.
4. Kirchhoff's current law is then applied to each node for which the voltage is not known. This produces a set of simultaneous equations.
5. These simultaneous equations are solved to determine each unknown node voltage.
6. If necessary, the node voltages are then used to calculate appropriate currents in the circuit.

To illustrate the use of this technique consider Figure 12.9(a), which shows a relatively simple circuit. Although none of the nodes is specifically labelled as the earth, we will choose the lower point in the circuit as our reference node. We then label the three remaining node voltages as V_1, V_2 and V_3 as shown in Figure 12.9(b). It is clear that V_1 is equal to E, since this is set by the voltage source, and this is marked on the diagram. The next step is to apply Kirchhoff's current law to each node for which the voltage is unknown. In this case only V_2 and V_3 are unknown, so we need only consider these two nodes.

Let us initially consider the node associated with V_2. Figure 12.9(c) labels the currents flowing into this node as I_A, I_B and I_C, and by applying Kirchhoff's current law we see that

$$I_A + I_B + I_C = 0$$

These currents can be easily determined from the circuit diagram. Each current is given by the voltages across the associated resistor, and in each case this voltage is simply the difference between two node voltages. Therefore

Figure 12.9 The application of nodal analysis

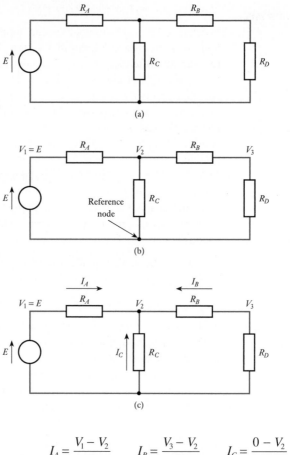

(a)

(b)

(c)

$$I_A = \frac{V_1 - V_2}{R_A} \qquad I_B = \frac{V_3 - V_2}{R_B} \qquad I_C = \frac{0 - V_2}{R_C}$$

Note that in each case we want the current flowing *into* the node associated with V_2, and therefore we subtract V_2 from the other node voltage in each case. Summing these currents then gives

$$\frac{V_1 - V_2}{R_A} + \frac{V_3 - V_2}{R_B} + \frac{0 - V_2}{R_C} = 0$$

A similar treatment of the node associated with V_3 gives

$$\frac{V_2 - V_3}{R_B} + \frac{0 - V_3}{R_D} = 0$$

We therefore have two equations that we can solve to find V_2 and V_3. From these values we can then calculate the various currents if necessary.

The circuit of Figure 12.9 contains a single voltage source, but nodal analysis can be applied to circuits containing multiple voltage sources or current sources. In a voltage source, the voltage across the device is known but the current flowing through it is not. In a current source, the current flowing through it is known but the voltage across it is not.

Example 12.8

The use of nodal analysis. Calculate the current I_1 in the following circuit.

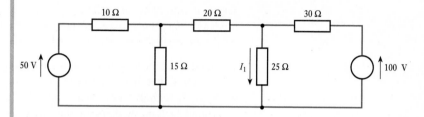

First we pick our reference node and label the various node voltages, assigning values where these are known.

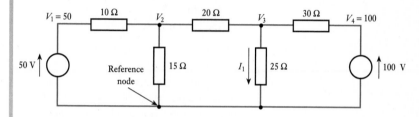

Next we sum the currents flowing into the nodes for which the node voltages are unknown. This gives

$$\frac{50 - V_2}{10} + \frac{V_3 - V_2}{20} + \frac{0 - V_2}{15} = 0$$

and

$$\frac{V_2 - V_3}{20} + \frac{100 - V_2}{30} + \frac{0 - V_3}{25} = 0$$

Solving these two equations (which is left as an exercise for the reader) gives

$$V_2 = 32.34 \text{ V}$$

$$V_3 = 40.14 \text{ V}$$

and the required current I_1 is given by

$$I_1 = \frac{V_3}{25 \ \Omega} = \frac{40.14 \text{ V}}{25 \ \Omega} = 1.6 \text{ A}$$

File 12D

Computer Simulation Exercise 12.4

Use circuit simulation to investigate the circuit of Example 12.8. Determine the voltages V_2 and V_3 and the current I_1 and confirm that these are as expected.

12.11   Mesh analysis

Mesh analysis, like nodal analysis, is a systematic means of obtaining a series of simultaneous equations describing the behaviour of a circuit. In this case, Kirchhoff's voltage law is applied to each mesh in the circuit. Again a series of steps are involved.

1. Identify the meshes in the circuit and assign a clockwise-flowing current to each. Label these currents I_1, I_2, I_3, etc.
2. Apply Kirchhoff's law to each mesh by summing the voltages clockwise around each mesh, equating the sum to zero. This produces a number of simultaneous equations (one for each loop).
3. Solve these simultaneous equations to determine the currents I_1, I_2, I_3, etc.
4. Use the values obtained for the various currents to compute voltages in the circuit as required.

This process is illustrated in Figure 12.10. The circuit of Figure 12.10(a) contains two meshes, which are labelled in Figure 12.10(b). Next we need to define the polarities of the various voltages in the circuit. This assignment is arbitrary at this stage, but it will enable us to interpret correctly the polarity of the voltages that we calculate. Figure 12.10(c) shows the way in

Figure 12.10   Mesh analysis

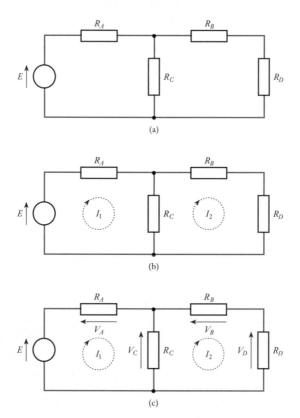

(a)

(b)

(c)

which the various voltages have been defined. Note that a positive current in one direction in a resistor produces a voltage drop in the other direction. Thus, in Figure 12.10(c), if I_1 is positive, V_A will also be positive.

Having defined our various voltages and currents, we are now in a position to write down our equations. We do this by summing the voltages clockwise around each loop and equating this sum to zero. For the first mesh, this gives

$$E - V_A - V_C = 0$$

$$E - I_1 R_A - (I_1 - I_2) R_C = 0$$

Note that only I_1 flows through R_A, so V_A is simply $I_1 R_A$. However, in R_C the current I_1 flows in one direction, while I_2 flows in the other direction. Consequently, the voltage across this resistor is $(I_1 - I_2) R_C$. Applying the same approach to the second loop gives

$$V_C - V_B - V_D = 0$$

$$(I_1 - I_2) R_C - I_2 R_B - I_2 R_D = 0$$

Hence we have two equations relating I_1 and I_2 to the circuit values. These simultaneous equations can be solved to obtain their values, which can then be used to calculate the various voltages.

As with nodal analysis, mesh analysis can be applied to circuits with any number of voltage or current sources.

Example 12.9	The use of mesh analysis. Calculate the voltage across the 10 Ω resistor in the following circuit.

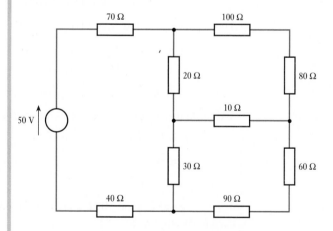

The circuit contains three meshes. To these we assign loop currents I_1, I_2 and I_3, as shown below. The diagram also defines the various voltages and, to aid explanation, assigns symbolic names to each resistor.

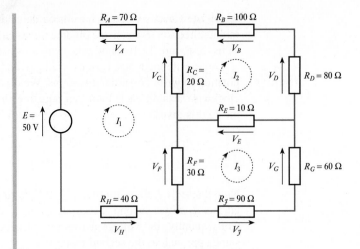

The next stage is to apply Kirchhoff's voltage law to each mesh. Normally, we would write down these equations directly using the component values and currents. However, to aid comprehension of the process, we will first write these symbolically. Considering the three loops in turn, this gives

$$E - V_A - V_C - V_F - V_H = 0$$

$$V_C - V_B - V_D + V_E = 0$$

$$V_F - V_E - V_G - V_J = 0$$

which gives the following simultaneous equations.

$$50 - 70I_1 - 20(I_1 - I_2) - 30(I_1 - I_3) - 40I_1 = 0$$

$$20(I_1 - I_2) - 100I_2 - 80I_2 + 10(I_3 - I_2) = 0$$

$$30(I_1 - I_3) - 10(I_3 - I_2) - 60I_3 - 90I_3 = 0$$

These can be rearranged to give

$$50 - 160I_1 + 20I_2 + 30I_3 = 0$$

$$20I_1 - 210I_2 + 10I_3 = 0$$

$$30I_1 + 10I_2 - 190I_3 = 0$$

and these three simultaneous equations may be solved to give

$$I_1 = 326 \text{ mA}$$

$$I_2 = 34 \text{ mA}$$

$$I_3 = 53 \text{ mA}$$

The voltage across the 10 Ω resistor is the product of its resistance and the current through it, so

$$V_E = R_E(I_3 - I_2)$$

$$= 10(0.053 - 0.034)$$

$$= 0.19 \text{ V}$$

Since the calculated voltage is positive, the polarity is as shown by the arrow, with the left-hand end of the resistor more positive than the right-hand end.

File 12E

Computer Simulation Exercise 12.5

Use circuit simulation to investigate the circuit of Example 12.9. Determine the three currents I_1, I_2 and I_3 and the voltage V_E and confirm that these are as expected.

The direction in which the currents are defined and the direction in which the voltages are summed are arbitrary. However, if you are consistent in the direction that you choose you are less likely to make mistakes. This is why the list of steps above specifies a clockwise direction in each case.

12.12 Solving simultaneous circuit equations

We have seen that both nodal analysis and mesh analysis produce a series of simultaneous equations that must be solved in order to determine the required voltages and currents. When considering simple circuits with few nodes or meshes, the number of equations generated may be small enough to be solved 'by hand' as illustrated above. However, with more complex circuits this approach becomes impractical.

A more attractive method of solving simultaneous equations is to express them in matrix form and to use the various tools of matrix algebra to obtain the solutions. For example, in Example 12.9 we obtained the following set of equations:

$$50 - 160I_1 + 20I_2 + 30I_3 = 0$$

$$20I_1 - 210I_2 + 10I_3 = 0$$

$$30I_1 + 10I_2 - 190I_3 = 0$$

These can be rearranged as

$$160I_1 - 20I_2 - 30I_3 = 50$$

$$20I_1 - 210I_2 + 10I_3 = 0$$

$$30I_1 + 10I_2 - 190I_3 = 0$$

and expressed in matrix form as

$$\begin{bmatrix} 160 & -20 & -30 \\ 20 & -210 & 10 \\ 30 & 10 & -190 \end{bmatrix} \begin{bmatrix} I_1 \\ I_2 \\ I_3 \end{bmatrix} = \begin{bmatrix} 50 \\ 0 \\ 0 \end{bmatrix}$$

This can be solved by hand using Cramer's rule or some other matrix algebra technique. Alternatively, automated tools can be used. Many scientific
calculators can solve such problems when they involve a handful of equations, while computer-based packages such as MATLAB or Mathcad can
solve large numbers of simultaneous equations if expressed in this form.

12.13   Choice of techniques

In this chapter, we have looked at a number of techniques that can be used
to analyse electrical circuits. This raises the question of how we know
which technique to use in a given situation. Unfortunately, there are no
simple rules to aid this choice, and it often comes down to the general form
of the circuit and which technique seems most appropriate. Techniques
such as nodal and mesh analysis will work in a wide range of situations,
but these are not always the simplest methods.

   To see how a range of techniques can be used, consider the circuit of
Figure 12.11(a). This circuit can be analysed by nodal or mesh analysis as
described earlier. Alternatively, the effects of each source could be investigated separately (using Ohm's law) and combined using the principle of

Figure 12.11   A simple circuit

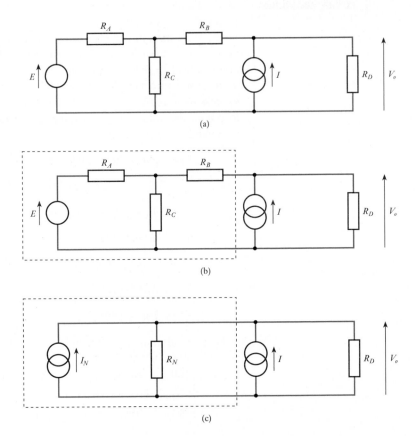

superposition. A third approach is to use Norton's theorem to simplify the circuit. Figure 12.11(b) shows the circuit with a group of components enclosed within a dotted box, and Figure 12.11(c) shows the effect of replacing these components by their Norton equivalent circuit. Determination of this equivalent circuit is relatively simple, as was seen in Example 12.3. The resultant circuit consists of two current sources and two resistors, which are all connected in parallel, and the output voltage is simply the sum of the currents produced by the sources times the effective resistance of the two resistors connected in parallel.

Whether the approach of Figure 12.11(c) is simpler than other techniques is open to debate, but for any given circuit it is likely that some approaches will be easier than others. Inevitably, your skill at picking the simplest method will increase with practice.

While the analysis of simple circuits is generally straightforward, more complex circuits can be very time-consuming. In these cases, we normally make use of computer-based network analysis tools. These often use nodal analysis and, if necessary, can handle circuits of great complexity. However, in many cases the manual techniques described in this chapter are completely adequate.

Key points

- An electric current is a flow of charge.

- A voltage source produces an e.m.f., which causes a current to flow in a circuit. An ideal voltage source has zero output resistance, but all real voltage sources have associated resistance.

- An ideal current source produces a constant current no matter what is connected to it. Such a source has an infinite output resistance.

- The current in a conductor is directly proportional to the voltage across it (this is Ohm's law). This voltage divided by the current gives the resistance of the conductor.

- The resistance of several resistors in series is given by the sum of their resistances.

- The resistance of several resistors in parallel is equal to the reciprocal of the sum of the reciprocals of their resistances.

- At any instant, the currents flowing into any node in a circuit sum to zero (Kirchhoff's current law).

- At any instant, the voltages around any loop in a circuit sum to zero (Kirchhoff's voltage law).

- Any two-terminal network of resistors and energy sources can be replaced by a series combination of a voltage source and a resistor (Thévenin's theorem).

- Any two-terminal network of resistors and energy sources can be replaced by a parallel combination of a current source and a resistor (Norton's theorem).

- In any linear network containing more than one energy source, the currents and voltages are equal to the sum of the voltages or currents that would be present if each source were considered separately (principle of superposition).

- Nodal analysis and mesh analysis each offer systematic methods of obtaining a set of simultaneous equations that can be solved to determine the voltages and currents in a circuit.

- When considering any particular circuit, a range of circuit analysis techniques can be used. The choice of technique should be based on an assessment of the nature of the circuit.

Exercises

12.1 Write down an equation relating current and charge.

12.2 What quantity of charge is transferred if a current of 5 A flows for 10 seconds?

12.3 What is the internal resistance of an ideal voltage source?

12.4 What is meant by a *controlled* voltage source?

12.5 What is the internal resistance of an ideal current source?

12.6 Determine the voltage V in each of the following circuits, being careful to note its polarity in each case.

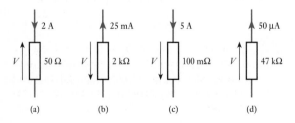

12.7 For each circuit in Exercise 12.6, determine the power dissipated in the resistor.

12.8 Estimate the resistance of a copper wire with a cross-sectional area of 1 mm^2 and a length of 1 m at 0 °C.

12.9 Determine the resistance of each of the following combinations.

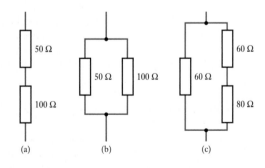

12.10 What resistance corresponds to 10 kΩ//10 kΩ?

12.11 Define the terms 'node', 'loop' and 'mesh'.

Exercises continued

12.12 Derive Thévenin and Norton equivalent circuits for the following arrangements.

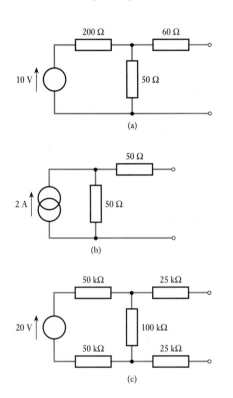

12.13 A two-terminal network is investigated by measuring the output voltage when connected to different loads. When a resistance of 12 Ω is connected across the output the output voltage is 16 V, and when a load of 48 Ω is connected the output voltage is 32 V. Use a graphical method to determine the Thévenin and Norton equivalent circuits of this arrangement.

12.14 Repeat Exercise 12.13 using a non-graphical approach.

12.15 Use the principle of superposition to determine the voltage V in each of the following circuits.

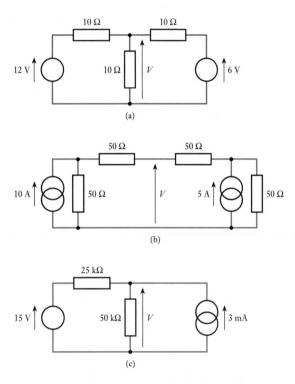

12.16 Derive an expression for the output voltage V_o of the following circuit in terms of the input voltages V_1, V_2 and V_3 and the component values.

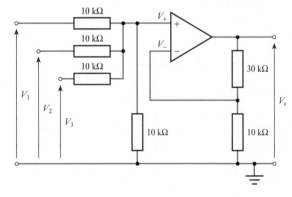

12.17 Simulate the circuit of Exercise 12.16 using one of the operational amplifiers supported by your simulation package. Apply appropriate input signals and hence confirm your answer to this exercise.

Exercises continued

12.18 Use nodal analysis to determine the voltage V in the following circuit.

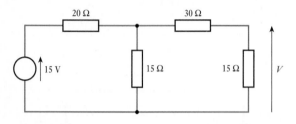

12.19 Simulate the circuit of Exercise 12.18 and hence confirm your answer to this exercise.

12.20 Use nodal analysis to determine the current I_1 in the following circuit.

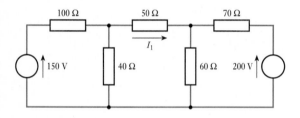

12.21 Simulate the circuit of Exercise 12.20 and hence confirm your answer to this exercise.

12.22 Use nodal analysis to determine the current I_1 in the following circuit.

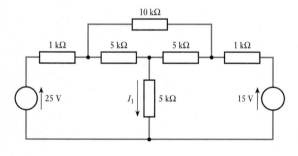

12.23 Simulate the circuit of Exercise 12.22 and hence confirm your answer to this exercise.

12.24 Use mesh analysis to determine the voltage V in the following circuit.

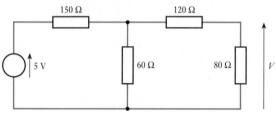

12.25 Simulate the circuit of Exercise 12.24 and hence confirm your answer to this exercise.

12.26 Use mesh analysis to determine the voltage V in the following circuit.

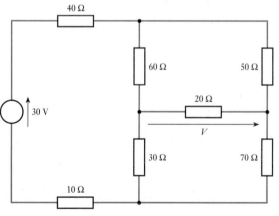

12.27 Simulate the circuit of Exercise 12.26 and hence confirm your answer to this exercise.

12.28 Use mesh analysis to determine the current I in the following circuit.

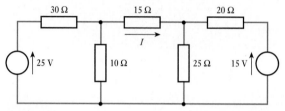

12.30 Use an appropriate form of analysis to determine the voltage V_o in the following circuit.

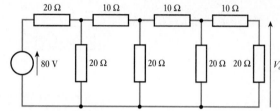

12.29 Simulate the circuit of Exercise 12.28 and hence confirm your answer to this exercise.

12.31 Simulate the circuit of Exercise 12.30 and hence confirm your answer to this exercise.

Chapter 13

Capacitance and Electric Fields

Objectives

When you have studied the material in this chapter you should be able to:

- describe the construction and form of a range of capacitors;
- explain the mechanism whereby charge is stored in a capacitor and the relationship between quantities such as charge, voltage, current and capacitance;
- define terms such as absolute permittivity and relative permittivity, and use these quantities to calculate the capacitance of a component from its dimensions;
- explain concepts such as electric field strength and electric flux density, and calculate the magnitudes of these quantities;
- determine the effective capacitance of a number of capacitors when connected in series or parallel;
- describe the relationship between the voltage and the current in a capacitor for both DC and AC signals;
- calculate the energy stored in a charged capacitor.

13.1 Introduction

In Chapter 12, we noted that an electric current represents a flow of electric charge. A capacitor is a component that can store electric charge and can therefore store energy. Capacitors are often used in association with alternating currents and voltages, and they are a key component in almost all electronic circuits.

13.2 Capacitors and capacitance

Capacitors consist of two conducting surfaces separated by an insulating layer called a **dielectric**. Figure 13.1(a) shows a simple capacitor consisting of two rectangular metal sheets separated by a uniform layer of dielectric. The gap between the conducting layers could be filled with air (since

Figure 13.1   Capacitors

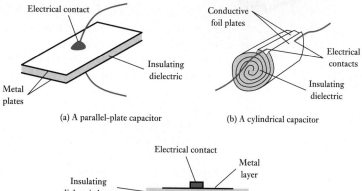

(a) A parallel–plate capacitor

(b) A cylindrical capacitor

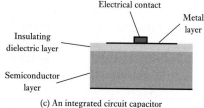

(c) An integrated circuit capacitor

Figure 13.2   A simple capacitor circuit

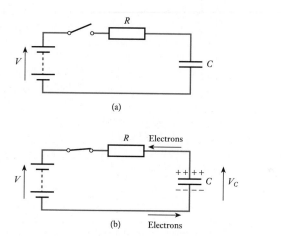

this is a good insulator) or with another dielectric material. In some cases the conductors are made of metal foil and the dielectric is flexible, allowing the arrangement to be rolled into a cylindrical shape as in Figure 13.1(b). In integrated circuits, a capacitor might be formed by depositing a layer of metal on to an insulating layer above a conducting semiconductor layer as shown in Figure 13.1(c). A wide range of construction methods are used to form capacitors, but in each case their basic operation is the same.

To understand the operation of a capacitor, consider the circuit of Figure 13.2(a). Here a battery, a resistor, a switch and a capacitor are connected in series. While the switch is open no current will flow around the circuit. However, when the switch is closed, the e.m.f. produced by the battery will attempt to drive a current around the network. Electrons flowing from the negative terminal of the battery will flow on to the lower plate of the capacitor, where they will repel electrons from the upper plate. As electrons are repelled from this plate, they leave a residual positive charge (in the form of a deficit of negatively charged electrons). The combined effect of these two processes is that electrons flow into the lower plate of the capacitor and flow out from the upper plate. Note that this represents a flow of conventional current in the opposite direction.

Since electrons flow *into* one side of the capacitor and flow *out of* the other, it might appear that current is flowing *through* it. However, this is an illusion. The dielectric between the plates of the capacitor is an insulator, and electrons do not actually cross this barrier. It should also be noted that this flow of electrons cannot last indefinitely. As electrons flow around the circuit they produce an increasing positive charge on one side of the capacitor and an increasing negative charge on the other. The result is an increasing **electric field** between the two plates. This produces a potential difference V_C across the capacitor, which opposes the e.m.f. of the battery. Eventually, the voltage across the capacitor is equal to that of the battery and the current falls to zero.

In this state, the capacitor is storing electrical charge and is therefore storing electrical energy. If the switch is opened at this point there is no path by which this charge can flow, and the capacitor will remain charged with a voltage of V_C across it. If now a resistor is connected across the charged capacitor, the stored energy will drive a current through the resistor, discharging the capacitor and releasing the stored energy. The capacitor therefore acts a little like a 'rechargeable battery', although the mechanism used to store the electrical energy is very different and the amount of energy stored is normally very small.

For a given capacitor, the charge stored q is directly proportional to the voltage across it V. The relationship between these two quantities is given by the capacitance C of the capacitor, such that

$$C = \frac{q}{V} \tag{13.1}$$

If the charge is measured in *coulombs* and the voltage in *volts*, then the capacitance has the units of **farads**.

| **Example 13.1** | A 10 μF capacitor has 10 V across it. What quantity of charge is stored in it? |

From Equation 13.1 we have

$$C = \frac{q}{V}$$

$$q = CV$$

$$= 10^{-5} \times 10$$

$$= 100 \ \mu C$$

Remember that C is the unit symbol for coulombs, while C is the symbol used for capacitance.

13.3   Capacitors and alternating voltages and currents

It is clear from the discussion above that a constant current (that is, a direct current) cannot flow through a capacitor. However, since the voltage across a capacitor is proportional to the charge on it, it follows that a changing voltage must correspond to a changing charge. Consequently, a changing voltage must correspond to a current flowing into or out of the capacitor. This is illustrated in Figure 13.3.

The voltage produced by the alternating voltage source in Figure 13.3 is constantly changing. When this voltage becomes more positive, this will produce a positive current *into* the top plate of the capacitor (a flow of electrons out of the top plate), making this plate more positive. When the source voltage becomes more negative, this will cause current to flow *out of* the top plate (a flow of electrons into this plate), making this plate more negative. Thus the alternating voltage produces an alternating current around the complete circuit.

It is important to remember at this point that the observed current does not represent electrons flowing from one plate to the other – these plates are separated by an insulator. However, the circuit behaves as if an alternating current is flowing through the capacitor. Perhaps one way of understanding this apparent paradox is to consider a mechanical analogy, as shown in Figure 13.4.

Figure 13.4(a) shows a section through a glass window that is fitted within a window frame. In this figure, the pressure on one side of the glass is higher than on the other, causing a deflection of the glass. In deflecting the glass in this way the air has done work on it, and energy is now stored in the glass (as it would be in a spring under tension). Figure 13.4(b) shows a similar position where the imbalance in air pressure is reversed. Again the

Figure 13.3   A capacitor and an alternating voltage

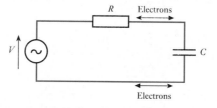

Figure 13.4   A mechanical analogy of a capacitor – a window

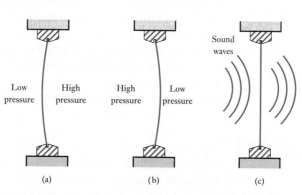

glass is deflected and is storing mechanical energy. In Figure 13.4(c), the average pressure on each side of the glass is equal, but sound waves are striking the glass on one side. Sound waves represent fluctuations in air pressure, and these will cause the window to vibrate backwards and forwards. This in turn will cause sound waves to be generated on the other side of the glass. We can see that a constant pressure difference across the window does not result in the passage of air through it. However, an alternating pressure difference (sound) is transmitted from one side to the other, even though no air passes through the window.

While a window is not a perfect analogy for a capacitor, it does illustrate quite well the way that a capacitor *blocks* direct currents but appears to *pass* alternating currents. However, it should be noted that, just as a window attenuates sounds that pass through it, so a capacitor impedes the flow of the current that appears to pass through it. In the case of a window, the attenuation will depend on the size of the window and the nature of the sound (its frequency range). Similarly, the effect of the capacitor will depend on its size (its capacitance) and the frequency of the signals present. We will return to look at this phenomenon in more detail when we look at alternating voltages and currents in Chapter 15.

13.4 The effect of a capacitor's dimensions on its capacitance

The capacitance of a capacitor is directly proportional to the area A of the conducting plates and inversely proportional to the distance d between them. Therefore $C \propto A/d$. The constant of proportionality of this relationship is the **permittivity** ε of the dielectric. The permittivity is normally expressed as the product of two terms: the **absolute permittivity** ε_0 and the **relative permittivity** ε_r of the dielectric used.

$$C = \frac{\varepsilon A}{d} \tag{13.2}$$

$$= \frac{\varepsilon_0 \varepsilon_r A}{d}$$

ε_0 is also referred to as the **permittivity of free space** and is the permittivity of a vacuum. It has a value of about 8.85 picofarads per metre (pF/m). ε_r represents the ratio of the permittivity of a material to that of a vacuum. Air has an ε_r very close to 1, while insulators have values from about 2 to a thousand or more. While capacitors can be produced using air as the dielectric material, much smaller components can be produced by using a material with a much higher relative permittivity.

Example 13.2

The conducting plates of a capacitor are 10×25 mm and have a separation of 7 μm. If the dielectric has a relative permittivity of 100, what will be the capacitance of the device?

From Equation 13.2 we have

$$C = \frac{\varepsilon_0 \varepsilon_r A}{d}$$

$$= \frac{8.85 \times 10^{-12} \times 100 \times 10 \times 10^{-3} \times 25 \times 10^{-3}}{7 \times 10^{-6}} = 31.6 \text{ nF}$$

Note that to achieve the same capacitance using air as the dielectric we would need a device 100×250 mm (assuming the same plate separation).

Capacitance is present not only between the plates of a capacitor but also between any two conductors that are separated by an insulator. Therefore, a small amount of capacitance exists between each of the conductors in electrical circuits (for example between each wire) and between the various elements in electrical components. These small, unintended capacitances are called stray capacitances and can have a very marked effect on circuit behaviour. The need to charge and discharge these stray capacitances limits the speed of operation of circuits and is a particular problem in high-speed circuits and those that use small signal currents.

13.5   Electric field strength and electric flux density

Electric charges of the same polarity repel each other, while those of opposite polarities attract. When charged particles experience a force as a result of their charge, we say that an electric field exists in that region. The magnitude of the force exerted on a charged particle is determined by the **electric field strength**, E, at that point in space. This quantity is defined as the force exerted on a unit charge at that point. When a voltage V exists between two points a distance d apart, the electric field strength is given by

$$E = \frac{V}{d} \tag{13.3}$$

and has units of volts per metre (V/m).

The charge stored in a capacitor produces a potential across the capacitor and an electric field across the dielectric material. This is illustrated in Figure 13.5, which shows a capacitor that has two plates each of area A, separated by a dielectric of thickness d. The capacitor holds a charge of Q and the potential across the capacitor is V. From Equation 13.3, we know that the electric field strength in the dielectric material is V/d volts per metre.

Figure 13.5   A charged capacitor

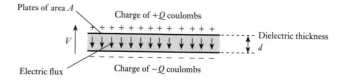

Plates of area A

Charge of $+Q$ coulombs

V

Dielectric thickness d

Electric flux

Charge of $-Q$ coulombs

| Example 13.3 | The conducting plates of a capacitor have a separation of 10 μm. If the potential across the capacitor is 100 V, what is the electric field strength in the dielectric? |

From Equation 13.3 we have

$$E = \frac{V}{d}$$

$$= \frac{100}{10^{-5}}$$

$$= 10^7 \text{ V/m}$$

All insulating materials have a maximum value for the field strength that they can withstand before they break down. This is termed their **dielectric strength** E_m. For this reason, all capacitors have a maximum operating voltage, which is related to the material used for the dielectric and its thickness. From Equation 13.2, it is clear that, to obtain the maximum capacitance for a device of a given plate size, we want to make the dielectric as thin as possible. Equation 13.3 shows that in doing this we increase the electrical stress on the insulating material, and in practice we need to compromise between physical size and breakdown voltage.

The force between positive and negative charges is often described in terms of an **electric flux** linking them. This is measured using the same units as electric charge (coulombs) and thus a charge of Q coulombs will produce a total electric flux of Q coulombs. We also define what is termed the **electric flux density**, D, as the amount of flux passing through a defined area perpendicular to the flux. In a capacitor, the size of the plates is always much greater than their separation, so 'edge effects' can be ignored and we can assume that all the flux produced by the stored charge passes through the area of the dielectric. Returning to Figure 13.5, we see that here a total flux of Q passes through an area of A, so the electric flux density is given by

$$D = \frac{Q}{A} \tag{13.4}$$

| Example 13.4 | The conducting plates of a capacitor have an area of 200 mm². If the charge on the capacitor is 15 μC, what is the electric flux density within the dielectric? |

From Equation 13.4 we have

$$D = \frac{Q}{A}$$

$$= \frac{15 \times 10^{-6}}{200 \times 10^{-6}}$$

$$= 75 \text{ mC/m}^2$$

By combining the results of Equations 13.1–13.4, it is relatively easy to show that

$$\varepsilon = \frac{D}{E} \qquad (13.5)$$

and thus the permittivity of the dielectric in a capacitor is equal to the ratio of the electric flux density to the electric field strength.

13.6 Capacitors in series and parallel

In Chapter 12, we considered the combined effect of putting several resistors in series and in parallel, and it is appropriate to look at combinations of capacitors in the same way. We will begin by looking at a parallel arrangement.

13.6.1 Capacitors in parallel

Consider a voltage V applied across two capacitors C_1 and C_2 connected in parallel, as shown in Figure 13.6(a). If we call the charge on these two capacitors Q_1 and Q_2, then

$$Q_1 = VC_1 \quad \text{and} \quad Q_2 = VC_2$$

If the two capacitors are now replaced by a single component with a capacitance of C such that this has the same capacitance as the parallel combination, then clearly the charge stored on C must be equal to $Q_1 + Q_2$. Therefore

$$\text{charge stored on } C = Q_1 + Q_2$$

$$VC = VC_1 + VC_2$$

$$C = C_1 + C_2$$

Therefore, the effective capacitance of two capacitors in parallel is equal to the sum of their capacitances. This result can clearly be extended to any number of components, and in general, for N capacitors in parallel:

$$C = C_1 + C_2 + \ldots + C_N \qquad (13.6)$$

Figure 13.6 Capacitors in parallel and in series

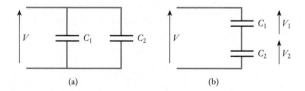

Example 13.5 Capacitors in parallel. What is the effective capacitance of this arrangement?

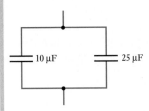

From Equation 13.6 we have

$$C = C_1 + C_2$$

$$= 10\ \mu F + 25\ \mu F$$

$$= 35\ \mu F$$

13.6.2 Capacitors in series

Consider a voltage V applied across two capacitors C_1 and C_2 connected in series, as shown in Figure 13.6(b). In this series arrangement, the only charge that can be delivered to the lower plate of C_1 is the charge supplied from the upper plate of C_2 and therefore the charge on each capacitor must be identical. We will call the charge on each capacitor Q.

If the two capacitors are now replaced by a single component with a capacitance of C such that this has the same capacitance as the series combination, then clearly the charge stored on C must also be equal to Q. From the diagram, it is clear that the applied voltage V is equal to $V_1 + V_2$ and therefore

$$V = V_1 + V_2$$

$$\frac{Q}{C} = \frac{Q}{C_1} + \frac{Q}{C_2}$$

$$\frac{1}{C} = \frac{1}{C_1} + \frac{1}{C_2}$$

Therefore, the effective capacitance of two capacitors in series is equal to the reciprocal of the sum of the reciprocals of their capacitances. This result can clearly be extended to any number of components, and in general, for N capacitors in series:

$$\frac{1}{C} = \frac{1}{C_1} + \frac{1}{C_2} + \ldots + \frac{1}{C_N} \tag{13.7}$$

Example 13.6 Capacitors in series. What is the effective capacitance of this arrangement?

10 μF   25 μF

From Equation 13.7 we have

$$\frac{1}{C} = \frac{1}{C_1} + \frac{1}{C_2}$$

$$= \frac{1}{10} + \frac{1}{25} = \frac{35}{250}$$

$$C = 7.14 \ \mu F$$

13.7 Relationship between voltage and current in a capacitor

While the voltage across a resistor is directly proportional to the current flowing though it, this is *not* the case in a capacitor. From Equation 13.1, we know that the voltage across a capacitor is directly related to the *charge* on the capacitor, and from Equation 12.2 we know that charge is given by the integral of the current with respect to time. Therefore, the voltage across a capacitor V is given by

$$V = \frac{Q}{C} = \frac{1}{C}\int I dt \tag{13.8}$$

One implication of this relationship is that the voltage across a capacitor cannot change instantaneously, since this would require an infinite current. The rate at which the voltage changes is controlled by the magnitude of the current.

Another way of looking at this relationship is to consider the current into the capacitor as a function of voltage, rather than the other way around. By differentiating our basic equation $Q = CV$, we obtain

$$\frac{dQ}{dt} = C\frac{dV}{dt}$$

and since dQ/dt is equal to current, this gives

$$I = C\frac{dV}{dt} \tag{13.9}$$

To investigate these relationships further, consider the circuit of Figure 13.7(a). If the capacitor is initially discharged, then the voltage across it will be zero (since $V = Q/C$). If now the switch is closed (at $t = 0$) then the voltage across the capacitor cannot change instantly, so initially $V_C = 0$. By applying Kirchhoff's voltage law around the circuit, it is clear

Figure 13.7 Relationship
between voltage and current in a
DC circuit

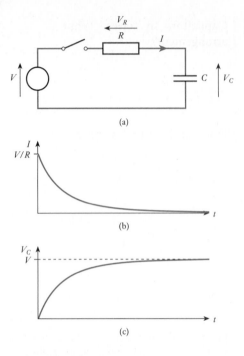

(a)

(b)

(c)

that $V = V_R + V_C$, and, if initially $V_C = 0$, then the entire supply voltage V will appear across the resistor. Therefore $V_R = V$, and the initial current flowing around the circuit, and thus into the capacitor, is given by $I = V/R$.

As the current flows into the capacitor its charge increases and V_C grows. This reduces the voltage across the resistor and hence the current. Therefore, as the voltage on the capacitor increases, the charging current decreases. The result is that the charging current is initially V/R but falls exponentially with time, and the voltage across the capacitor is initially zero but rises exponentially with time. Eventually, the voltage on the capacitor is virtually equal to the supply voltage and the charging current becomes negligible. This behaviour is shown in Figures 13.7(b) and 13.7(c).

File 13A

Computer Simulation Exercise 13.1

Simulate the circuit of Figure 13.7(a) with $V = 1$ V, $R = 1$ MΩ and $C = 200$ nF. Include in your circuit a switch that closes at $t = 0$ and another switch (not shown in Figure 13.7(a)) that opens at $t = 0$. This second switch should be connected directly across C to ensure that the capacitor is initially discharged. Use transient simulation to investigate the behaviour of the circuit during the first second after the switches change. Plot V_C and I against time on separate graphs and confirm that the circuit behaves as expected. Experiment with different values of the circuit components and note the effects on the voltage and current graphs.

It is clear from the above that the charging current is determined by R and the voltage across it. Therefore, increasing R will increase the time needed to charge up the capacitor, while reducing R will speed up the process. It is also clear that increasing the capacitance C will also increase the time taken for it to charge to a given voltage (since the voltage on a capacitor is inversely proportional to its capacitance for a given charge). Thus the time taken for the capacitor to charge to a particular voltage increases with both C and R. This leads to the concept of the **time constant** for the circuit, which is equal to the product CR. The time constant is given the symbol T (upper case Greek letter *tau*). It can be shown that the charging rate of the circuit of Figure 13.7(a) is determined by the value of the time constant of the circuit rather than the actual values of C and R. We will return to this topic in Chapter 18 when we look in more detail at transient behaviour.

File 13A

Computer Simulation Exercise 13.2

Repeat Computer Simulation Exercise 13.1 noting the effect of different component values. Begin with the same values as in the previous exercise and then change the values of C and R while keeping their product constant. Again plot V_C and I against time on separate graphs and confirm that the characteristics are unchanged. Hence confirm that the characteristics are determined by the time constant CR rather than the actual values of C and R.

13.8 Sinusoidal voltages and currents

So far we have considered the relationship between voltage and current in a DC circuit containing a capacitor. It is also important to look at the situation when using sinusoidal quantities.

Consider the arrangement of Figure 13.8(a), where an alternating voltage is applied across a capacitor. Figure 13.8(b) shows the sinusoidal voltage waveform across the capacitor, which in turn dictates the charge on the capacitor and the current into it. From Equation 13.9, we know that the current into a capacitor is given by $C\, dV/dt$, so the current is directly proportional to the *time differential* of the voltage. Since the differential of a sine wave is a cosine wave, we obtain a current waveform as shown in Figure 13.8(c). The current waveform is phase-shifted with respect to the voltage waveform by 90° (or $\pi/2$ radians). It is also clear that the current waveform *leads* the voltage waveform. This is a very important property of capacitors, and we will return to look at the mathematics of this relationship in Chapter 15.

Figure 13.8 Capacitors and alternating quantities

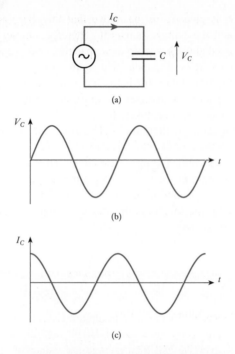

(a)

(b)

(c)

File 13B

Computer Simulation Exercise 13.3

Simulate the circuit of Figure 13.8(a) using any value of capacitor. Use a sinusoidal voltage source to apply a voltage of 1 V peak at 1 Hz and use transient analysis to display the voltage across the capacitor, and the current through it, over a period of several seconds. Note the phase relationship between the two waveforms and hence confirm that the current leads the voltages by 90° (or $\pi/2$ radians). Note the effect of varying the capacitor value, and the frequency used.

Another implication of the fact that the current into a capacitor is equal to $C\,dV/dt$ is that the magnitude of the current is determined by the rate at which the voltage changes. In a sinusoidal waveform, this rate of change is related to its frequency. This **frequency dependence** will be investigated in more detail in Chapter 15.

13.9   Energy stored in a charged capacitor

We noted earlier that capacitors store energy, and it is now time to quantify this effect. To move a charge Q through a potential difference V requires an amount of energy QV. As we progressively charge up a capacitor, we can consider that we are repeatedly adding small amounts of charge ΔQ by moving them through the voltage on the capacitor V. The energy needed to do this is clearly $V \times \Delta Q$. Since $Q = CV$, it follows that ΔQ is equivalent to

$C\Delta V$ (since C is constant) and hence the energy needed to add an amount of charge ΔQ is $V \times C\Delta V$ or, rearranging, $CV\Delta V$. If we now take an uncharged capacitor and calculate the energy needed to add successive charges to raise the voltage to V, this is

$$E = \int_0^V CV dV = \frac{1}{2}CV^2 \qquad (13.10)$$

Alternatively, since $V = Q/C$, the energy can be written as

$$E = \frac{1}{2}CV^2 = \frac{1}{2}C\left(\frac{Q}{C}\right)^2 = \frac{1}{2}\frac{Q^2}{C}$$

Thus the energy stored within a charged capacitor is $\frac{1}{2}CV^2$ or $\frac{1}{2}Q^2/C$. The unit of energy is the joule (J).

Example 13.7

Calculate the energy stored in a 10 μF capacitor when it is charged to 100 V.

From Equation 13.10

$$E = \frac{1}{2}CV^2$$

$$= \frac{1}{2} \times 10^{-5} \times 100^2$$

$$= 50 \text{ mJ}$$

13.10 Circuit symbols

Although we have used a single circuit symbol to represent capacitors, other symbols are sometimes used to differentiate between different types of device. Figure 13.9 shows a range of symbols that are commonly used. Figure 13.9(a) shows the standard symbol for a fixed capacitor and Figure 13.9(b) the symbol for a variable capacitor. Some devices, such as **electrolytic capacitors**, use construction methods that mean that they can only be used with voltages of a single polarity. In this case, the symbol is modified to indicate how the component should be connected. This can be done by adding a '+' sign, as in Figure 13.9(c), or by using a modified symbol as in Figure 13.9(d). In this last figure, the upper terminal of the component is the positive terminal.

Figure 13.9 Circuit symbols for capacitors

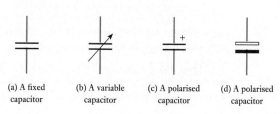

(a) A fixed capacitor   (b) A variable capacitor   (c) A polarised capacitor   (d) A polarised capacitor

Key points

- A capacitor consists of two conducting plates separated by an insulating dielectric.

- Electrons flowing around a circuit can produce a positive charge on one side of the capacitor and an equal negative charge on the other. This results in an electric field between the two plates.

- The charge stored on a capacitor is directly proportional to the voltage across it.

$$C = \frac{q}{V}$$

- A capacitor *blocks* direct currents but appears to *pass* alternating currents.

- The capacitance of a parallel-plate capacitor is proportional to the surface area of the plates and inversely proportional to their separation. The constant of proportionality is the permittivity of the dielectric.

$$C = \frac{\varepsilon A}{d} = \frac{\varepsilon_0 \varepsilon_r A}{d}$$

- The charge on the plates of the capacitor produces an electric field with a strength $E = V/d$.

- The stored charge produces an electric flux in the dielectric. The flux density $D = Q/A$.

- The capacitance of several capacitors in parallel is given by the sum of their individual capacitances.

- The capacitance of several capacitors in series is given by the reciprocal of the sum of the reciprocals of the individual capacitances.

- The voltage on a capacitor is given by

$$V = \frac{1}{C} \int I \, dt$$

- The current into a capacitor is given by

$$I = C \frac{dV}{dt}$$

- When a capacitor is char ged through a resistor, the charging rate is determined by the time constant CR.

- When capacitors are used with sinusoidal signals, the current leads the voltage by 90° ($\pi/2$ radians).

- The energy stored in a charged capacitor is $\frac{1}{2}CV^2$ or $\frac{1}{2}Q^2/C$.

Exercises

13.1 Explain what is meant by a dielectric.

13.2 If electrons represent negative charge in a capacitor, what constitutes positive charge?

13.3 If the two plates of a capacitor are insulated from each other, why does it appear that under some circumstances a current flows between them?

13.4 Why does the presence of charge on the plates of a capacitor represent the storage of energy?

13.5 How is the voltage across a capacitor related to the stored charge?

13.6 A 22 µF capacitor holds 1 mC of stored charge. What voltage is seen across its terminals?

13.7 A capacitor has a voltage of 25 V across it when it holds 500 µC of charge. What is its capacitance?

13.8 Why does a capacitor appear to pass AC signals while blocking DC signals?

13.9 How is the capacitance of a parallel-plate capacitor related to its dimensions?

13.10 The conducting plates of a capacitor are 5×15 mm and have a separation of 10 µm. What would be the capacitance of such a device if the space between the plates were filled with air?

13.11 What would be the capacitance of the device described in Exercise 13.10 if the space between the plates were filled with a dielectric with a relative permittivity of 200?

13.12 What is meant by stray capacitance, and why is this sometimes a problem?

13.13 Explain what is meant by an electric field and by electric field strength.

13.14 The plates of a capacitor have 250 V across them and have a separation of 15 µm. What is the electric field strength in the dielectric?

13.15 What is meant by dielectric strength?

13.16 Explain what is meant by electric flux and by electric flux density.

13.17 The plates of a capacitor are 15×35 mm and store a charge of 35 µC. Calculate the electric flux density in the dielectric.

13.18 Determine the effective capacitance of each of the following arrangements?

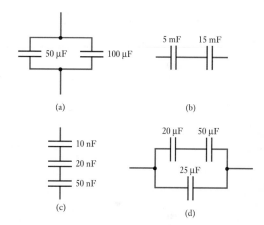

13.19 How is voltage related to current in a capacitor?

13.20 Repeat Computer Simulation Exercise 13.1 with $V = 5$ V, $R = 100$ kΩ and $C = 1$ µF. Plot the voltage across the capacitor as a function of time and hence estimate the time taken for the capacitor voltage to reach 2.5 V.

13.21 Explain what is meant by a time constant. What is the time constant of the circuit in Exercise 13.20?

13.22 The circuit of Exercise 13.20 is modified by changing R to 10 kΩ. What value should be chosen for C so that the time taken for the capacitor to charge to 2.5 V is unchanged?

13.23 Confirm your answer to Exercise 13.22 using computer simulation.

Exercises continued

13.24 Describe the relationship between the voltage across a capacitor and the current if the voltage is sinusoidal.

13.25 Give an expression for the energy stored in a charged capacitor.

13.26 A 5 mF capacitor is charged to 15 V. What is the energy stored in the capacitor?

13.27 A 50 μF capacitor contains 1.25 mC of charge. What energy is stored in the capacitor?

Chapter 14

Inductance and Magnetic Fields

Objectives

When you have studied the material in this chapter you should be able to:

- **explain the meaning and significance of terms such as magnetic field strength, magnetic flux, permeability, reluctance and inductance;**
- **outline the basic principles of electromagnetism and apply these to simple calculations of magnetic circuits;**
- **describe the mechanisms of self-induction and mutual induction;**
- **estimate the inductance of simple inductors from a knowledge of their physical construction;**
- **describe the relationship between the current and voltage in an inductor for both DC and AC signals;**
- **calculate the energy stored in an inductor in terms of its inductance and its current;**
- **describe the operation and characteristics of transformers;**
- **explain the operation of a range of inductive sensors.**

14.1 Introduction

We noted in Chapter 13 that capacitors store energy by producing an electric field within a piece of dielectric material. Inductors also store energy, but in this case it is stored within a *magnetic* field. In order to understand the operation and characteristics of inductors, and related components such as transformers, first we need to look at *electromagnetism*.

14.2 Electromagnetism

A wire carrying an electrical current causes a **magnetomotive force** (**m.m.f.**), F, which produces a **magnetic field** about it, as shown in Figure 14.1(a). One can think of an m.m.f. as being similar in some ways to an e.m.f. in an electric circuit. The presence of an e.m.f. results in an electric field and in the production of an electric current. Similarly, in magnetic

Figure 14.1 The magnetic effects of an electric current in a wire

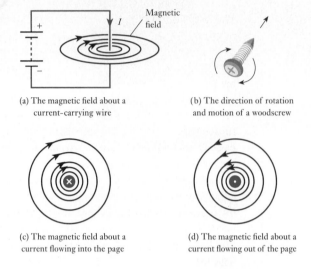

(a) The magnetic field about a current-carrying wire

(b) The direction of rotation and motion of a woodscrew

(c) The magnetic field about a current flowing into the page

(d) The magnetic field about a current flowing out of the page

circuits, the presence of an m.m.f. results in a magnetic field and the production of magnetic flux. The m.m.f. has units of amperes and for a single wire F is simply equal to the current I.

The magnitude of the field is defined by the **magnetic field strength**, H, which in this arrangement is given by

$$H = \frac{I}{l} \tag{14.1}$$

where I is the current flowing in the wire and l is the length of the magnetic circuit. The units of H are amperes per metre. The length of the circuit increases as the circumference of the circles increases, and hence the field gets weaker as we move further from the wire. Since the circumference of a circle is linearly related to its radius (being equal to $2\pi r$), the field strength is directly proportional to the current I and inversely proportional to the distance from the wire.

Example 14.1

A straight wire carries a current of 5 A. What is the magnetic field strength, H, at a distance of 100 mm from the wire?

Since the field about a straight wire is symmetrical, the length of the magnetic path at a distance r from the wire is given by the circumference of a circle of this radius. When $r = 100$ mm, the circumference is equal to $2\pi r = 0.628$ m. Therefore, from Equation 14.1

$$\text{magnetic field strength, } H = \frac{I}{l}$$

$$= \frac{5}{0.628}$$

$$= 7.96 \text{ A/m}$$

The direction of the electric field is determined by the direction of the current in the wire. For a long straight wire the electric field is circular about its axis, and one way of remembering the direction of the magnetic field is to visualise a woodscrew lying along the axis of the wire. In this arrangement, the rotation of the screw bears the same relationship to the direction of motion of the screw as the direction of the magnetic field has to the flow of current in the wire. This is shown in Figure 14.1(b). If we imagine a wire running perpendicular through this page, then a current flowing into the page would produce a clockwise magnetic field, while one flowing out of the page would result in an anticlockwise field, as shown in Figures 14.1(c) and 14.1(d). The direction of current flow in these figures is indicated by a cross to show current into the page and a dot to show current coming out of the page. To remember this notation, you may find it useful to visualise the head or the point of the screw of Figure 14.1(b).

The magnetic field produces a **magnetic flux** that flows in the same direction as the field. Magnetic flux is given the symbol Φ, and the unit of flux is the **weber** (Wb).

The strength of the flux at a particular location is measured in terms of the **magnetic flux density**, B, which is the flux per unit area of cross-section. Therefore

$$B = \frac{\Phi}{A} \qquad (14.2)$$

The unit of flux density is the tesla (T), which is equal to 1 Wb/m^2.

The flux density at a point depends on the strength of the field at that point, but it is also greatly affected by the material present. If a current-carrying wire is surrounded by air, this will result in a relatively small amount of magnetic flux as shown in Figure 14.2(a). However, if the wire is surrounded by a ferromagnetic ring, the flux within the ring will be orders of magnitude greater, as illustrated in Figure 14.2(b).

Magnetic flux density is related to the field strength by the expression

$$B = \mu H \qquad (14.3)$$

where μ is the **permeability** of the material through which the field passes. One can think of the permeability of a material as a measure of the ease with which a magnetic flux can pass through it. This expression is often rewritten as

$$B = \mu_0 \mu_r H \qquad (14.4)$$

Figure 14.2 Magnetic flux associated with a current-carrying wire

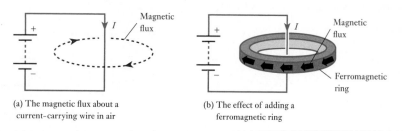

(a) The magnetic flux about a current-carrying wire in air

(b) The effect of adding a ferromagnetic ring

Figure 14.3   The magnetic field
in a coil

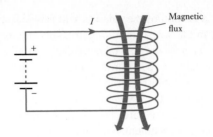

where μ_0 is the permeability of free space, and μ_r is the relative permeability of the material present. μ_0 is a constant with a value of $4\pi \times 10^{-7}$ H/m. μ_r is the ratio of the flux density produced in a material to that produced in a vacuum. For air and most non-magnetic materials, $\mu_r = 1$ and $B = \mu_0 H$. For ferromagnetic materials, μ_r may have a value of 1000 or more. Unfortunately, for ferromagnetic materials μ_r varies considerably with the magnetic field strength.

When a current-carrying wire is formed into a coil, as shown in Figure 14.3, the magnetic field is concentrated within the coil, and it increases as more and more turns are added. The m.m.f. is now given by the product of the current I and the number of turns of the coil N, so that

$$F = IN \tag{14.5}$$

For this reason, the m.m.f. is often expressed in *ampere-turns*, although formally its units are amperes, since the number of turns is dimensionless.

In a long coil with many turns, most of the magnetic flux passes through the centre of the coil. Therefore, it follows from Equations 14.1 and 14.5 that the magnetic field strength produced by such a coil is given by

$$H = \frac{IN}{l} \tag{14.6}$$

where l is the length of the flux path as before.

As discussed earlier, the flux density produced as a result of a magnetic field is determined by the permeability of the material present. Therefore, the introduction of a ferromagnetic material in a coil will dramatically increase the flux density. Figure 14.4 shows examples of arrangements that use such materials in coils. The first shows an iron bar placed within a

Figure 14.4   The use of
ferromagnetic materials in coils

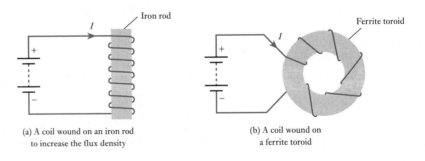

(a) A coil wound on an iron rod
to increase the flux density

(b) A coil wound on
a ferrite toroid

linear coil to increase its flux density. The second shows a coil wound on a ferrite toroid (a ring with a circular cross-section).

| **Example 14.2** | A coil is formed by winding 500 turns of wire on to a non-magnetic toroid that has a mean circumference of 400 mm and a cross-sectional area of 300 mm². If the current in the coil is 6 A, calculate: |

(a) the magnetomotive force;
(b) the magnetic field strength within the coil;
(c) the flux density in the coil;
(d) the total flux.

How would these quantities be affected if the toroid were replaced by one of similar dimensions but constructed of a magnetic material with a μ_r of 100?

(a) The magnetomotive force is given by the 'ampere-turns' of the coil and is therefore

$$F = IN$$

$$= 6 \times 500$$

$$= 3000 \text{ ampere-turns}$$

(b) The magnetic field strength is given by the m.m.f. divided by the length of the magnetic path. In this case, the length of the magnetic path is the mean circumference of the coil, so

$$H = \frac{IN}{l}$$

$$= \frac{3000}{0.4}$$

$$= 7500 \text{ A/m}$$

(c) For a non-magnetic material $B = \mu_0 H$, so

$$B = \mu_0 H$$

$$= 4\pi \times 10^{-7} \times 7500$$

$$= 9.42 \text{ mT}$$

(d) The total flux can be deduced from Equation 14.2, from which it is clear that $\Phi = BA$. Hence

$$\Phi = BA$$

$$= 9.42 \times 10^{-3} \times 300 \times 10^{-6}$$

$$= 2.83 \text{ } \mu\text{Wb}$$

If the toroid were replaced by a material with a μ_r of 100, this would have no effect on (a) and (b) but would increase (c) and (d) by a factor of 100.

14.3   Reluctance

As we know, in electric circuits, when an electromotive force is applied across a resistive component a current is produced. The ratio of the voltage to the resultant current is termed the *resistance* of the component and is a measure of how the component opposes the flow of electricity.

A directly equivalent concept exists in magnetic circuits. Here a magnetomotive force produces a magnetic flux, and the ratio of one to the other is termed the **reluctance**, S, of the magnetic circuit. In this case, the reluctance is a measure of how the circuit opposes the flow of *magnetic flux*. Just as resistance is equal to V/I, so the reluctance is given by the m.m.f. (F) divided by the flux (Φ) and hence

$$S = \frac{F}{\Phi} \tag{14.7}$$

The units of reluctance are amperes per weber (A/Wb).

14.4   Inductance

A changing magnetic flux induces an electrical voltage (an e.m.f.) in any conductor within the field. The magnitude of the effect is given by **Faraday's law**, which states that:

The magnitude of the e.m.f. induced in a circuit is proportional to the rate of change of the magnetic flux linking the circuit.

Also of importance is **Lenz's law**, which states that:

The direction of the e.m.f. is such that it tends to produce a current that opposes the change of flux responsible for inducing that e.m.f.

When a circuit forms a single loop, the e.m.f. induced by changes in the magnetic flux associated with that circuit is simply given by the rate of change of the flux. When a circuit contains many loops, then the resulting e.m.f. is the sum of the e.m.f.s produced by each loop. Therefore, if a coil of N turns experiences a change in magnetic flux, then the induced voltage V is given by

$$V = N\frac{\mathrm{d}\Phi}{\mathrm{d}t} \tag{14.8}$$

where $\mathrm{d}\Phi/\mathrm{d}t$ is the rate of change of flux in Wb/s.

This property, whereby an e.m.f. is induced into a wire as a result of a change in magnetic flux, is referred to as **inductance**.

14.5   Self-inductance

We have seen that a current flowing in a coil (or in a single wire) produces a magnetic flux about it, and that changes in the current will cause changes in the magnetic flux. We have also seen that, when the magnetic flux associated with a circuit changes, this induces an e.m.f. in that circuit which opposes the changing flux. It follows, therefore, that, when the current in a coil changes, an e.m.f. is induced in that coil which tends to oppose the change in the current. This process is known as **self-inductance**.

The voltage produced across the inductor as a result of changes in the current is given by the expression

$$V = L\frac{\mathrm{d}I}{\mathrm{d}t} \tag{14.9}$$

where L is the inductance of the coil. The unit of inductance is the **henry** (symbol H), which can be defined as the inductance of a circuit when an e.m.f. of 1 V is induced by a change in the current of 1 A/s.

14.5.1   Notation

It should be noted that some textbooks assign a negative polarity to the voltages of Equations 14.8 and 14.9 to reflect the fact that the induced voltage *opposes* the change in flux or current. This notation reflects the implications of Lenz's law. However, either polarity can be used provided that the calculated quantity is applied appropriately, and in this text we will use the *positive* notation since this is consistent with the treatment of voltages across resistors and capacitors.

| Example 14.3 | The current in a **10 mH inductor changes at a constant rate of 3 A/s.** **What voltage is induced across this coil?** |

From Equation 14.9

$$V = L\frac{\mathrm{d}I}{\mathrm{d}t}$$

$$= 10 \times 10^{-3} \times 3$$

$$= 30\,\mathrm{mV}$$

14.6   Inductors

Circuit elements that are designed to provide inductance are called **inductors**. Typical components for use in electronic circuits will have an inductance of the order of microhenries or millihenries, although large components may have an inductance of the order of henries.

Small-value inductors can be produced using air-filled coils, but larger devices normally use ferromagnetic materials. As we noted earlier, the presence of a ferromagnetic material dramatically increases the flux density in a coil and consequently also increases the rate of change of flux. There-fore, adding a ferromagnetic core to a coil greatly increases its inductance. Inductor cores may take many forms, including rods, as in Figure 14.4(a), or rings, as in Figure 14.4(b). Small inductor cores are often made from iron oxides called **ferrites**, which have very high permeability. Larger com-ponents are often based on laminated steel cores.

Unfortunately, the permeability of ferromagnetic materials decreases with increasing magnetic field strength, making inductors non-linear. Air does not suffer from this problem, so air-filled inductors are linear. For this reason, air-filled devices may be used in certain applications even though they may be physically larger than components using ferromagnetic cores.

14.6.1 Calculating the inductance of a coil

The inductance of a coil is determined by its dimensions and by the material around which it is formed. Although it is fairly straightforward to calculate the inductance of simple forms from first principles, designers often use standard formulae. Here we will look at a couple of examples, as shown in Figure 14.5.

Figure 14.5(a) shows a simple, helical, air-filled coil of length l and cross-sectional area A. The characteristics of this arrangement vary with the dimensions, but provided that the length is much greater than the diameter, the inductance of this coil is given by the expression

$$L = \frac{\mu_0 A N^2}{l} \qquad (14.10)$$

Figure 14.5(b) shows a coil wound around a toroid that has a mean cir-cumference of l and a cross-sectional area of A. The inductance of this arrangement is given by

$$L = \frac{\mu_0 \mu_r A N^2}{l} \qquad (14.11)$$

Figure 14.5 Examples of standard inductor formats

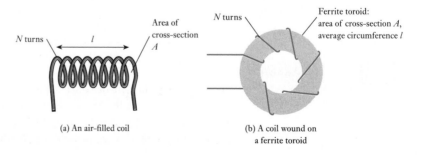

(a) An air-filled coil

(b) A coil wound on a ferrite toroid

where μ_r is the relative permeability of the material used for the toroid. If this is a non-magnetic material then μ_r will be equal to 1, and the inductance becomes

$$L = \frac{\mu_0 A N^2}{l} \tag{14.12}$$

which is the same as for the long air-filled coil described earlier (although the meaning of l is slightly different). Although these two examples have very similar equations, other coil arrangements will have different characteristics.

In these two examples, and in many other inductors, the inductance increases as the square of the number of turns.

Example 14.4 Calculate the inductance of a helical, air-filled coil 200 mm in length, with a cross-sectional area of 30 mm² and having 400 turns.

From Equation 14.10

$$L = \frac{\mu_0 A N^2}{l}$$

$$= \frac{4\pi \times 10^{-7} \times 30 \times 10^{-6} \times 400^2}{200 \times 10^{-3}}$$

$$= 30 \text{ μH}$$

14.6.2 Equivalent circuit of an inductor

So far we have considered inductors as idealised components. In practice, all inductors are made from wires (or other conductors) and therefore all real components will have resistance. We can model a real component as an **ideal inductor** (that is, one that has inductance but no resistance) in series with a resistor that represents its internal resistance. This is shown in Figure 14.6.

Figure 14.6 An equivalent circuit of a real inductor

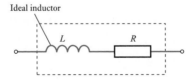

Ideal inductor

14.6.3 Stray inductance

While circuit designers will often use inductors to introduce inductance into circuits, the various conductors in *all* circuits introduce **stray inductance** that is often unwanted. We have seen that even a straight wire exhibits inductance, and though this is usually small (perhaps 1 nH per mm length of wire) the combined effects of these small amounts of inductance can dramatically affect circuit operation – particularly in high-speed circuits. In such cases, great care must be taken to reduce both stray inductance and stray capacitance (as discussed in Chapter 13).

14.7 Inductors in series and parallel

When several inductors are connected together, their effective inductance is computed in the same way as when resistors are combined, *provided that they are not linked magnetically*. Therefore, when inductors are connected in series their inductances add. Similarly, when inductors are connected in parallel their combined inductance is given by the reciprocal of the sum of the reciprocals of the individual inductances. This is shown in Figure 14.7.

Figure 14.7   Inductors in series and parallel

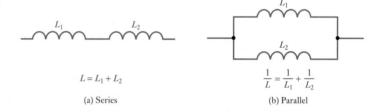

$$L = L_1 + L_2$$

(a) Series

$$\frac{1}{L} = \frac{1}{L_1} + \frac{1}{L_2}$$

(b) Parallel

Example 14.5

Calculate the inductance of:

(a) a 10 H and a 20 H inductor in series;
(b) a 10 H and a 20 H inductor in parallel.

(a) Inductances in series add

$$L = L_1 + L_2$$

$$= 10\,H + 20\,H$$

$$= 30\,H$$

(b) Inductances in parallel sum as their reciprocals

$$\frac{1}{L} = \frac{1}{L_1} + \frac{1}{L_2}$$

$$= \frac{1}{10} + \frac{1}{20}$$

$$= \frac{30}{200}$$

$$L = 6.67\,H$$

14.8 Relationship between voltage and current in an inductor

From Equation 14.9, we know that the relationship between the voltage across an inductor and the current through it is given by

$$V = L\frac{\mathrm{d}I}{\mathrm{d}t}$$

This implies that when a constant current is passed through an inductor ($\mathrm{d}I/\mathrm{d}t = 0$) the voltage across it is zero. However, when the current changes a voltage is produced that tends to oppose this change in current. Another implication of the equation is that the current through an inductor cannot change instantaneously, since this would correspond to $\mathrm{d}I/\mathrm{d}t = \infty$ and would produce an infinite induced voltage opposing the change in current. Thus inductors tend to stabilise the *current* flowing through them. (You may recall that in capacitors the voltage cannot change instantaneously, so capacitors tend to stabilise the *voltage* across them.)

The relationship between the voltage and the current in an inductor is illustrated in Figure 14.8. In the circuit of Figure 14.8(a), the switch is initially open and no current flows in the circuit. If now the switch is closed (at $t = 0$), then the current through the inductor cannot change instantly, so initially $I = 0$, and consequently $V_R = 0$. By applying Kirchhoff's voltage law around the circuit, it is clear that $V = V_R + V_L$, and if initially $V_R = 0$, then the entire supply voltage V will appear across the inductor, and $V_L = V$.

The voltage across the inductor dictates the initial rate of change of the current (since $V_L = L\ \mathrm{d}I/\mathrm{d}t$) and hence the current steadily increases. As I grows, the voltage across the resistor grows and V_L falls, reducing $\mathrm{d}I/\mathrm{d}t$.

Figure 14.8 Relationship between voltage and current in an inductor

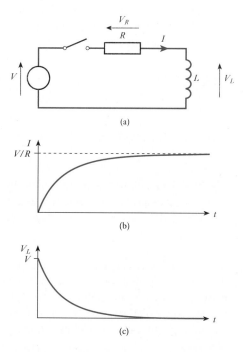

(a)

(b)

(c)

Therefore, the rate of increase of the current decreases with time. Gradually, the voltage across the inductor tends to zero and all the applied voltage appears across the resistor. This produces a steady-state current of V/R. The result is that the current is initially zero but increases exponentially with time, and the voltage across the inductor is initially V but falls exponentially with time. This behaviour is shown in Figures 14.8(b) and 14.8(c). You might like to compare these curves with the corresponding results produced for a capacitor in Figure 13.7.

File 14A

Computer Simulation Exercise 14.1

Simulate the circuit of Figure 14.8(a) with $V = 1$ V, $R = 1$ Ω and $L = 1$ H. Include in your circuit a switch that closes at $t = 0$. Use transient simulation to investigate the behaviour of the circuit during the first 5 s after the switch changes. Plot V_L and I against time on separate graphs and confirm that the circuit behaves as expected. Experiment with different values of the circuit components and note the effects on the voltage and current graphs.

In Chapter 13, we noted that the time taken for a capacitor to charge increases with both the capacitance C and the series resistance R, and we defined a term called the time constant, equal to the product CR, which determines the charging time. In the inductor circuit discussed above, the rate at which the circuit approaches its steady-state condition increases with the inductance L but *decreases* with the value of R. The reason for this effect will become clear in Chapter 18, but for the moment we will simply note that circuits of this type have a time constant (T) equal to L/R.

File 14A

Computer Simulation Exercise 14.2

Repeat Computer Simulation Exercise 14.1 noting the effect of different component values. Begin with the same values as in the previous simulation exercise and then change the values of L and R while keeping the ratio L/R constant. Again plot V_L and I against time on separate graphs and confirm that the characteristics are unchanged. Hence confirm that the characteristics are determined by the time constant L/R rather than the actual values of L and R.

It is interesting to consider what happens in the circuit of Figure 14.8(a) if the switch is opened some time after being closed. From Figure 14.8(b), we know that the current stabilises at a value of V/R. If the switch is now opened, this would suggest that the current would instantly go to zero. This would imply that dI/dt would be infinite and that an infinite voltage would be produced across the coil. In practice, the very high induced voltage appears across the switch and causes 'arcing' at the switch contacts. This maintains the current for a short time after the switch is operated and reduces the rate of change of current. This phenomenon is used to advantage in some situations such as in automotive ignition coils. However, arcing across switches can cause severe damage to the contacts and also generates electrical interference. For this reason, when it is necessary to switch inductive loads, we normally add circuitry to reduce the rate of change of the current. This circuitry may be as simple as a capacitor placed across the switch.

So far in this section we have assumed the use of an ideal inductor and have ignored the effects of any internal resistance. In Section 14.6, we noted that an inductor with resistance can be modelled as an ideal inductor in series with a resistor. In Chapter 15, we will look at the characteristics of circuits containing elements of various types (resistive, inductive and capacitive), so we will leave the effects of internal resistance until then.

14.9 Sinusoidal voltages and currents

Having looked at the relationship between voltage and current in a DC circuit containing an inductor, it is now time to turn our attention to circuits using sinusoidal quantities.

Consider the arrangement of Figure 14.9(a), where an alternating current is passed through an inductor. Figure 14.9(c) shows the sinusoidal current

Figure 14.9 Inductors and alternating quantities

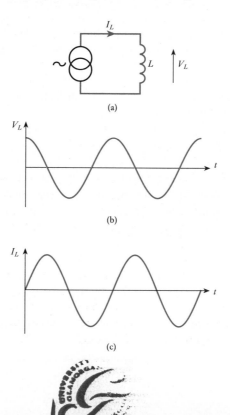

(a)

(b)

(c)

waveform in the inductor, which in turn dictates the voltage across the inductor. From Equation 14.9 we know that the voltage across an inductor is given by $L \, dI/dt$, so the voltage is directly proportional to the *time differential* of the current. Since the differential of a sine wave is a cosine wave, we obtain a voltage waveform as shown in Figure 14.9(b). The current waveform is phase-shifted with respect to the voltage waveform by 90° (or $\pi/2$ radians). It is also clear that the current waveform *lags* the voltage waveform. You might like to compare this result with that shown in Figure 13.8 for a capacitor. You will note that in a capacitor the current *leads* the voltage, while in an inductor the current *lags* the voltage. We will return to the analysis of sinusoidal waveforms in Chapter 15.

File 14B

Computer Simulation Exercise 14.3

Simulate the circuit of Figure 14.9(a) using any value of inductor. Use a sinusoidal current source to produce a current of 1 A peak at 1 Hz and use transient analysis to display the voltage across the inductor, and the current through it, over a period of several seconds. Note the phase relationship between the two waveforms and hence confirm that the current lags the voltages by 90° (or $\pi/2$ radians). Note the effect of varying the inductor value, and the frequency used.

14.10 Energy storage in an inductor

Inductors store energy within a magnetic field. The amount of energy stored in this way can be determined by considering an initially un-energised inductor of inductance L, in which a current is gradually increased from zero to I amperes. If the rate of change of the current at a given time is di/dt, then the instantaneous voltage across the inductor (v) will be given by

$$v = L\frac{di}{dt}$$

In a small amount of time dt, the amount of energy added to the magnetic field is equal to the product of the instantaneous voltage (v), the instantaneous current (i) and the time interval (dt).

$$\text{Energy added} = vidt$$

$$= L\frac{di}{dt}idt$$

$$= Lidi$$

Therefore, the energy added to the magnetic field as the current increases from zero to I is given by

$$\text{stored energy} = L\int_0^I i\,dt$$

$$\text{stored energy} = \frac{1}{2}LI^2 \tag{14.13}$$

| Example 14.6 | What energy is stored in an inductor of 10 mH when a current of 5 A is passing through it? |

From Equation 14.3

$$\text{stored energy} = \frac{1}{2}LI^2$$

$$= \frac{1}{2}\times 10^{-2}\times 5^2$$

$$= 125 \text{ mJ}$$

14.11 Mutual inductance

If two conductors are linked magnetically, then a changing current in one of these will produce a changing magnetic flux associated with the other and will result in an induced voltage in this second conductor. This is the principle of **mutual inductance**.

Mutual inductance is quantified in a similar way to self-inductance, such that, if a current I_1 flows in one circuit, the voltage induced in a second circuit V_2 is given by

$$V_2 = M\frac{dI_1}{dt} \tag{14.14}$$

where M is the mutual inductance between the two circuits. The unit of mutual inductance is the henry, as for self-inductance. Here, a henry would be defined as the mutual inductance between two circuits when an e.m.f. of 1 V is induced in one by a change in the current of 1 A/s in the other. The mutual inductance between two circuits is determined by their individual inductances and the magnetic linkage between them.

Often our interest is in the interaction of coils, as in a **transformer**. Here a changing current in one coil (the primary) is used to induce a changing current in a second coil (the secondary). Figure 14.10 shows arrangements of two coils that are linked magnetically. In Figure 14.10(a), the two coils are loosely coupled with a relatively small part of the flux of the first coil linking with the second. Such an arrangement would have a relatively low mutual inductance. The degree of coupling between circuits is described by their **coupling coefficient**, which defines the fraction of the flux of one coil that links with the other. A value of 1 represents total flux linkage, while a

Figure 14.10 Mutual
inductance between two coils

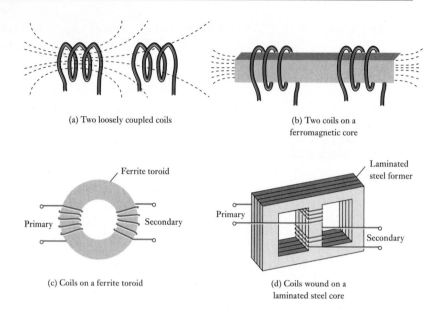

(a) Two loosely coupled coils

(b) Two coils on a
ferromagnetic core

(c) Coils on a ferrite toroid

(d) Coils wound on a
laminated steel core

value of 0 represents no linkage. The coupling between the two coils can
be increased in a number of ways, such as by moving the coils closer
together, by wrapping one coil around the other, or by adding a **ferromagnetic core** as in Figure 14.10(b). Excellent coupling is achieved by wrapping coils around a continuous ferromagnetic loop as in Figures 14.10(c)
and 14.10(d). In these examples, the cores increase the inductance of the
coils and increase the flux linkage between them.

14.12   Transformers

The basic form of a transformer is illustrated in Figure 14.11(a). Two coils,
a primary and a secondary, are wound on to a ferromagnetic core or former
in an attempt to get a coupling coefficient as close as possible to unity. In
practice, many transformers are very efficient and for the benefit of this
discussion we will assume that all the flux from the primary coil links with
the secondary. That is, we will assume an ideal transformer with a coupling
coefficient of 1.

Figure 14.11 A transformer

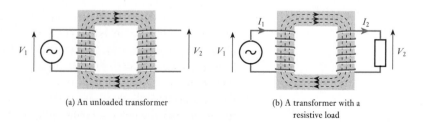

(a) An unloaded transformer

(b) A transformer with a
resistive load

If an alternating voltage V_1 is applied to the primary, this will produce an alternating current I_1, which in turn will produce an alternating magnetic field. Since the variation in the magnetic flux associated with the primary coil is the same as that associated with the secondary, the voltage induced *in each turn* of the primary and the secondary will be the same. Let us call this V_T. Now, if the number of turns in the primary is N_1, then the voltage induced across the primary will be N_1V_T. Similarly, if the number of turns in the secondary is N_2, then the voltage across the secondary will be N_2V_T. Therefore, the ratio of the output voltage V_2 to the input voltage V_1 is given by

$$\frac{V_2}{V_1} = \frac{N_2V_T}{N_1V_T}$$

and thus

$$\frac{V_2}{V_1} = \frac{N_2}{N_1} \tag{14.15}$$

Thus the transformer works as a voltage amplifier with a gain determined by the ratio of the number of turns in the secondary to that in the primary. N_2/N_1 is often called the **turns ratio** of the transformer.

However, there are, several points to note about this arrangement. The first is that this voltage amplification clearly applies only to alternating voltages – a constant voltage applied to the primary will not produce a changing magnetic flux and consequently no output voltage will be induced. Second, it must be remembered that this 'amplifier' has no energy source other than the input signal (that is, it is a passive amplifier) and consequently the power delivered at the output cannot be greater than that absorbed at the input. This second point is illustrated in Figure 14.11(b), where a resistive load has been added to our transformer. The addition of a load means that a current will now flow in the secondary circuit. This current will itself produce magnetic flux, and the nature of induction means that this flux will oppose that generated by the primary circuit. Consequently, the current flowing in the secondary coil tends to reduce the voltage in that coil. The overall effect of this mechanism is that when the secondary is open-circuit, or when the output current is very small, the output voltage is as predicted by Equation 14.15, but as the output current increases the output voltage falls.

The efficiency of modern transformers is very high and therefore the power delivered at the output is almost the same as that absorbed at the input. For an ideal transformer

$$V_1I_1 = V_2I_2 \tag{14.16}$$

If the secondary of a transformer has many more turns than the primary we have a **step-up transformer**, which provides an output voltage that is much higher than the input voltage, but it can deliver a smaller output current. If the secondary has fewer turns than the primary we have a **step-down transformer**, which provides a smaller output voltage but can supply a greater current. Step-down transformers are often used in power supplies for low-voltage electronic equipment, where they produce an output voltage of a few volts from the supply voltage. An additional advantage of this arrangement is that the transformer provides **electrical isolation** from the supply lines, since there is no electrical connection between the primary and the secondary circuits.

14.13   Circuit symbols

We have looked at several forms of inductor and transformer, and some of these may be indicated through the use of different circuit symbols. Figure 14.12 shows various symbols and identifies their distinguishing characteristics. Figure 14.12(f) shows a transformer with two secondary coils. This figure also illustrates what is termed the **dot notation** for indicating the polarity of coil windings. Current flowing *into* each winding at the connection indicated by the dot will produce magnetomotive forces in the same direction within the core. Reversing the connections to a coil will invert the corresponding voltage waveform. The dot notation allows the required connections to be indicated on the circuit diagram.

Figure 14.12   Circuit symbols for inductors and transformers

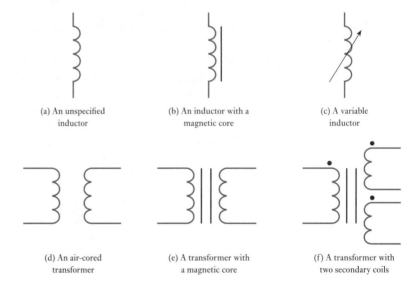

(a) An unspecified inductor

(b) An inductor with a magnetic core

(c) A variable inductor

(d) An air-cored transformer

(e) A transformer with a magnetic core

(f) A transformer with two secondary coils

14.14   The use of inductance in sensors

Inductors and transformers are used in a wide range of electrical and electronic systems, and we shall be meeting several such applications in later chapters. However, at this point, it might be useful to look at a couple of situations where inductance is used as a means of measuring physical quantities. The first of these we have already encountered in Chapter 3.

14.14.1 Inductive proximity sensors

We looked briefly at inductive proximity sensors in Section 3.6 and looked at some real devices in Figure 3.7. The essential elements of such a sensor are shown in Figure 14.13. The device is basically a coil wrapped around a ferromagnetic rod. The arrangement is used as a sensor by combining it with a ferromagnetic plate (attached to the object to be sensed) and a circuit to measure the self-inductance of the coil. When the plate is close to the coil it increases its self-inductance, allowing its presence to be detected. The sensor can be used to measure the separation between the coil and the plate but is more often used in a binary mode to sense its presence or absence.

Figure 14.13 An inductive proximity sensor

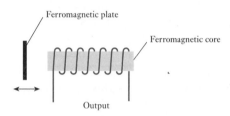

14.14.2 Linear variable differential transformers (LVDTs)

An LVDT consists of three coils wound around a hollow, non-magnetic tube, as shown in Figure 14.14. The centre coil forms the primary of the transformer and is exited by an alternating voltage. The remaining coils form identical secondaries, positioned symmetrically either side of the primary. The two secondary coils are connected in series in such a way that their output voltages are out of phase (note the position of the dots in Figure 14.14) and therefore cancel. If a sinusoidal signal is applied to the primary coil, the symmetry of the arrangement means that the two secondary coils produce identical signals that cancel each other, and the output is zero. This assembly is turned into a useful sensor by the addition of a movable 'slug' of ferromagnetic material inside the tube. The material increases the mutual inductance between the primary and the secondary coils and thus increases

Figure 14.14 A linear variable differential transformer (LVDT)

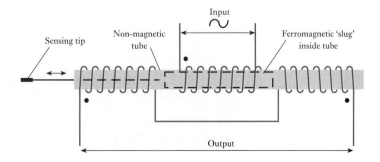

the magnitude of the voltages induced in the secondary coils. If the slug is positioned centrally with respect to the coils, it will affect both coils equally and the output voltages will still cancel. However, if the slug is moved slightly to one side or the other, it will increase the coupling to one and decrease the coupling to the other. The arrangement will now be out of balance, and an output voltage will be produced. The greater the displacement of the slug from its central position the greater the resulting output signal. The output is in the form of an alternating voltage where the magnitude represents the offset from the central position and the phase represents the direction in which the slug is displaced. A simple circuit can be used to convert this alternating signal into a more convenient DC signal if required.

LVDTs can be constructed with ranges from a few metres down to a fraction of a millimetre. They typically have a resolution of about 0.1 percent of their full range and have good linearity. Unlike resistive potentiometers, they do not require a frictional contact and so can have a very low operating force and long life.

Key points

■ Inductors store energy within a magnetic field.

■ A wire carrying an electrical current causes a magnetomotive force (m.m.f.), which produces a magnetic field about it.

■ The magnetic field strength, H, is proportional to the current and inversely proportional to the length of the magnetic circuit.

■ The magnetic field produces a magnetic flux, Φ, which flows in the same direction.

■ The flux density is determined by the field strength and the permeability of the material present.

■ When a current-carrying wire is formed into a coil, the magnetic field is concentrated. The m.m.f. increases with the number of turns of the coil.

■ A changing magnetic flux induces an electrical voltage in any conductors within the field.

■ The direction of the induced e.m.f. is such that it opposes the change of flux.

■ When the current in a coil changes, an e.m.f. is induced in that coil which tends to oppose the change in the current. This is self-inductance.

■ The induced voltage is proportional to the rate of change of the current in the coil.

■ Inductors can be made by coiling wire in air, but much greater inductance is produced if the coil is wound around a ferromagnetic core.

- All real inductors have some resistance.

- When inductors are connected in series their inductances add. When inductors are connected in parallel the resultant inductance is the reciprocal of the sum of the reciprocals of the individual inductances.

- The current in an inductor cannot change instantly.

- When using sinusoidal signals the current lags the voltage by 90° (or π/2 radians).

- The energy stored in an inductor is equal to $\frac{1}{2}LI^2$.

- When two conductors are linked magnetically, a changing current in one will induce a voltage in the other. This is mutual induction.

- When a transformer is used with alternating signals, the voltage gain is determined by the turns ratio.

- Several forms of sensor make use of variations in inductance.

Exercises

14.1 Explain what is meant by a magnetomotive force (m.m.f.).

14.2 Describe the field produced by a current flowing in a straight wire.

14.3 A straight wire carries a current of 3 A. What is the magnetic field at a distance of 1 m from the wire? What is the direction of this field?

14.4 What factors determine the flux density at a particular point in space adjacent to a current-carrying wire?

14.5 Explain what is meant by the permeability of free space? What are its value and units?

14.6 Give an expression for the magnetomotive force produced by a coil of N turns that is passing a current of I amperes?

14.7 A coil is formed by wrapping wire around a wooden toroid. The cross-sectional area of the coil is 400 mm², the number of turns is 600, and the mean circumference of the toroid is 900 mm. If the current in the coil is 5 A, calculate the magnetomotive force, the magnetic field strength in the coil, the flux density in the coil and the total flux.

14.8 If the toroid in Exercise 14.7 were to be replaced by a magnetic toroid with a relative permeability of 500, what effect would this have on the values calculated?

14.9 If an m.m.f. of 15 ampere-turns produces a total flux of 5 mWb, what is the reluctance of the magnetic circuit?

14.10 State Faraday's law and Lenz's law.

14.11 Explain what is meant by inductance.

14.12 Explain what is meant by self-inductance.

14.13 How is the voltage induced in a conductor related to the rate of change of the current within it?

14.14 Define the henry as it applies to the measurement of self-inductance.

14.15 The current in an inductor changes at a constant rate of 50 mA/s, and there is a voltage across it of 150 μV. What is its inductance?

Exercises continued

14.16 Why does the presence of a ferromagnetic core increase the inductance of an inductor?

14.17 Calculate the inductance of a helical, air-filled coil 500 mm in length, with a cross-sectional area of 40 mm^2 and having 600 turns.

14.18 Calculate the inductance of a coil wound on a magnetic toroid of 300 mm mean circumference and 100 mm^2 cross-sectional area, if there are 250 turns on the coil and the relative permeability of the toroid is 800.

14.19 Calculate the effective inductance of the following arrangements.

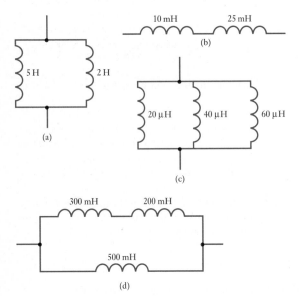

14.20 Describe the relationship between voltage and current in an inductor.

14.21 Why is it not possible for the current in an inductor to change instantaneously?

14.22 Repeat Computer Simulation Exercise 14.1 with $V = 15$ V, $R = 5$ Ω and $L = 10$ H. Plot the current through the inductor as a function of time and hence estimate the time taken for the inductor current to reach 2 A.

14.23 Explain what is meant by a time constant. What is the time constant of the circuit in Exercise 14.22?

14.24 The circuit of Exercise 14.22 is modified by changing R to 10 Ω. What value should be chosen for L so that the time taken for the inductor current to reach 2 A is unchanged?

14.25 Confirm your answer to Exercise 14.24 using computer simulation.

14.26 Discuss the implications of induced voltages when switching inductive circuits.

14.27 How do real inductors differ from ideal inductors?

14.28 What is the relationship between the sinusoidal current in an inductor and the voltage across it?

14.29 What is the energy stored in an inductor of 2 mH when a current of 7 A is passing through it?

14.30 Explain what is meant by mutual inductance.

14.31 Define the henry as it applies to the measurement of mutual inductance.

14.32 What is meant by a coupling coefficient?

14.33 What is meant by the turns ratio of a transformer?

14.34 A transformer has a turns ratio of 10. A sinusoidal voltage of 5 V peak is applied to the primary coil, with the secondary coil open-circuit. What voltage would you expect to appear across the secondary coil?

14.35 Explain the dot notation used when representing transformers in circuit diagrams.

14.36 Describe the operation of an inductive proximity sensor.

14.37 Describe the construction and operation of an LVDT.

Chapter 15

Alternating Voltages and Currents

Objectives

When you have studied the material in this chapter you should be able to:

- describe the relationship between sinusoidal voltages and currents in resistors, capacitors and inductors;
- explain the meaning of terms such as reactance and impedance, and calculate values for these quantities for individual components and simple circuits;
- use phasor diagrams to determine the relationship between voltages and currents in a circuit;
- analyse circuits containing resistors, capacitors and inductors to determine the associated voltages and currents;
- explain the use of complex numbers in the description and analysis of circuit behaviour;
- use complex notation to calculate the voltages and currents in AC circuits.

15.1 Introduction

From our earlier consideration of sine waves in Chapter 11, we know that a sinusoidal voltage waveform can be described by the equation

$$v = V_p \sin(\omega t + \phi)$$

where V_P is the **peak voltage** of the waveform, ω is the **angular frequency** and ϕ represents its **phase angle**.

The angular frequency of the waveform is related to its natural frequency by the expression

$$\omega = 2\pi f$$

It follows that the period of the waveform, T, is given by

$$T = \frac{1}{f} = \frac{2\pi}{\omega}$$

Figure 15.1　Sinusoidal voltage
waveforms

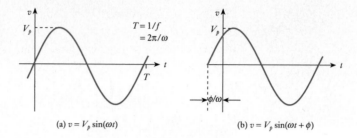

(a) $v = V_p \sin(\omega t)$　　　　　(b) $v = V_p \sin(\omega t + \phi)$

If the phase angle ϕ is expressed in radians, then the corresponding time delay t is given by

$$t = \frac{\phi}{\omega}$$

These relationships are illustrated in Figure 15.1, which shows two voltage waveforms of similar magnitude and frequency but with different phase angles.

15.2　Relationship between voltage and current

In earlier chapters, we looked at the relationship between voltage and current in a range of components. It is interesting to compare the voltages across a resistor, an inductor and a capacitor when a similar sinusoidal current is passed through each. In this case, we will use a current given by

$$i = I_P \sin(\omega t)$$

15.2.1　Resistors

In a resistor, the relationship between the voltage and the current is given by Ohm's law, and we know that

$$v_R = iR$$

If $i = I_P \sin(\omega t)$, then

$$v_R = I_P R \sin(\omega t) \tag{15.1}$$

15.2.2　Inductors

In an inductor, the voltage is related to the current by the expression

$$v_L = L\frac{di}{dt}$$

If $i = I_P \sin(\omega t)$, then

$$v_L = L\frac{d(I_P \sin(\omega t))}{dt} = \omega L I_P \cos(\omega t) \tag{15.2}$$

15.2.3 Capacitors

In a capacitor, the voltage is related to the current by the expression

$$v_C = \frac{1}{C}\int i\,dt$$

If $i = I_P\sin(\omega t)$, then

$$v_C = \frac{1}{C}\int I_P\sin(\omega t) = -\frac{I_P}{\omega C}\cos(\omega t) \tag{15.3}$$

Figure 15.2 shows the corresponding voltages across a resistor, an inductor and a capacitor when the same sinusoidal current is passed through each. In this figure, the magnitudes of the various traces are unimportant, since these will depend on component values. The dotted line shows the current waveform and acts as a reference for the other traces. The voltage across the resistor v_R is in phase with the applied current, as indicated by Equation 15.1. The voltage across v_L is given as a cosine wave in Equation 15.2 and is therefore phase-shifted by 90° with respect to the current waveform (and therefore with respect to v_R). The voltage across the capacitor v_C is also given as a cosine wave in Equation 15.3, but here the magnitude is negative, inverting the waveform as shown in Figure 15.2. Inverting a sinusoidal waveform has the same effect as shifting its phase by 180°. Thus the voltage across the resistor is *in phase* with the current; the voltage across the inductor *leads* the current by 90°, and the voltage across the capacitor *lags* the current by 90°. This is consistent with the discussion of inductors and capacitors in earlier chapters.

While it is relatively easy to remember that there is a phase difference of 90° between the current and the voltage in an inductor and a capacitor, it is important to remember which leads which in each case. One way of remembering this is to use the simple mnemonic

C I V I L

In C, I leads V; V leads I in L.

In other words, in a capacitor the current leads the voltage, while the voltage leads the current in an inductor.

Figure 15.2 Relationship between the voltage across a resistor, a capacitor and an inductor

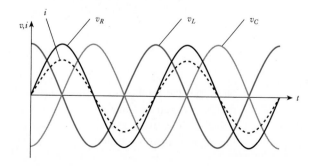

15.3   Reactance of inductors and capacitors

Equations 15.1 to 15.3 show the relationship between current and voltage for a resistor, an inductor and a capacitor. Let us ignore, for the moment, the phase relationship between the current and the voltage and consider instead the relationship between the magnitudes of these quantities. In each case, we will consider the relationship between the peak voltage and the peak current of each component.

15.3.1   Resistance

From Equation 15.1, the ratio of the peak magnitude of the voltage to the peak magnitude of the current is given by

$$\frac{\text{peak value of voltage}}{\text{peak value of current}} = \frac{\text{peak value of } (I_P R \sin(\omega t))}{\text{peak value of } (I_P \sin(\omega t))} = \frac{I_P R}{I_P} = R$$

15.3.2   Inductance

From Equation 15.2, the ratio of the peak magnitude of the voltage to the peak magnitude of the current is given by

$$\frac{\text{peak value of voltage}}{\text{peak value of current}} = \frac{\text{peak value of } (\omega L I_P \cos(\omega t))}{\text{peak value of } (I_P \sin(\omega t))}$$

$$= \frac{\omega L I_P}{I_P} = \omega L$$

15.3.3   Capacitance

From Equation 15.3, the ratio of the peak magnitude of the voltage to the peak magnitude of the current is given by

$$\frac{\text{peak value of voltage}}{\text{peak value of current}} = \frac{\text{peak value of } \left(-\dfrac{I_P}{\omega C}\cos(\omega t)\right)}{\text{peak value of } (I_P \sin(\omega t))}$$

$$= \frac{\dfrac{I_P}{\omega C}}{I_P} = \frac{1}{\omega C}$$

Note that the three expressions derived above would have been identical had we chosen to compare the r.m.s. value of the voltage with the r.m.s. value of the current in each case, since this would have simply multiplied the upper and lower term in each expression by $1/\sqrt{2}$ (0.707).

This ratio of the voltage to the current, ignoring any phase shift, is a measure of how the component opposes the flow of electricity. In the case of a resistor, we already know that this is termed its resistance. In the case of inductors and capacitors, this quantity is termed its **reactance**, which is given the symbol X. Therefore

$$\text{reactance of an inductor, } X_L = \omega L \qquad\qquad (15.4)$$

$$\text{reactance of a capacitor, } X_C = \frac{1}{\omega C} \qquad\qquad (15.5)$$

Since reactance represents the ratio of voltage to current it has the units of ohms.

Example 15.1   Calculate the reactance of an inductor of 1 mH at an angular frequency of 1000 rad/s.

$$\text{reactance, } X_L = \omega L$$
$$= 1000 \times 10^{-3}$$
$$= 1\ \Omega$$

Example 15.2   Calculate the reactance of a capacitor of 2 μF at a frequency of 50 Hz.

At a frequency of 50 Hz, the angular frequency is given by

$$\omega = 2\pi f$$
$$= 2 \times \pi \times 50$$
$$= 314.2 \text{ rad/s}$$

$$\text{reactance, } X_C = \frac{1}{\omega C}$$

$$= \frac{1}{314.2 \times 2 \times 10^{-6}}$$

$$= 1.59 \text{ k}\Omega$$

The reactance of a component can be used to calculate the voltage across it from a knowledge of the current through it, and vice versa, in much the same way as we use resistance for a resistor. Therefore, for an inductor

$$V = IX_L$$

and for a capacitor

$$V = IX_C$$

Note that these relationships are true whether V and I represent r.m.s., peak or peak-to-peak values, provided that the same measure is used for both.

Example 15.3   A sinusoidal voltage of 5 V peak and 100 Hz is applied across an inductor of 25 mH. What will be the peak current in the inductor?

At this frequency, the reactance of the inductor is given by

$$X_L = \omega L$$

$$= 2\pi f L$$

$$= 2 \times \pi \times 100 \times 25 \times 10^{-3}$$

$$= 15.7 \; \Omega$$

Therefore

$$I_L = \frac{V_L}{X_L}$$

$$= \frac{5}{15.7}$$

$$= 2 \times \pi \times 100 \times 25 \times 10^{-3}$$

$$= 318 \; \text{mA}$$

Example 15.4   A sinusoidal current of 2 A r.m.s. at 25 rad/s flows through a capacitor of 10 mF. What voltage will appear across the capacitor?

At this frequency, the reactance of the capacitor is given by

$$X_C = \frac{1}{\omega C}$$

$$= \frac{1}{25 \times 10 \times 10^{-3}}$$

$$= 4 \; \Omega$$

Therefore

$$V_C = I_C X_C$$

$$= 2 \times 4$$

$$= 8 \; \text{V r.m.s}$$

When describing sinusoidal quantities, we very often use r.m.s. values (for reasons discussed in Chapter 11). However, since the calculation of currents and voltages is essentially the same whether we are using r.m.s., peak or peak-to-peak quantities, in the remaining examples in this chapter we will simply give magnitudes in volts and amps, ignoring the form of the measurement.

15.4 Phasor diagrams

We have seen that sinusoidal signals are characterised by their *magnitude*, their *frequency* and their *phase*. In many cases, the voltages and currents at different points in a system are driven by a single source (such as the AC supply voltage) such that they all have a common frequency. However, the magnitudes of the signals at different points will be different, and, as we have seen above, the phase relationship between these signals may also be different. We therefore are often faced with the problem of combining or comparing signals of the same frequency but of differing magnitude and phase. A useful tool in this area is the **phasor diagram**, which allows us to represent both the magnitude and the phase of a signal in a single diagram.

Figure 15.3(a) shows a single phasor representing a sinusoidal voltage. The length of the phasor, L, represents the magnitude of the voltage, while the angle ϕ represents its phase angle with respect to some reference waveform. The end of the phasor is marked with an arrowhead to indicate its direction and also to make the end visible should it coincide with one of the axes or another phasor. Phase angles are traditionally measured anticlockwise from the right-pointing horizontal axis.

The use of phasors clearly indicates the phase relationship between signals of the same frequency. For example, Figure 15.3(b) shows the three voltage waveforms v_R, v_L and v_C of Figure 15.2 represented by phasors $\mathbf{V_R}$, $\mathbf{V_L}$ and $\mathbf{V_C}$. Here the current waveform is taken as the reference phase, and therefore $\mathbf{V_R}$ has a phase angle of zero, $\mathbf{V_L}$ has a phase angle of +90° (+π/2 rad), and $\mathbf{V_C}$ is shown with a phase angle of −90° (−π/2 rad). Conventionally, the name of a phasor is written in bold (for example $\mathbf{V_R}$), the magnitude of the phasor is written in italics (for example V_R) and the instantaneous value of the sinusoidal quantity it represents is written in lower case italics (for example v_R).

Phasor diagrams can be used to represent the addition or subtraction of signals of the same frequency but different phase. Consider for example Figure 15.4(a), which shows two phasors, **A** and **B**. Since the two sinusoids have different phase angles, if we add them together the magnitude of the resulting signal will *not* be equal to the arithmetic sum of the magnitudes of **A** and **B**. A phasor diagram can be used to determine the effect of adding

Figure 15.3 Phasor diagrams

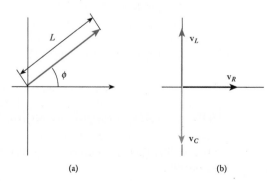

(a)   (b)

Figure 15.4 Representing the addition or subtraction of waveforms using phasors

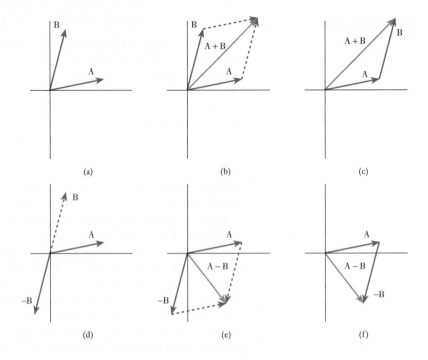

the two signals, using techniques similar to those used in **vector analysis** to add vectors. Figure 15.4(b) shows the technique of 'completing the parallelogram' used to compute the effect of adding **A** and **B**. It can be seen that the diagram gives the magnitude and the phase of the resulting waveform. Another way of combining phasors is illustrated in Figure 15.4(c), where the phasors **A** and **B** are added by drawing **B** from the end of **A**. This process can be seen as adding **B** to **A** and can be repeated to combine any number of phasors. The *subtraction* of one sinusoid from another is equivalent to *adding* a sinusoid of the opposite polarity. This in turn is equivalent to adding a signal that is phase-shifted by 180°. Figure 15.4(d) shows the phasors **A** and **B** from Figure 15.4(a) and also shows the phasor −**B**. It can be seen that −**B** is represented by a line of equal magnitude to **B** but pointing in the opposite direction. These two vectors can again be combined by completing the parallelogram (as in Figure 15.4(e)) or by drawing one on the end of the other (as in Figure 15.4(f)).

Fortunately, the nature of circuit components means that we are normally concerned with phasors that are at right angles to each other (as illustrated in Figure 15.3(b)). This makes the use of phasor diagrams much simpler.

15.4.1   Phasor analysis of an *RL* circuit

Consider the circuit of Figure 15.5, which shows a circuit containing a resistor and an inductor in series with a sinusoidal voltage source. The

Figure 15.5 Phasor analysis of an *RL* circuit

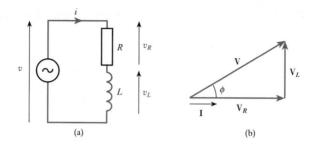

(a)   (b)

reference phase for the diagram is the current signal *i* and consequently the voltage across the resistor (represented by $\mathbf{V}_R$ in the phasor diagram) has zero phase (since this is in phase with the current), while the voltage across the inductor (represented by $\mathbf{V}_L$) has a phase angle of 90° (or $\pi/2$ radians). The voltage across the voltage source *v* is represented by the phasor $\mathbf{V}$, and from the circuit diagram it is clear that $v = v_L + v_R$. Thus in the phasor diagram $\mathbf{V} = \mathbf{V}_R + \mathbf{V}_L$; the diagram shows that the voltage across the circuit is not in phase with the current and that *v* leads *i* by a phase angle of ϕ.

Phasor diagrams can be used in connection with values for resistance and reactance to determine the voltages and currents within a circuit, and the phase relationships between them.

Example 15.5   A sinusoidal current of 5 A at 50 Hz flows through a series combination of a resistor of 10 Ω and an inductor of 25 mH. Determine:

(a) the voltage across the combination;
(b) the phase angle between this voltage and the current.

(a) The voltage across the resistor is given by

$$V_R = IR$$

$$= 5 \times 10$$

$$= 50 \text{ V}$$

At a frequency of 50 Hz, the reactance of the inductor is given by

$$X_L = 2\pi f L$$

$$= 2 \times \pi \times 50 \times 0.025$$

$$= 7.85 \ \Omega$$

Therefore, the magnitude of the voltage across the inductor is given by

$$V_L = IX_L$$

$$= 5 \times 7.85$$

$$= 39.3 \text{ V}$$

These can be combined using a phasor diagram as follows:

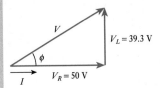

From the diagram, the magnitude of the voltage across the combination is given by

$$V = \sqrt{(V_R^2 + V_L^2)}$$

$$= \sqrt{(50^2 + 39.3^2)}$$

$$= 63.6 \text{ V}$$

(b) The phase angle is given by

$$\phi = \tan^{-1}\frac{V_L}{V_R}$$

$$= \tan^{-1}\frac{39.3}{50}$$

$$= 38.2°$$

Therefore, the voltage leads the current by 38.2°.

File 15A

Computer Simulation Exercise 15.1

Simulate the circuit of Example 15.5 using a sinusoidal current source. Use transient analysis to investigate the magnitude and the phase of the voltage across the resistor/inductor combination and compare these with the values calculated above.

15.4.2 Phasor analysis of an *RC* circuit

A similar approach can be used with circuits involving resistors and capacitors, as shown in Figure 15.6. Again the reference phase is that of the current *i*. Since in a capacitor the voltage lags the current by 90°, the voltage across the capacitor has a phase angle of −90°, and the phasor is drawn vertically downwards. The resultant phase angle ϕ is now negative, showing that the voltage across the combination lags the current through it.

As before, we can use the phasor diagram to determine the relationships between currents and voltages in the circuit.

Figure 15.6 Phasor analysis of an *RC* circuit

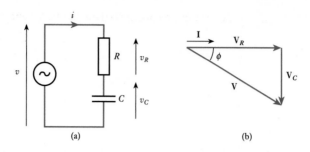

(a)   (b)

Example 15.6

A sinusoidal voltage of 10 V at 1 kHz is applied across a series combination of a resistor of 10 kΩ and a capacitor of 30 nF. Determine:

(a) the current flowing through the combination;
(b) the phase angle between this current and the applied voltage.

(a) In this example, the current in the circuit is unknown – we will call this current *I*. The magnitude of the voltage across the resistor is given by

$$V_R = IR = I \times 10^4 \text{ V}$$

At a frequency of 1 kHz, the reactance of the capacitor is given by

$$X_C = \frac{1}{2\pi fC}$$

$$= \frac{1}{2 \times \pi \times 10^3 \times 3 \times 10^{-8}}$$

$$= 5.3 \text{ k}\Omega$$

Therefore, the magnitude of the voltage across the capacitor is given by

$$V_C = IX_C = I \times 5.3 \times 10^3 \text{ V}$$

These can be combined using a phasor diagram as follows:

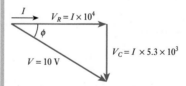

From the diagram, we see that

$$V^2 = V_R^2 + V_C^2$$

$$10^2 = (I \times 10^4)^2 + (I \times 5.3 \times 10^3)^2$$

$$= I^2 \times 1.28 \times 10^8$$

Which can be solved to give

$$I = 884 \text{ μA}$$

(b) The phase angle is given by

$$\phi = \tan^{-1}\frac{V_C}{V_R}$$

$$= \tan^{-1}\frac{I \times 5.3 \times 10^3}{I \times 10^4}$$

$$= -27.9°$$

The phase angle is negative by inspection of the diagram. This means that the voltage lags the current by nearly 28° or, alternatively, the current leads the voltage by this amount.

File 15B

Computer Simulation Exercise 15.2

Simulate the circuit of Example 15.6 using a sinusoidal voltage source. Use transient analysis to investigate the magnitude and the phase of the current flowing in the resistor/capacitor combination and compare these with the values calculated above.

15.4.3   Phasor analysis of *RLC* circuits

The techniques described above can be used with circuits containing any combination of resistors, inductors and capacitors. For example, the circuit of Figure 15.7(a) contains two resistors, one inductor and one capacitor, and Figure 15.7(b) shows the corresponding phasor diagram. It is interesting to note that summing the voltages across the four components in a different order *should* give us the same overall voltage. Figure 15.7(c) illustrates this and shows that this does indeed give us the same voltage (both in magnitude and phase angle).

Figure 15.7   Phasor analysis of an *RLC* circuit

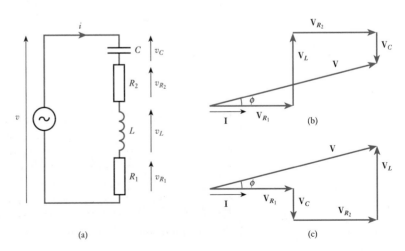

15.4.4 Phasor analysis of parallel circuits

The phasor diagrams shown above relate to series combinations of components. In such circuits, the current is the same throughout the network and we are normally interested in the voltages across each component. Our phasor diagram therefore shows the relationship between the various voltages in the circuit. In a parallel network, the voltage across each component is the same and it is the currents that are of interest. We can also use phasor diagrams to represent currents as shown in Figure 15.8. In these circuits, the applied voltage v is taken as the reference phase, and therefore the current though the resistor i_R has a phase angle of zero. Since the current through an inductor *lags* the applied voltage, the phasor $\mathbf{I}_L$ is shown vertically downwards. Similarly, since the current in a capacitor *leads* the applied voltage, $\mathbf{I}_C$ is shown vertically upwards. These directions are clearly opposite to those in voltage phasor diagrams.

As with series arrangements, phasor diagrams can be used with parallel circuits containing any number of components, and calculations can be performed in a similar manner.

Figure 15.8 Phasor diagrams of parallel networks

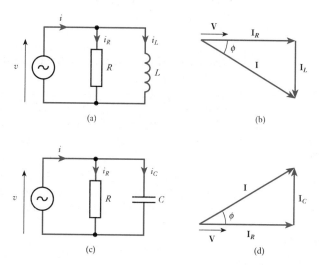

15.5 Impedance

In circuits containing only resistive elements, the current is related to the applied voltage by the *resistance* of the arrangement. In circuits containing reactive components, the relationship between the current and the applied voltage can be described by the **impedance Z** of the arrangement, which represents the effect of the circuit not only on the *magnitude* of the current but also on its *phase*. Impedance can be used in reactive circuits in a manner similar to the way that resistance is used in resistive circuits, but it should be remembered that impedance changes with frequency.

Figure 15.9   A series *RL* network

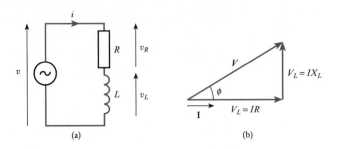

(a)   (b)

Consider Figure 15.9, which shows a series *RL* network as in Figure 15.5. From the phasor diagram of Figure 15.9(b), it is clear that the magnitude of the voltage across the *RL* combination, *V*, is given by

$$V = \sqrt{(V_R{}^2 + V_L{}^2)}$$

$$= \sqrt{(IR)^2 + (IX_L)^2}$$

$$= I\sqrt{R^2 + X_L^2}$$

$$= IZ$$

where $Z = \sqrt{R^2 + X_L{}^2}$. As before, we use letters in italics to represent the magnitude of a quantity that has both magnitude and phase, so $Z = |\mathbf{Z}|$ is the magnitude of the impedance of the *RL* combination. Similar consideration of the *RC* circuit of Figure 15.6 shows that the magnitude of the impedance of this arrangement is given by $Z = \sqrt{R^2 + X_C{}^2}$. Thus in each case the magnitude of the impedance is given by the expression

$$Z = \sqrt{R^2 + X^2} \tag{15.6}$$

The impedance of a circuit has not only a magnitude (*Z*) but also a phase angle (ϕ), which represents the phase between the voltage and the current. From Figure 15.9, we can see that the phase angle ϕ is given by

$$\phi = \tan^{-1}\frac{V_L}{V_R} = \tan^{-1}\frac{IX_L}{IR} = \tan^{-1}\frac{X_L}{R}$$

and a similar analysis can be performed for a series *RC* circuit, which produces the result that $\phi = \tan^{-1} X_C/R$. This leads to the general observation that

$$\phi = \tan^{-1}\frac{X}{R} \tag{15.7}$$

where the sign of the resulting phase angle can be found by inspection of the phasor diagram.

Figure 15.9(b) shows a phasor representation of circuit *voltages*. However, a similar diagram could be used to combine *resistive* and *reactive* components to determine their combined impedance. This is shown in Figure 15.10, where Figure 15.10(a) shows the impedance of a series *RL*

Figure 15.10 A graphical representation of impedance

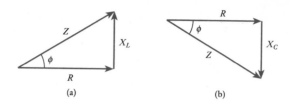

(a)   (b)

combination, and Figure 15.10(b) shows the impedance of a series *RC* arrangement. The impedance (**Z**) can be expressed in a **rectangular form** as a resistive plus a reactive component, or in a **polar form** by giving the magnitude *Z* and the angle ϕ.

The technique demonstrated in Figure 15.10 can be extended to compute the impedance of other combinations of resistors, inductors and capacitors by adding the various circuit elements in turn. However, it is often more convenient to use what is termed 'complex notation' to represent impedances, so we will turn our attention to this approach.

15.6	Complex notation

Readers who are familiar with complex mathematics will have noticed several similarities between the phasor diagrams discussed above and **Argand diagrams** used to represent complex quantities. If you are not familiar with this topic, you are advised to read Appendix D before continuing with this section, since this gives a brief introduction to complex numbers.

The distinction between real and imaginary numbers in complex arithmetic is in many ways similar to the difference between the resistive and the reactive components of impedance. When a sinusoidal current is passed through a resistor, the resulting voltage is in phase with the current and for this reason we consider the impedance of a resistor to be *real*. When a similar current is passed though an inductor or a capacitor, the voltage produced leads or lags the current with a phase angle of 90°. For this reason, we consider reactive elements to represent an *imaginary* impedance. Because the phase shift produced by inductors is opposite to that produced by capacitors, we assign different polarities to the imaginary impedances associated with these elements. By convention, inductors are taken to have *positive* impedance and capacitors are assumed to have *negative* impedance. The *magnitude* of the impedance of capacitors and inductors is determined by the reactance of the component.

Therefore, the impedance of resistors, inductors and capacitors is given by:

resistors: $\quad \mathbf{Z_R} \;=\; R$

inductors: $\quad \mathbf{Z_L} \;=\; jX_L \;=\; j\omega L$

capacitors: $\quad \mathbf{Z_C} \;=\; -jX_C \;=\; -j\dfrac{1}{\omega C} = \dfrac{1}{j\omega C}$

Figure 15.11   Complex impedances

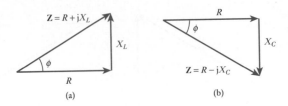

Figure 15.11 shows the complex impedances of the arrangements of Figure 15.10.

An attractive characteristic of complex impedances is that they can be used for sinusoidal signals in a similar manner to the way resistances are used for DC circuits.

15.6.1   Series combinations

For a series combination of impedances $\mathbf{Z}_1, \mathbf{Z}_2, \ldots, \mathbf{Z}_N$, the effective total impedance is equal to the sum of the individual impedances.

$$\mathbf{Z} = \mathbf{Z}_1 + \mathbf{Z}_2 + \ldots + \mathbf{Z}_N \tag{15.8}$$

This is illustrated in Figure 15.12(a).

Figure 15.12   The impedance of series and parallel arrangements

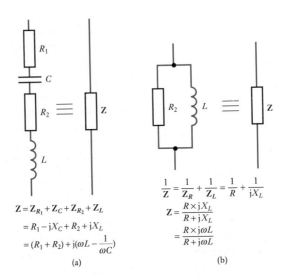

15.6.2   Parallel combinations

For a parallel combination of impedances $\mathbf{Z}_1, \mathbf{Z}_2, \ldots, \mathbf{Z}_N$, the effective total impedance is equal to the reciprocal of the sum of the reciprocals of the individual impedances.

$$\frac{1}{\mathbf{Z}} = \frac{1}{\mathbf{Z}_1} + \frac{1}{\mathbf{Z}_2} + \ldots + \frac{1}{\mathbf{Z}_N} \qquad (15.9)$$

This is illustrated in Figure 15.12(b).

Example 15.7 Determine the complex impedance of the following series arrangement at a frequency of 50 Hz.

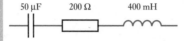

50 μF   200 Ω   400 mH

A frequency of 50 Hz corresponds to an angular frequency of

$$\omega = 2\pi f$$
$$= 2 \times \pi \times 50$$
$$= 314 \text{ rad/s}$$

Therefore

$$\mathbf{Z} = \mathbf{Z}_C + \mathbf{Z}_R + \mathbf{Z}_L$$
$$= R + j(X_L - X_C)$$
$$= R + j\left(\omega L - \frac{1}{\omega C} \right)$$
$$= 200 + j\left(314 \times 400 \times 10^{-3} - \frac{1}{314 \times 50 \times 10^{-6}} \right)$$
$$= 200 + j62 \ \Omega$$

Note that *at this frequency* the impedance of the arrangement is equivalent to a resistor of 200 Ω in series with an inductor of $X_L = 62$ Ω. Since $X_L = \omega L$, this equivalent inductance $L = X_L/\omega = 62/314 = 197$ mH. Therefore, at this single frequency, the circuit above is equivalent to

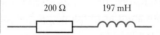

200 Ω   197 mH

15.6.3 Expressing complex quantities

We can express complex quantities in a number of ways, the most common forms being the rectangular form $(a + jb)$, the polar form $(r\angle\theta)$ and the **exponential form** $(re^{j\theta})$. If you are unfamiliar with these forms, and with arithmetic based on them, you are advised to read Appendix D before continuing with this section.

If we wish to add (or subtract) complex quantities (for example, two impedances), this is easiest to achieve if they are both expressed in a rectangular form, since

$$(a + jb) + (c + jd) = (a + c) + j(b + d)$$

However, if we wish to multiply (or divide) complex quantities (for example, multiplying a sinusoidal current by a complex impedance), this is easier using either polar or exponential forms. For example, using a polar form

$$\frac{A\angle\alpha}{B\angle\beta} = \frac{A}{B}\angle(\alpha - \beta)$$

Fortunately, conversion between the various forms is straightforward (see Appendix D), and we will often change the form of a quantity in order to simplify the arithmetic.

Sinusoidal voltages and currents of the same frequency are often described in polar form by their magnitude and their phase angle. Since phase angles are relative, we need to define a reference phase and this will often be the input voltage or the current in a circuit. For example, if a circuit has a sinusoidal input voltage with a magnitude of 20 V, and we choose this signal to define our reference phase, then we can describe this signal in polar form as $20\angle0$ V. If we determine that another signal in the circuit is $5\angle30°$, then this will have a magnitude of 5 V and a phase angle of 30° with respect to the reference (input) waveform. Positive values for this phase angle mean that the signal *leads* the reference waveform, while negative values indicate a phase *lag*.

15.6.4   Using complex impedance

Complex impedances can be used with sinusoidal signals in a similar manner to the way resistances are used in DC circuits. Consider, for example, the circuit of Figure 15.13, where we wish to determine the current i. If this were a purely resistive circuit, then the current would be given by $i = v/R$. So, in this case the current is given by $i = v/\mathbf{Z}$, where $\mathbf{Z}$ is the complex impedance of the circuit. In this circuit, the driving voltage is equal to 100 sin 250t, so the angular frequency ω is equal to 250. Thus the impedance $\mathbf{Z}$ of the circuit is given by

Figure 15.13   Use of impedance in a simple *RC* circuit

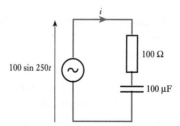

100 sin 250t

100 Ω

100 μF

$$Z = R - jX_C$$

$$= R - j\frac{1}{\omega C}$$

$$= 100 - j\frac{1}{250 \times 10^{-4}}$$

$$= 100 - j40$$

The current is given by $v/\mathbf{Z}$, and this is much easier to compute if we express each quantity in its polar form. If we define the reference phase to be that of the input voltage, then $v = 100\angle 0$. The polar form of $\mathbf{Z}$ is given by

$$\mathbf{Z} = 100 - j40$$

$$|\mathbf{Z}| = \sqrt{100^2 + 40^2}$$

$$= 107.7$$

$$\angle\mathbf{Z} = \tan^{-1}\frac{-40}{100}$$

$$= -21.8°$$

$$\mathbf{Z} = 107.7\angle -21.8°$$

Therefore

$$i = \frac{v}{\mathbf{Z}}$$

$$= \frac{100\angle 0}{107.7\angle -21.8}$$

$$= 0.93\angle 21.8°$$

Another example of the use of complex impedance is demonstrated by the circuit of Figure 15.14(a), which contains two resistors, a capacitor and an inductor. To analyse this circuit, we first replace each component with its

Figure 15.14 The use of impedance in circuit analysis

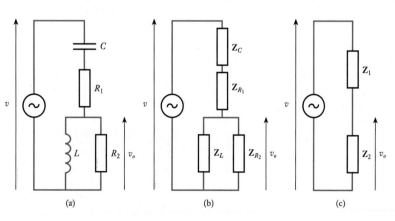

impedance as shown in Figure 15.14(b). Z_C and Z_{R1} are in series and may be combined to form a single impedance Z_1 as discussed in the last section. Similarly, Z_L and Z_{R2} are in parallel and may be combined to form a single impedance Z_2. This reduces our circuit to that shown in Figure 15.14(c), from which the output voltage can be determined using our standard formula for a potential divider (with the resistances replaced by impedances).

Example 15.8

Calculate the output voltage v_o in the circuit of Figure 15.14(a) if $C = 200\ \mu\text{F}$, $R_1 = 5\ \Omega$, $L = 50\ \text{mH}$, $R_2 = 50\ \Omega$ and the input voltage $v = 10 \sin 500t$.

Since $v = 10 \sin 500t$, $\omega = 500$ rad/s. We will take the input voltage as the reference phase, so v corresponds to $10\angle 0$.

Our first task is to calculate the impedances corresponding to Z_1 and Z_2 in Figure 15.14(c).

$$Z_1 = R_1 - jX_C$$

$$= R_1 - j\frac{1}{\omega C}$$

$$= 5 - j\frac{1}{500 \times 200 \times 10^{-6}}$$

$$= 5 - j10\ \Omega$$

$$\frac{1}{Z_2} = \frac{1}{R_2} + \frac{1}{jX_L}$$

$$Z_2 = \frac{jX_L R_2}{R_2 + jX_L}$$

$$= \frac{jX_L R_2 (R_2 - jX_L)}{(R_2 + jX_L)(R_2 - jX_L)}$$

$$= \frac{R_2 X_L^2 + jR_2^2 X_L}{R_2^2 + X_L^2}$$

$$= \frac{R_2 \omega^2 L^2 + jR_2^2 \omega L}{R_2^2 + \omega^2 L^2}$$

$$= \frac{(50 \times 500^2 \times 0.05^2) + (j \times 50^2 \times 500 \times 0.05)}{50^2 + (500^2 \times 0.05^2)}$$

$$= \frac{31{,}250 + j62{,}500}{3125}$$

$$= 10 + j20\ \Omega$$

From Figure 15.14(c), it is clear that

$$v_o = v \times \frac{\mathbf{Z}_2}{\mathbf{Z}_1 + \mathbf{Z}_2}$$

$$= v \times \frac{10 + j20}{(5 - j10) + (10 + j20)}$$

$$= v \times \frac{10 + j20}{15 + j10}$$

Division and multiplication are often easier using the polar form, which gives

$$v_o = v \times \frac{22.4\angle 63.4°}{18.0\angle 33.7°}$$

$$= v \times 1.24\angle 29.7°$$

$$= 10\angle 0 \times 1.24\angle 29.7°$$

$$= 12.4\angle 29.7°$$

Therefore, since the frequency of v_o is equal to that of v,

$$v_o = 12.4 \sin(500t + 29.7°)$$

and the output voltage leads the input voltage by 29.7°.

File 15C

Computer Simulation Exercise 15.3

Simulate the circuit of Example 15.8 using a sinusoidal voltage source for v. Use transient analysis to investigate the magnitude and the phase of the output voltage v_o and compare these with the values calculated above.

Impedances can be used in place of resistances in calculations involving Ohm's law, and Kirchhoff's voltage and current laws, allowing the various circuit analysis techniques described in Chapter 12 to be applied to alternating voltages and currents.

Key points

■ A sinusoidal voltage waveform can be described by the equation

$$v = V_p \sin(\omega t + \phi)$$

where V_p is the peak voltage of the waveform, ω is the angular frequency and ϕ represents its phase angle.

■ When a sinusoidal current flows through a resistor, an inductor and a capacitor:

- the voltage across the resistor is *in phase* with the current;
- the voltage across the inductor *leads* the current by 90°;
- the voltage across the capacitor *lags* the current by 90°.

■ The magnitude of the voltage across an inductor or a capacitor is determined by its reactance, where

- the reactance of an inductor, $X_L = \omega L$
- the reactance of a capacitor, $X_C = \dfrac{1}{\omega C}$

■ The *magnitudes* of the voltages across resistors, inductors and capacitors are given by:

- resistor: $V = IR$
- inductor: $V = IX_L$
- capacitor: $V = IX_C$

■ Phasor diagrams can be used to represent both the magnitude and the phase of quantities of the same frequency.

■ Representing alternating voltages by phasors allows signals of different phase to be added or subtracted easily.

■ The relationship between the current and the voltage in a circuit containing reactive components is described by its impedance **Z**.

■ Complex notation simplifies calculations of currents and voltages in circuits that have reactive elements. Using this notation, the impedance of resistors, inductors and capacitors is given by:

- resistors: $\mathbf{Z_R} = R$
- inductors: $\mathbf{Z_L} = j\omega L$
- capacitors: $\mathbf{Z_C} = -j\dfrac{1}{\omega C} = \dfrac{1}{j\omega C}$

■ Complex quantities may be expressed in a number of ways, the most common forms being the *rectangular form* $(a + jb)$, the *polar form* $(r\angle\theta)$ and the *exponential form* $(re^{j\theta})$.

■ Complex impedances can be used with sinusoidal signals in a similar manner to the way resistances are used in DC circuits.

Exercises

15.1 A signal v is described by the expression $v = 15 \sin 100t$. What is the angular frequency of this signal, and what is its peak magnitude?

15.2 A signal v is described by the expression $v = 25 \sin 250t$. What is the frequency of this signal (in Hz), and what is its r.m.s. magnitude?

15.3 Give an expression for a sinusoidal signal with a peak voltage of 20 V and an angular frequency of 300 rad/s.

15.4 Give an expression for a sinusoidal signal with an r.m.s. voltage of 14.14 V and a frequency of 50 Hz.

15.5 Give an expression relating the voltage across an inductor to the current through it.

15.6 Give an expression relating the voltage across a capacitor to the current through it.

15.7 If a sinusoidal current is passed through a resistor, what is the phase relationship between this current and the voltage across the component?

15.8 If a sinusoidal current is passed through a capacitor, what is the phase relationship between this current and the voltage across the component?

15.9 If a sinusoidal current is passed through an inductor, what is the phase relationship between this current and the voltage across the component?

15.10 Explain what is meant by the term 'reactance'.

15.11 What is the reactance of a resistor?

15.12 What is the reactance of an inductor?

15.13 What is the reactance of a capacitor?

15.14 Calculate the reactance of an inductor of 20 mH at a frequency of 100 Hz, being sure to include the units in your answer.

15.15 Calculate the reactance of a capacitor of 10 nF at an angular frequency of 500 rad/s, being sure to include the units in your answer.

15.16 A sinusoidal voltage of 15 V r.m.s. at 250 Hz is applied across a 50 μF capacitor. What will be the current in the capacitor?

15.17 A sinusoidal current of 2 mA peak at 100 rad/s flows through an inductor of 25 mH. What voltage will appear across the inductor?

15.18 Explain briefly the use of a phasor diagram.

15.19 What is the significance of the length and direction of a phasor?

15.20 Estimate the magnitude and phase of $(\mathbf{A} + \mathbf{B})$ and $(\mathbf{A} - \mathbf{B})$ in the following phasor diagram.

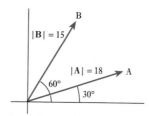

15.21 A voltage is formed by summing two sinusoidal waveforms of the same frequency. The first has a magnitude of 20 V and is taken as the reference phase (that is, its phase angle is taken as 0°). The second has a magnitude of 10 V and leads the first waveform by 45°. Draw a phasor diagram of this arrangement and hence estimate the magnitude and phase of the resultant signal.

15.22 A sinusoidal current of 3 A at 100 Hz flows through a series combination of a resistor of 25 Ω and an inductor of 75 mH. Use a phasor diagram to determine the voltage across the combination and the phase angle between this voltage and the current.

15.23 Use circuit simulation to confirm your results for Exercise 15.22. In doing this, you might find it useful to start with your circuit for Computer Simulation Exercise 15.1.

15.24 A sinusoidal voltage of 12 V at 500 Hz is applied across a series combination of a resistor of 5 kΩ and a capacitor of 100 nF. Use a phasor diagram to determine the current through the combination and the phase angle between this current and the applied voltage.

15.25 Use circuit simulation to confirm your results for Exercise 15.24. In doing this, you might find it useful to start with your circuit for Computer Simulation Exercise 15.2.

15.26 Use a phasor diagram to determine the magnitude and phase angle of the impedance formed by the series combination of a resistance of 25 Ω and a capacitance of 10 μF, at a frequency of 300 Hz.

15.27 If $x = 5 + j7$ and $y = 8 - j10$, evaluate $(x + y)$, $(x - y)$, $(x \times y)$ and $(x \div y)$.

15.28 What is the complex impedance of a resistor of 1 kΩ at a frequency of 1 kHz?

15.29 What is the complex impedance of a capacitor of 1 μF at a frequency of 1 kHz?

15.30 What is the complex impedance of an inductor of 1 mH at a frequency of 1 kHz?

15.31 Determine the complex impedance of the following arrangements at a frequency of 200 Hz.

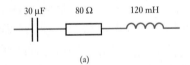

(a)

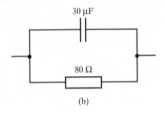

(b)

15.32 Express $x = 20 + j30$ in polar form and in exponential form.

15.33 Express $y = 25\angle{-40°}$ in rectangular form and in exponential form.

15.34 A voltage $v = 60 \sin 314t$ is applied across a series combination of a 10 Ω resistor and an inductance of 50 mH. Determine the magnitude and phase of the resulting current.

15.35 Use circuit simulation to confirm your results for Exercise 15.34.

15.36 A current of $i = 0.5 \sin 377t$ is passed through a parallel combination of a resistance of 1 kΩ and a capacitance of 5 μF. Determine the magnitude and phase of the resulting voltage across the combination.

15.37 Use circuit simulation to confirm your results for Exercise 15.36.

Chapter 16

Power in AC Circuits

Objectives

When you have studied the material in this chapter you should be able to:

- **explain concepts such as apparent power, active power, reactive power and power factor;**
- **calculate the power dissipated in circuits containing resistors, capacitors and inductors when these are used with AC signals;**
- **discuss the importance of the power factor in determining the efficiency of power utilisation and distribution;**
- **determine the power factor of a given circuit arrangement and propose appropriate additional components to achieve power factor correction if necessary;**
- **select an appropriate load impedance to achieve maximum power transfer;**
- **describe the measurement of power in both single-phase and three-phase arrangements.**

16.1 Introduction

The instantaneous power dissipated in a resistor is given by the product of the instantaneous voltage and the instantaneous current. In a DC circuit, this produces the well-known relationship

$$P = VI$$

while in an AC circuit

$$p = vi$$

where v and i are the instantaneous values of an alternating waveform, and p is the instantaneous power dissipation.

In a purely resistive circuit v and i are in phase, and calculation of p is straightforward. However, in circuits containing reactive elements, there will normally be a phase difference between v and i, and calculating the power is slightly more complicated. Here we will begin by looking at the power dissipated in resistive loads before going on to consider inductive, capacitive and mixed loads.

16.2　Power dissipation in resistive components

If a sinusoidal voltage $v = V_P \sin \omega t$ is applied across a resistance R, then the current i through the resistance will be

$$i = \frac{v}{R}$$

$$= \frac{V_P \sin \omega t}{R}$$

$$= I_P \sin \omega t$$

where $I_P = V_P/R$.

The resultant power p is given by

$$p = vi$$

$$= V_P \sin \omega t \times I_P \sin \omega t$$

$$= V_P I_P (\sin^2 \omega t)$$

$$= V_P I_P \left(\frac{1 - \cos 2\omega t}{2} \right)$$

We see that p varies at a frequency *double* that of v and i. The relationship between v, i and p is shown in Figure 16.1. Since the average value of a cosine function is zero, the average value of $(1 - \cos 2\omega t)$ is 1, and the average value of p is $\frac{1}{2}V_P I_P$. Therefore, the average power P is equal to

$$P = \frac{1}{2}V_P I_P$$

$$= \frac{V_P}{\sqrt{2}} \times \frac{I_P}{\sqrt{2}}$$

$$= VI$$

where V and I are the r.m.s. voltage and current. This is consistent with our discussion of r.m.s. values in Chapter 11.

Figure 16.1　Relationship between voltage, current and power in a resistor

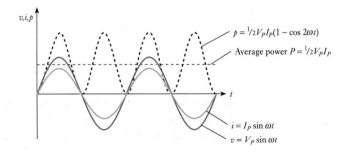

$p = \frac{1}{2}V_P I_P(1 - \cos 2\omega t)$

Average power $P = \frac{1}{2}V_P I_P$

$i = I_P \sin \omega t$

$v = V_P \sin \omega t$

16.3 Power in capacitors

From our discussions in Chapter 13, we know that the current in a capacitor leads the voltage by 90°. Therefore, if $v = V_p \sin \omega t$, then the resultant current will be given by $i = I_P \cos \omega t$, and the power p will be

$$p = vi$$

$$= V_P \sin \omega t \times I_P \cos \omega t$$

$$= V_P I_P (\sin \omega t \times \cos \omega t)$$

$$= V_P I_P \left(\frac{\sin 2\omega t}{2} \right)$$

The relationship between v, i and p is shown in Figure 16.2. Again p varies at a frequency *double* that of v and i, but since the average value of a sine function is zero, the average value of p is also zero. Therefore, power flows *into* the capacitor during part of the cycle and then flows *out* of the capacitor again. Thus the capacitor stores energy for part of the cycle and returns it to the circuit again, with the average power dissipated P in the capacitor being zero.

Figure 16.2 Relationship between voltage, current and power in a capacitor

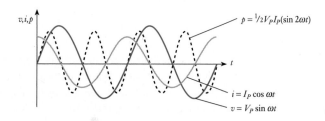

16.4 Power in inductors

From Chapter 14, we know that the current in an inductor lags the voltage by 90°. Therefore, if $v = V_p \sin \omega t$, the resultant current will be given by $i = -I_P \cos \omega t$. Therefore, the power p is given by

$$p = vi$$

$$= V_P \sin \omega t \times -I_P \cos \omega t$$

$$= -V_P I_P (\sin \omega t \times \cos \omega t)$$

$$= -V_P I_P \left(\frac{\sin 2\omega t}{2} \right)$$

The situation is very similar to that of the capacitor, and again the power dissipated in the inductor is zero. The relationship between v, i and p is shown in Figure 16.3.

Figure 16.3 Relationship between voltage, current and power in an inductor

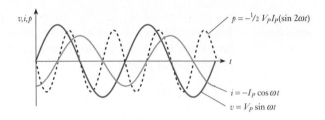

$p = -\frac{1}{2}V_P I_P(\sin 2\omega t)$

$i = -I_P \cos \omega t$

$v = V_P \sin \omega t$

16.5 Power in circuits with resistance and reactance

When a sinusoidal voltage $v = V_P \sin \omega t$ is applied across an impedance that contains both resistive and inductive elements, the resulting current will be of the general form $i = I_P \sin(\omega t - \phi)$. Therefore, the instantaneous power p is given by

$$p = vi$$

$$= V_P \sin \omega t \times I_P \sin(\omega t - \phi)$$

$$= \frac{1}{2}V_P I_P \{\cos \phi - \cos(2\omega t - \phi)\}$$

$$p = \frac{1}{2}V_P I_P \cos \phi - \frac{1}{2}V_P I_P \cos(2\omega t - \phi) \tag{16.1}$$

It can be seen that the expression for p has two components. The second is a function of $2\omega t$ and consequently this component oscillates at twice the frequency of v. The average value of this term over a complete cycle, or over a long period of time, is zero. This represents the energy that is stored in the reactive elements and is then returned to the circuit within each voltage cycle. The first component is independent of time and therefore has a constant value. This represents the power dissipated in the resistive elements in the circuit. Consequently, the average power dissipated is given by

$$P = \frac{1}{2}V_P I_P (\cos \phi)$$

$$= \frac{V_P}{\sqrt{2}} \times \frac{I_P}{\sqrt{2}} \times (\cos \phi)$$

$$P = VI \cos \phi \tag{16.2}$$

where V and I are r.m.s. values of the voltage and current. This average power dissipation P is called the **active power** and is measured in watts.

If one were to measure the r.m.s. voltage V across a load and the r.m.s. current through the load I, and multiply these two quantities together, one would obtain a quantity equal to VI, which is termed the **apparent power**. One could imagine that an inexperienced engineer, unfamiliar with the effects of phase angle, would take such measurements and perform such a calculation in order to calculate the power dissipated in the load. From the discussion above, we know that this product does *not* give the dissipated power, but it is still a quantity of interest and the apparent power is given the symbol S. To avoid confusion with dissipated power, S is given the units of volt amperes (VA).

From Equation 16.2, we know that $P = VI \cos \phi$, and therefore

$$P = S \cos \phi$$

In other words, the active power is equal to the apparent power times the cosine of the phase angle. This cosine term is referred to as the **power factor**, which is defined as the ratio of the active to the apparent power. Thus

$$\frac{\text{active power (in watts)}}{\text{apparent power (in volt amperes)}} = \text{power factor} \qquad (16.3)$$

From the above

$$\text{power factor} = \frac{P}{S} = \cos \phi \qquad (16.4)$$

Example 16.1 | The voltage across a component is measured as **50 V r.m.s.** and the current through it is **5 A r.m.s.** If the current leads the voltage by **30°**, calculate:

(a) the apparent power;
(b) the power factor;
(c) the active power.

(a) The apparent power is

$$\text{apparent power } S = VI$$
$$= 50 \times 5$$
$$= 250 \text{ VA}$$

(b) The power factor is

$$\text{power factor} = \cos \phi$$
$$= \cos 30°$$
$$= 0.866$$

(c) The active power is

$$\text{active power } P = S \cos \phi$$
$$= 250 \times 0.866$$
$$= 216.5 \text{ W}$$

File 16A

16.6   Active and reactive power

From Equation 16.1, it is clear that, when a load has both resistive and reactive elements, the resultant power will have two components. The first is that which is *dissipated* in the resistive element of the load, which we have described as the active power. The second element is not dissipated but is *stored* and *returned* by the reactive elements in the circuit. We refer to this as the **reactive power** in the circuit, which is given the symbol Q.

While reactive power is not dissipated in the load, its presence does have an effect on the rest of the system. The need to supply power to the reactive elements during part of the supply cycle, and to accept power from them during other parts of the cycle, increases the current that must be supplied by the power source and also increases the losses due to resistance in power cables.

In order to quantify this effect, let us consider the situation where a sinusoidal voltage V is applied to a complex load given by $\mathbf{Z} = R + jX$ as shown in Figure 16.4(a). A voltage phasor diagram of this arrangement is shown in Figure 16.4(b), and this is redrawn in Figure 16.4(c) expressing V_R and V_X in terms of the applied voltage V and the phase angle ϕ. If we take the *magnitudes* of the various elements of this phasor diagram and multiply each by I, we obtain a triangle of a similar shape, as shown in Figure 16.4(d). If V and I represent the r.m.s. voltage and current, then the hypotenuse of this triangle is of length VI, which we identified earlier as being the apparent power S; the base of the triangle is of length $VI \cos \phi$, which we identified as being the active power P; and the vertical line is of length

Figure 16.4 Active and
reactive power

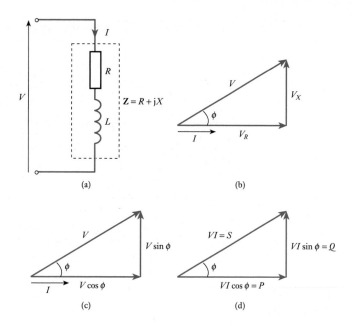

(a)   (b)

(c)   (d)

$VI \sin \phi$, which represents the reactive power Q in the circuit. Reactive
power is given the units of **volt amperes reactive** or **var** (to differentiate it
from active power, with which it is dimensionally equivalent). For obvious
reasons, Figure 16.4(d) is referred to as a **power triangle**. Therefore

$$\text{active power } P = VI \cos \phi \text{ W} \tag{16.5}$$

$$\text{reactive power } Q = VI \sin \phi \text{ var} \tag{16.6}$$

$$\text{apparent power } S = VI \text{ VA} \tag{16.7}$$

$$S^2 = P^2 + Q^2 \tag{16.8}$$

| **Example 16.2** | A 2 kVA motor operates from a 240 V supply at 50 Hz and has a power factor of 0.75. Determine the apparent power, the active power, the reactive power and the current in the motor. |

The apparent power S of the motor is 2000 VA, since this is the rating of
the motor. The power factor ($\cos \phi$) is 0.75. Therefore, the active power in
the motor is

$$\text{active power } P = S \cos \phi$$

$$= 2000 \times 0.75$$

$$= 1500 \text{ W}$$

Since $\cos \phi = 0.75$, it follows that $\sin \phi = \sqrt{1 - \cos^2 \phi} = 0.661$. Therefore

$$\text{reactive power } Q = S \sin \phi$$

$$= 2000 \times 0.6614$$

$$= 1323 \text{ var}$$

The current is given by the apparent power divided by the voltage

$$\text{current } I = \frac{S}{V}$$

$$= \frac{2000}{240}$$

$$= 8.33\text{A}$$

16.7 Power factor correction

The power factor is of particular importance in high-power applications. A given power supply or generator normally has a maximum output voltage and a maximum output current, and in general we wish to maximise the useful power that we can obtain from such a source. If the load connected to the supply has a power factor of unity, then the available power will be maximised, and all the current supplied by the generator will be used to produce active power. If the power factor is less than 1, the power available at the load will be less than this maximum value, and reactive currents will result in increased losses in the associated cables.

Inductive loads are said to have *lagging* power factors, since the current lags the applied voltage, while capacitive loads are said to have *leading* power factors. Many high-power devices, such as motors, are inductive, and a typical AC motor will have a lagging power factor of 0.9 or less. Consequently, the total load applied to national power distribution grids tends to have a power factor of 0.8–0.9 lagging. This leads to major inefficiencies in the power generation and distribution process, and power companies therefore charge fees that penalise industrial users who introduce a poor power factor.

The problems associated with inductive loads can be tackled by adding additional components to bring the power factor closer to unity. A capacitor of an appropriate size in parallel with a *lagging* load can improve the load factor by 'cancelling out' the inductive element in the load's impedance. The effect is that the reactive currents associated with the inductance now flow in and out of the capacitor, rather than to and from the power supply. This not only reduces the current flowing into and out of the power supply but also reduces the resistive losses in the cables. A similar effect can also be obtained by adding a capacitor in series with the load, although a parallel arrangement is more common because it does not alter the voltage on the load.

Example 16.3

A capacitor is to be added in parallel with the motor of Example 16.2 to increase its power factor to 1.0. Calculate the value of the required capacitor, and calculate the active power, the apparent power and the current after power factor correction.

In Example 16.2, we determined that for this motor:

$$\text{apparent power } S = 2000 \text{ VA}$$

$$\text{active power } P = 1500 \text{ W}$$

$$\text{current } I = 8.33 \text{ A}$$

$$\text{reactive power } Q = 1323 \text{ var}$$

The capacitor is required to cancel the *lagging* reactive power. We therefore need to add a capacitive element with a *leading* reactive power Q_C of -1323 var.

Now, just as $P = V^2/R$, so $Q = V^2/X$. Since capacitive reactive power is negative

$$Q_C = -\frac{240^2}{X_C} = -1323 \text{ var}$$

$$X_C = \frac{240^2}{1323} = 43.54 \ \Omega$$

$X_C = 1/\omega C$ which is equal to $1/2\pi fC$. Therefore

$$\frac{1}{2\pi fC} = 43.54$$

$$C = \frac{1}{43.54 \times 2 \times \pi \times f}$$

$$= \frac{1}{43.54 \times 2 \times 3.142 \times 50}$$

$$= 73 \ \mu\text{F}$$

The power factor correction does not affect the active power in the motor, and P is therefore unchanged at 1500 W. However, since the power factor is now 1, the apparent power is now $S = P = 1500$. The current is now given by

$$\text{current } I = \frac{S}{V} = \frac{1500}{240} = 6.25 \text{ A}$$

Thus the apparent power is reduced from 2000 VA to 1500 VA as a result of the addition of the capacitor, while the current drops from 8.33 A to 6.25 A. The active power dissipated by the motor remains unchanged at 1500 W.

In Example 16.3, we chose a capacitor to increase the power factor to unity, but this is not always appropriate. High-voltage capacitors suitable for this purpose are expensive, and it may be more cost-effective to increase the power factor by a more modest amount, perhaps up to about 0.9.

16.8 Power transfer

In Section 6.5, we observed that maximum power transfer occurs in DC circuits when the load resistance is equal to the source (or output) resistance. Our discussion of AC circuits in the last chapter might suggest that for AC circuits maximum power transfer would occur when the load impedance is equal to the source impedance, but this is *not* the case.

In this chapter, we have seen that, for a given value of apparent power, the active power is greatest when the output current is in phase with the output voltage. This corresponds to a power factor of unity. It would therefore seem reasonable to assume that, in the case of a source and load arrangement, maximum power will be delivered to the load when the output current is in phase with the output voltage. In order to achieve this, the total reactance of the output circuit must be zero. Therefore, if the source has an output reactance of $+jX$, then the load must have a reactance of $-jX$. In order to maximise power transfer, the resistance of the source and load must be equal, as in the DC case. Therefore, if the output impedance $\mathbf{Z}_o = R + jX$, then for maximum power transfer the load impedance $\mathbf{Z}_L = R - jX$. In other words, the load impedance should be the **complex conjugate** of the output impedance, and $\mathbf{Z}_L = \mathbf{Z}_o{}^*$. This is illustrated in Figure 16.5. If you are unfamiliar with the concept of complex conjugates, see Appendix D. This relationship is the AC form of the **maximum power transfer theorem** and is also the general form. Clearly, the DC form derived in Chapter 6 is simply a special case of this theorem where the reactance is zero.

While this relationship has been derived empirically, it can be obtained analytically by deriving an expression for the output power and differentiating it to find its maximum value. However, the above explanation probably gives a greater understanding of the nature of this relationship.

Figure 16.5 The condition for maximum power transfer

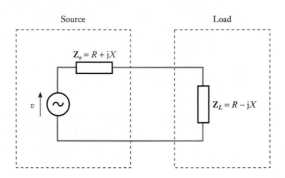

Example 16.4 An amplifier has an output impedance Z_o of $50 - j20\ \Omega$. What value of load impedance will permit maximum power transfer?

For maximum power transfer, the load impedance $Z_L = Z_o{}^*$. Therefore

$$Z_L = Z_o{}^*$$
$$= (50 - j20)^*$$
$$= 50 + j20\ \Omega$$

16.9 Three-phase systems

So far, our description of AC signals has been restricted to **single-phase** arrangements. This is the most common arrangement and is that used in conventional domestic supplies, where power is delivered using a single pair of cables (often with the addition of an earth conductor). While this arrangement works well for heating and lighting applications, there are certain situations (in particular when using large electric motors) when single-phase arrangements are unsatisfactory. In such situations, it is common to use a three-phase supply, which provides power using three alternating waveforms, each differing in phase by 120°. The three **phases** are given the labels red, yellow and blue, which are normally abbreviated to R, Y and B. The relationship between these phases is shown in Figure 16.6.

Three-phase power can be supplied using three or four conductors. Where three are used each provides one of the phases, and loads are connected between the conductors, as shown in Figure 16.7(a). In a four-line system, the additional wire is a neutral conductor. Loads may then be connected between each phase and neutral, as shown in Figure 16.7(b).

Three-phase arrangements are common in high-power industrial applications, particularly where electrical machines are involved.

Figure 16.6 Voltage waveforms of a three-phase arrangement

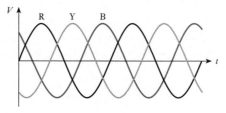

Figure 16.7 Three-phase connections

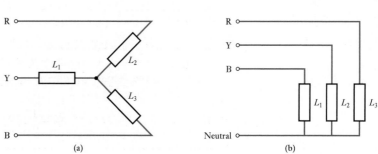

(a)   (b)

16.10 Power measurement

When using sinusoidal signals, the power dissipated in a load is determined not only by the r.m.s. values of the voltage and current but also by the phase angle between the voltage and current waveforms (which determines the power factor). Consequently, it is not possible to calculate the power simply by taking independent measurements of the voltage and the current.

In single-phase AC circuits, power is normally measured using an **electrodynamic wattmeter**. This device passes the load current through a series of low-resistance field coils and places the load voltage across a high-resistance armature coil. The resulting deflection is directly related to the product of the instantaneous current and voltage and hence to the instantaneous power in the load. The inertia of the coil smooths out the line frequencies and produces a reading that is proportional to the average value of the power. The device can therefore be directly calibrated in watts.

In three-phase circuits, it is necessary to sum the power taken from each phase in order to measure the total power consumption. In a three-wire system, because of the interaction between the phases, the readings from two wattmeters can be used to measure the total power. Summing the readings from the two meters gives the total power, whether or not the same power is taken from each phase. In a four-line arrangement, it may be necessary to use three wattmeters to measure the power taken from each of the phases. However, if it is known that the system is balanced (that is, that equal power is taken from each phase), then a single wattmeter can be used. This is connected to any one of the phases and its reading multiplied by three to get the total power.

Key points

- In both DC and AC circuits, the instantaneous power in a resistor is given by the product of the instantaneous voltage and the instantaneous current.

- In purely resistive circuits, the current is in phase with the voltage and therefore calculating power is straightforward: the average power is equal to VI, where V and I are r.m.s. values.

- In a capacitor, the current leads the voltage by 90° and the average power dissipation is zero.

- In an inductor, the current lags the voltage by 90° and again the average power dissipation is zero.

- In circuits that have both resistive and inductive elements, the average power $P = VI \cos \phi$, where ϕ is the phase angle between the current and the voltage waveforms. P is termed the active power, which has units of watts (W).

- The product of the r.m.s. voltage V and the r.m.s. current I is termed the apparent power S. This has units of volt amperes (VA).

- The ratio of the active power to the apparent power is the power factor.

$$\text{power factor} = \frac{P}{S} = \cos \phi$$

- The power stored and returned to the system by reactive elements is termed the reactive power Q. This has units of volt amperes reactive (var).

- The efficiency of power utilisation and distribution can be increased by increasing the power factor – a process called power factor correction. Since most high-power loads have an inductive element, correction is normally achieved by adding a capacitor in parallel with the load.

- Maximum power transfer is achieved when the load impedance is equal to the complex conjugate of the source (output) impedance. $\mathbf{Z}_L = \mathbf{Z}_o{}^*$.

- High-power applications often make use of three-phase supplies.

- Power can be measured directly using a wattmeter, which multiplies instantaneous values of voltage and current to measure power directly.

Exercises

16.1 A sinusoidal voltage $v = 10 \sin 377t$ is applied to a resistor of 50 Ω. Calculate the average power dissipated in it.

16.2 The voltage of Exercise 16.1 is applied across a capacitor of 1 µF. Derive an expression for the current in the capacitor and calculate the average power dissipated in it.

16.3 The voltage of Exercise 16.1 is applied across an inductor of 1 mH. Derive an expression for the current in the inductor and the average power dissipated in it.

16.4 The voltage across a component is 100 V r.m.s. and the current is 7 A r.m.s. If the current lags the voltage by 60°, calculate the apparent power, the power factor and the active power.

16.5 Explain the difference between the units of watts, VA and var.

16.6 A sinusoidal voltage of 100 V r.m.s. at 50 Hz is applied across a series combination of a 40 Ω resistor and an inductor of 100 mH. Determine the r.m.s. current, the apparent power, the power factor, the active power and the reactive power.

16.7 A machine operates on a 250 V supply at 60 Hz; it is rated at 500 VA and has a power factor of 0.8. Determine the apparent power, the active power, the reactive power and the current in the machine.

16.8 Explain what is meant by power factor correction and explain why this is of importance in high-power systems.

16.9 Calculate the value of capacitor required to be added in parallel with the machine of Exercise 16.7 to achieve a power factor of 1.0.

16.10 Calculate the value of capacitor required to be added in parallel with the machine of Exercise 16.7 to achieve a power factor of 0.9.

Exercises continued

16.11 A sinusoidal signal of 20 V peak at 50 Hz, is applied to a load consisting of a 10 Ω resistor and a 16 mH inductor connected in series. Calculate the power factor of this arrangement and the active power dissipated in the load.

16.12 Simulate the arrangement of Exercise 16.11 and plot the voltage and the current. Estimate the phase difference between these two waveforms and hence confirm the value you calculated for the power factor. Plot the product of the voltage and the current and estimate its average value. Hence confirm your calculated value for the active power of the circuit.

16.13 Determine the value of capacitor needed to be added in series with the circuit of Exercise 16.11 to produce a power factor of 1.0. Calculate the active power that would be dissipated in the circuit with the addition of such a capacitor.

16.14 Simulate the arrangement of Exercise 16.13 and confirm your predictions for its behaviour.

16.15 State the maximum power transfer theorem as it applies to AC systems.

16.16 What load should be used with a source having an output impedance of $30 + j20$ in order to achieve maximum power transfer?

16.17 Explain the difference between three- and four-conductor arrangements of three-phase power supplies.

16.18 Explain why it is not possible to calculate the power dissipated in an AC network by multiplying the readings of a voltmeter and an ammeter.

16.19 Explain how it is possible to measure power directly in a single-phase system.

University of Glamorgan
Learning Resources Centre -
Treforest
Self Issue Receipt (TR1)

**Customer name: MR NAIM
AKBAR**
Customer ID: **********001

Title: Electrical & electronic systems

ID: 7312819395
**Due: 23:59 Monday, 23 January
2012**

Total items: 1
16/01/2012 21:37

Thank you for using the Self-
Service system
Diolch yn fawr

Chapter 17

Frequency Characteristics of AC Circuits

Objectives

When you have studied the material in this chapter you should be able to:

- design simple high-pass and low-pass networks using *RC* or *RL* circuits;
- explain how the gain and phase shift produced by these circuits varies with frequency;
- predict the effects of combining a number of high-pass or low-pass stages and outline the characteristics of the resulting arrangement;
- describe the characteristics of simple circuits containing resistors, inductors and capacitors, and calculate the resonant frequency and bandwidth of such circuits;
- discuss the operation and characteristics of a range of passive and active filters;
- explain the importance of stray capacitance and stray inductance in determining the frequency characteristics of electronic circuits.

17.1 Introduction

In Chapter 5, we noted that all systems have limits to the range of frequencies over which they will operate, and in Chapter 6 we looked briefly at the bandwidth and the frequency response of amplifiers. Having now studied the AC behaviour of circuit components, we are in a position to consider frequency response in more detail.

While the properties of a pure resistance are not affected by the frequency of the signal concerned, this is not true of reactive components. The reactance of both inductors and capacitors is dependent on frequency, and we know that

$$X_L = \omega L$$

$$X_C = \frac{1}{\omega C}$$

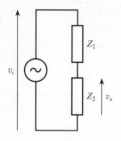

Figure 17.1 A potential divider circuit

where $\omega = 2\pi f$. Therefore, the characteristics of any circuit that includes capacitors or inductors will change with frequency. However, the situation is more complex than this because, as we noted in Chapters 13 and 14, all real circuits have both stray capacitance and stray inductance. Inevitably, therefore, the characteristics of all circuits will change with frequency.

In order to understand the nature of these frequency-related effects, we will start by looking at very simple circuits containing resistors and capacitors, or resistors and inductors. In Chapter 15, we looked at circuits involving impedances, including the potential divider arrangement shown in Figure 17.1. From our earlier consideration of the circuit, we know that the output voltage of this circuit is given by

$$v_o = v_i \times \frac{Z_2}{Z_1 + Z_2}$$

Another way of describing the behaviour of this circuit is to give an expression for the output voltage divided by the input voltage. In this case, this gives

$$\frac{v_o}{v_i} = \frac{Z_2}{Z_1 + Z_2} \tag{17.1}$$

We have previously described this ratio as the voltage gain of the circuit, but it is also referred to as its **transfer function**. We will now use this expression to analyse the behaviour of simple RC and RL circuits.

17.2 A high-pass RC network

Consider the circuit of Figure 17.2(a), which shows a potential divider, formed from a capacitor and a resistor. This circuit is shown redrawn in Figure 17.2(b), which is electrically identical. Applying Equation 17.1, we see that

$$\frac{v_o}{v_i} = \frac{Z_R}{Z_R + Z_C} = \frac{R}{R - j\dfrac{1}{\omega C}} = \frac{1}{1 - j\dfrac{1}{\omega CR}} \tag{17.2}$$

Figure 17.2 A simple RC network

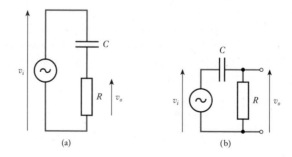

(a)　　　(b)

At high frequencies, ω is large and the value of $1/j\omega CR$ is small compared with 1. Therefore, the denominator of the expression is close to unity and the voltage gain is approximately 1.

However, at lower frequencies the magnitude of $1/\omega CR$ becomes more significant and the gain of the network decreases. Since the denominator of the expression for the gain has both real and imaginary parts, the magnitude of the voltage gain is given by:

$$| \text{voltage gain} | = \frac{1}{\sqrt{1^2 + \left(\dfrac{1}{\omega CR}\right)^2}}$$

When the value of $1/\omega CR$ is equal to 1, this gives

$$| \text{voltage gain} | = \frac{1}{\sqrt{1 + 1}} = \frac{1}{\sqrt{2}} = 0.707$$

Since power gain is proportional to the square of the voltage gain, this is a halving of the power gain (or a fall of 3 dB) compared with the gain at high frequencies. Therefore, as discussed in Chapter 6, this corresponds to the cut-off frequency of the circuit. If the angular frequency corresponding to this cut-off frequency is given the symbol ω_c, then $1/\omega_c CR$ is equal to 1, and

$$\omega_c = \frac{1}{CR} = \frac{1}{\mathsf{T}} \text{ rad/s} \tag{17.3}$$

where $\mathsf{T} = CR$ is the time constant of the capacitor–resistor combination that produces the cut-off frequency.

Since it is often more convenient to deal with *cyclic* frequencies (which are measured in hertz) rather than *angular* frequencies (which are measured in radians per second) we can use the relationship $\omega = 2\pi f$ to calculate the corresponding cyclic cut-off frequency f_c.

$$f_c = \frac{\omega_c}{2\pi} = \frac{1}{2\pi CR} \text{ Hz} \tag{17.4}$$

Example 17.1   **Calculate the time constant T, the angular cut-off frequency ω_c and the cyclic cut-off frequency f_c of the following arrangement.**

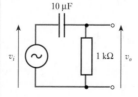

From above

$$\mathsf{T} = CR = 10 \times 10^{-6} \times 1 \times 10^3 = 0.01 \text{ s}$$

$$\omega_c = \frac{1}{T} = \frac{1}{0.01} = 100 \text{ rad/s}$$

$$f_c = \frac{\omega_c}{2\pi} = \frac{100}{2\pi} = 15.9 \text{ Hz}$$

If we substitute for ω (where $\omega = 2\pi f$) and CR (where $CR = 1/2\pi f_c$) in Equation 17.2, we obtain an expression for the gain of the circuit in terms of the signal frequency f and the cut-off frequency f_c:

$$\frac{v_o}{v_i} = \frac{1}{1 - j\dfrac{1}{\omega CR}} = \frac{1}{1 - j\dfrac{1}{(2\pi f)\left(\dfrac{1}{2\pi f_c}\right)}} = \frac{1}{1 - j\dfrac{f_c}{f}} \qquad (17.5)$$

This is a general expression for the voltage gain of this form of CR network.

From Equation 17.5, it is clear that the voltage gain is a function of the signal frequency f and that the magnitude of the gain varies with frequency. Since the gain has an imaginary component, it is also clear that the circuit produces a **phase shift** that changes with frequency. To investigate how these two quantities change with frequency, let us consider the gain of the circuit in different frequency ranges.

17.2.1   When $f \gg f_c$

When the signal frequency f is much greater than the cut-off frequency f_c, then in Equation 17.5 f_c/f is much less than unity, and the voltage gain is approximately equal to 1. Here the imaginary part of the gain is negligible and the gain of the circuit is effectively real. Hence the phase shift produced is negligible. This situation is shown in the phasor diagram of Figure 17.3(a).

Figure 17.3   Phasor diagrams of the gain of the circuit of Figure 17.2 at different frequencies

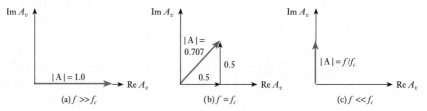

17.2.2   When $f = f_c$

When the signal frequency f is equal to the cut-off frequency f_c, then Equation 17.5 becomes

$$\frac{v_o}{v_i} = \frac{1}{1 - j\dfrac{f_c}{f}} = \frac{1}{1 - j}$$

Multiplying the numerator and the denominator by $(1 + j)$ gives

$$\frac{v_o}{v_i} = \frac{(1 + j)}{(1 - j)(1 + j)} = \frac{(1 + j)}{2} = 0.5 + 0.5j$$

This is illustrated in the phasor diagram of Figure 17.3(b), which shows that the magnitude of the gain at the cut-off frequency is 0.707. This is consistent with our earlier analysis, which predicted that the gain at the cut-off frequency should be $1/\sqrt{2}$ (or 0.707) times the mid-band gain. In this case, the mid-band gain is the gain some way above the cut-off frequency, which we have just shown to be 1. The phasor diagram also shows that at this frequency the phase angle of the gain is $+45°$. This shows that the output voltage *leads* the input voltage by $45°$. The gain is therefore $0.707\angle 45°$.

17.2.3   When $f \ll f_c$

The third region of interest is where the signal frequency is well below the cut-off frequency. Here f_c/f is much greater than 1, and Equation 17.5 becomes

$$\frac{v_o}{v_i} = \frac{1}{1 - j\dfrac{f_c}{f}} \approx \frac{1}{-j\dfrac{f_c}{f}} = j\frac{f}{f_c}$$

The 'j' signifies that the gain is imaginary, as shown in the phasor diagram of Figure 17.3(c). The magnitude of the gain is simply f/f_c and the phase shift is $+90°$, the '+' sign meaning that the output voltage *leads* the input voltage by $90°$.

Since f_c is a constant for a given circuit, in this region the voltage gain is linearly related to frequency. If the frequency is halved the voltage gain will be halved. Therefore, the gain falls by a factor of 0.5 for every octave drop in frequency (an **octave** is a doubling or halving of frequency and is equivalent to an octave jump on a piano or other musical instrument). A fall in voltage gain by a factor of 0.5 is equivalent to a change in gain of -6 dB. Therefore, the rate of change of gain can be expressed as 6 dB per octave. An alternative way of expressing the rate of change of gain is to specify the change of gain for a decade change in frequency (a **decade**, as its name suggests, is a change in frequency of a factor of 10). If the frequency falls to 0.1 of its previous value, the voltage gain will also drop to 0.1 of its previous value. This represents a change in gain of -20 dB. Thus the rate of change of gain is 20 dB per decade.

Example 17.2   Determine the frequencies corresponding to:

(a) an octave above 1 kHz;
(b) three octaves above 10 Hz;
(c) an octave below 100 Hz;

(d) a decade above 20 Hz;
(e) three decades below 1 MHz;
(f) two decades above 50 Hz.

(a) an octave above 1 kHz = $1000 \times 2 = 2$ kHz
(b) three octaves above 10 Hz = $10 \times 2 \times 2 \times 2 = 80$ Hz
(c) an octave below 100 Hz = $100 \div 2 = 50$ Hz
(d) a decade above 20 Hz = $20 \times 10 = 200$ Hz
(e) three decades below 1 MHz = $1,000,000 \div 10 \div 10 \div 10 = 1$ kHz
(f) two decades above 50 Hz = $50 \times 10 \times 10 = 5$ kHz

17.2.4 Frequency response of the high-pass *RC* network

Figure 17.4 shows the gain and phase response of the circuit of Figure 17.2 for frequencies above and below the cut-off frequency. It can be seen that, at frequencies much greater than the cut-off frequency, the magnitude of the gain tends to a straight line corresponding to a gain of 0 dB (that is, a gain of 1). Therefore, this line (shown dotted in Figure 17.4) forms an **asymptote** to the response. At frequencies much less than the cut-off frequency, the response tends to a straight line drawn at a slope of 6 dB per octave (20 dB per decade) change in frequency. This line forms a second asymptote to the response and is also shown dotted on Figure 17.4. The two asymptotes intersect at the cut-off frequency. At frequencies considerably above or below the cut-off frequency, the gain response tends towards these two asymptotes. Near the cut-off frequency, the gain deviates from the two straight lines and is 3 dB below their intersection at the cut-off frequency.

Figure 17.4 Gain and phase responses (or Bode diagram) for the high-pass *RC* network

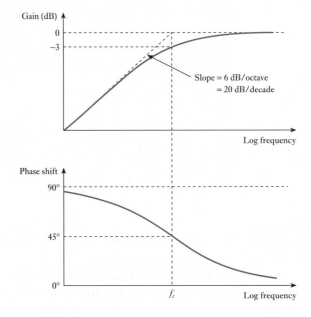

Figure 17.4 also shows the variation of phase with frequency of the *RC* network. At frequencies well above the cut-off frequency, the network produces very little phase shift and its effects may generally be ignored. However, as the frequency decreases the phase shift produced by the arrangement increases, reaching 45° at the cut-off frequency and increasing to 90° at very low frequencies.

Asymptotic diagrams of gain and phase of the form shown in Figure 17.4 are referred to as **Bode diagrams** (or sometimes **Bode plots**). These plot logarithmic gain (usually in dB) and phase against logarithmic frequency. Such diagrams are easy to plot and give a useful picture of the characteristic of the circuit. We will look at the Bode diagrams for a range of other circuits in this chapter and then consider how they may be easily drawn and used.

It can be seen that the *RC* network passes signals of some frequencies with little effect but that signals of other frequencies are attenuated and are subjected to a phase shift. Thus signals that have a frequency near or below the cut-off frequency (or which have components that are near or below the cut-off frequency) are subjected to **amplitude** and **phase distortion** by this arrangement. The network has the characteristics of a **high-pass filter**, since it allows high-frequency signals to pass but filters out low-frequency signals. We shall look at filters in more detail later in this chapter.

File 17A

Computer Simulation Exercise 17.1

Calculate the cut-off frequency of the circuit of Figure 17.2 if $R = 1\ \text{k}\Omega$ and $C = 1\ \mu\text{F}$. Simulate the circuit using these component values and perform an AC sweep to measure the response over a range from 1 Hz to 1 MHz. Plot the gain (in dB) and the phase of the output over this frequency range, estimate the cut-off frequency from these plots and compare this with the predicted value. Measure the phase shift at the estimated cut-off frequency and compare this with the value predicted above. Repeat this exercise for different values of *R* and *C*.

17.3   A low-pass *RC* network

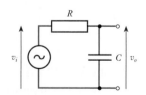

Figure 17.5   A low-pass *RC* network

The circuit of Figure 17.5 shows an *RC* arrangement similar to the earlier circuit but with the positions of the resistor and the capacitor reversed. Applying Equation 17.1 produces

$$\frac{v_o}{v_i} = \frac{\mathbf{Z}_C}{\mathbf{Z}_R + \mathbf{Z}_C} = \frac{-j\dfrac{1}{\omega C}}{R - j\dfrac{1}{\omega C}} = \frac{1}{1 + j\omega CR} \qquad (17.6)$$

Comparing this expression with that of Equation 17.2 shows that it has a very different frequency characteristic. At low frequencies, ω is small and the value of $j\omega CR$ is small compared with 1. Therefore, the denominator of the expression is close to unity and the voltage gain is approximately 1. At high frequencies, the magnitude of ωCR becomes more significant and the gain of the network decreases. We therefore have a **low-pass filter** arrangement.

A similar analysis to that in the last section will show that the magnitude of the voltage gain is now given by

$$| \text{voltage gain} | = \frac{1}{\sqrt{1 + (\omega CR)^2}}$$

When the value of ωCR is equal to 1, this gives

$$| \text{voltage gain} | = \frac{1}{\sqrt{1+1}} = \frac{1}{\sqrt{2}} = 0.707$$

and again this corresponds to a cut-off frequency. The angular frequency of the cut-off ω_c corresponds to the condition that $\omega CR = 1$, therefore

$$\omega_c = \frac{1}{CR} = \frac{1}{T} \text{ rad/s} \tag{17.7}$$

as before. Therefore the expression for the cut-off frequency is identical to that in the previous circuit.

Example 17.3 | Calculate the time constant T, the angular cut-off frequency ω_c and the cyclic cut-off frequency f_c of the following arrangement.

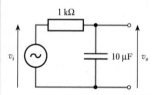

From above

$$T = CR = 10 \times 10^{-6} \times 1 \times 10^3 = 0.01 \text{ s}$$

$$\omega_c = \frac{1}{T} = \frac{1}{0.01} = 100 \text{ rad/s}$$

$$f_c = \frac{\omega_c}{2\pi} = \frac{100}{2\pi} = 15.9 \text{ Hz}$$

While the cut-off frequency of this circuit is identical to that of the previous arrangement, you should note that in the circuit of Figure 17.2 the cut-off attenuates low-frequency signals and is therefore a **low-frequency cut-off**. However, in the circuit of Figure 17.5 high frequencies are attenuated, so this circuit has a **high-frequency cut-off**.

Substituting into Equation 17.6 gives

$$\frac{v_o}{v_i} = \frac{1}{1 + j\omega CR} = \frac{1}{1 + j\dfrac{\omega}{\omega_c}} = \frac{1}{1 + j\dfrac{f}{f_c}}$$

(17.8)

You might like to compare this with the expression for a high-pass network in Equation 17.5. As before, we can investigate the behaviour of this arrangement in different frequency ranges.

17.3.1 When $f \ll f_c$

When the signal frequency f is much lower than the cut-off frequency f_c, then in Equation 17.8 f/f_c is much less than unity, and the voltage gain is approximately equal to 1. The imaginary part of the gain is negligible and the gain of the circuit is effectively real. This situation is shown in the phasor diagram of Figure 17.6(a).

Figure 17.6 Phasor diagrams of the gain of the low-pass network at different frequencies

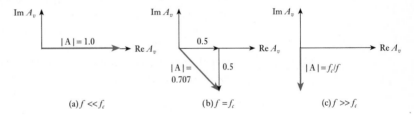

(a) $f \ll f_c$ (b) $f = f_c$ (c) $f \gg f_c$

17.3.2 When $f = f_c$

When the signal frequency f is equal to the cut-off frequency f_c, then Equation 17.8 becomes

$$\frac{v_o}{v_i} = \frac{1}{1 + j\dfrac{f}{f_c}} = \frac{1}{1 + j}$$

Multiplying the numerator and the denominator by $(1 - j)$ gives

$$\frac{v_o}{v_i} = \frac{(1 - j)}{(1 + j)(1 - j)} = \frac{(1 - j)}{2} = 0.5 - 0.5j$$

This is illustrated in the phasor diagram of Figure 17.6(b), which shows that the magnitude of the gain at the cut-off frequency is 0.707 and the phase angle of the gain is $-45°$. This shows that the output voltage *lags* the input voltage by 45°. The gain is therefore $0.707\angle-45°$.

17.3.3   When $f \gg f_c$

At high frequencies f/f_c is much greater than 1, and Equation 17.8 becomes

$$\frac{v_o}{v_i} = \frac{1}{1 + j\dfrac{f}{f_c}} \approx \frac{1}{j\dfrac{f}{f_c}} = -j\frac{f_c}{f}$$

The 'j' signifies that the gain is imaginary, and the minus sign indicates that the output lags the input. This is shown in the phasor diagram of Figure 17.6(c). The magnitude of the gain is simply f_c/f and, since f_c is a constant, the voltage gain is inversely proportional to frequency. If the frequency is halved, the voltage gain will be doubled. Therefore, the rate of change of gain can be expressed as −6 dB/octave or −20 dB/decade.

17.3.4   Frequency response of the low-pass *RC* network

Figure 17.7 shows the gain and phase response (or Bode diagram) of the low-pass network for frequencies above and below the cut-off frequency. The magnitude response is very similar in form to that of the high-pass network shown in Figure 17.4, with the frequency scale reversed. The phase response is a similar shape to that in Figure 17.4, but here the phase goes from 0° to −90° as the frequency is increased, rather than from +90° to 0° as in the previous arrangement. From the figure it is clear that this is a low-pass filter arrangement.

Figure 17.7   Gain and phase responses (or Bode diagram) for the low-pass *RC* network

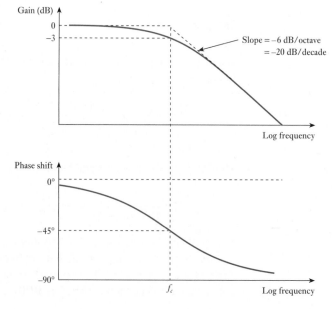

Computer Simulation Exercise 17.2

Calculate the cut-off frequency of the circuit of Figure 17.5 if $R = 1 \text{ k}\Omega$ and $C = 1 \text{ μF}$. Simulate the circuit using these component values and perform an AC sweep to measure the response over a range from 1 Hz to 1 MHz. Plot the gain (in dB) and the phase of the output over this frequency range, estimate the cut-off frequency from these plots and compare this with the predicted value. Measure the phase shift at the estimated cut-off frequency and compare this with the value predicted above. Repeat this exercise for different values of R and C.

17.4 A low-pass RL network

High-pass and low-pass arrangements may also be formed using combinations of resistors and inductors. Consider for example the circuit of Figure 17.8. This shows a circuit similar to that of Figure 17.2, but with the capacitor replaced by an inductor. If we apply a similar analysis to that used above, we obtain

$$\frac{v_o}{v_i} = \frac{\mathbf{Z}_R}{\mathbf{Z}_R + \mathbf{Z}_L} = \frac{R}{R + \mathrm{j}\omega L} = \frac{1}{1 + \mathrm{j}\omega \dfrac{L}{R}} \tag{17.9}$$

A similar analysis to that in the last section will show that the magnitude of the voltage gain is now given by

$$|\text{ voltage gain }| = \frac{1}{\sqrt{1 + \left(\omega \dfrac{L}{R}\right)^2}}$$

When the value of $\omega L/R$ is equal to 1, this gives

$$|\text{ voltage gain }| = \frac{1}{\sqrt{1 + 1}} = \frac{1}{\sqrt{2}} = 0.707$$

and this corresponds to a cut-off frequency. The angular frequency of the cut-off ω_c corresponds to the condition that $\omega L/R = 1$, therefore

Figure 17.8 A low-pass RL network

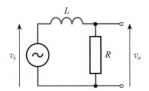

$$\omega_c = \frac{R}{L} = \frac{1}{T} \text{ rad/s} \tag{17.10}$$

As before, T is the time constant of the circuit, and in this case T is equal to L/R.

Example 17.4 Calculate the time constant T, the angular cut-off frequency ω_c and the cyclic cut-off frequency f_c of the following arrangement.

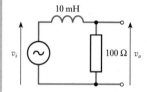

From above

$$T = \frac{L}{R} = \frac{10 \times 10^{-3}}{100} = 10^{-4} \text{ s}$$

$$\omega_c = \frac{1}{T} = \frac{1}{1 \times 10^{-4}} = 10^4 \text{ rad/s}$$

$$f_c = \frac{\omega_c}{2\pi} = \frac{1 \times 10^4}{2\pi} = 1.59 \text{ kHz}$$

Substituting into Equation 17.9, we have

$$\frac{v_o}{v_i} = \frac{1}{1 + j\omega\dfrac{L}{R}} = \frac{1}{1 + j\dfrac{\omega}{\omega_c}} = \frac{1}{1 + j\dfrac{f}{f_c}} \tag{17.11}$$

This expression is identical to that of Equation 17.8, and thus the frequency behaviour of this circuit is identical to that of the circuit of Figure 17.5.

File 17C

Computer Simulation Exercise 17.3

Calculate the cut-off frequency of the circuit of Figure 17.8 if $R = 10 \ \Omega$ and $L = 5$ mH. Simulate the circuit using these component values and perform an AC sweep to measure the response over a range from 1 Hz to 1 MHz. Plot the gain (in dB) and the phase of the output over this frequency range, estimate the cut-off frequency from these plots and compare this with the predicted value. Measure the phase shift at the estimated cut-off frequency and compare this with the value predicted above. Repeat this exercise for different values of R and L.

17.5 A high-pass *RL* network

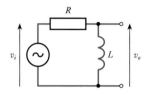

Figure 17.9 A high-pass *RL* network

Interchanging the components of Figure 17.8 gives the circuit of Figure 17.9. Analysing this as before, we obtain

$$\frac{v_o}{v_i} = \frac{\mathbf{Z_L}}{\mathbf{Z_R} + \mathbf{Z_L}} = \frac{j\omega L}{R + j\omega L} = \frac{1}{1 + \dfrac{R}{j\omega L}} = \frac{1}{1 - j\dfrac{R}{\omega L}} \qquad (17.12)$$

If we substitute $\omega_c = R/L$ as before, this gives

$$\frac{v_o}{v_i} = \frac{1}{1 - j\dfrac{R}{\omega L}} = \frac{1}{1 - j\dfrac{\omega_c}{\omega}} = \frac{1}{1 - j\dfrac{f_c}{f}} \qquad (17.13)$$

This expression is identical to that of Equation 17.5, and thus the frequency behaviour of this circuit is identical to that of the circuit of Figure 17.2.

File 17D

Computer Simulation Exercise 17.4

Calculate the cut-off frequency of the circuit of Figure 17.9 if $R = 10 \ \Omega$ and $L = 5$ mH. Simulate the circuit using these component values and perform an AC sweep to measure the response over a range from 1 Hz to 1 MHz. Plot the gain (in dB) and the phase of the output over this frequency range, estimate the cut-off frequency from these plots and compare this with the predicted value. Measure the phase shift at the estimated cut-off frequency and compare this with the value predicted above. Repeat this exercise for different values of R and L.

17.6 A comparison of *RC* and *RL* networks

From the above it is clear that *RC* and *RL* circuits have many similarities. The behaviour of the circuits we have considered is summarised in Figure 17.10. Each of the circuits has a cut-off frequency, and in each case this frequency is determined by the time constant T of the circuit. In the *RC* circuits $T = CR$, and in the *RL* circuits $T = L/R$. In each case the angular cut-off frequency is then given by $\omega_c = 1/T$ and the cyclic cut-off frequency by $f_c = \omega_c/2\pi$.

Two of the circuits of Figure 17.10 have upper cut-off frequencies (low-pass circuits), and two have lower cut-off frequencies (high-pass circuits). Transposing the components in a particular circuit will change it from a high-pass to a low-pass circuit, and vice versa. Replacing a capacitor by an inductor, or replacing an inductor by a capacitor, will also change it from a high-pass to a low-pass circuit, and vice versa.

Figure 17.10 A comparison of *RC* and *RL* networks

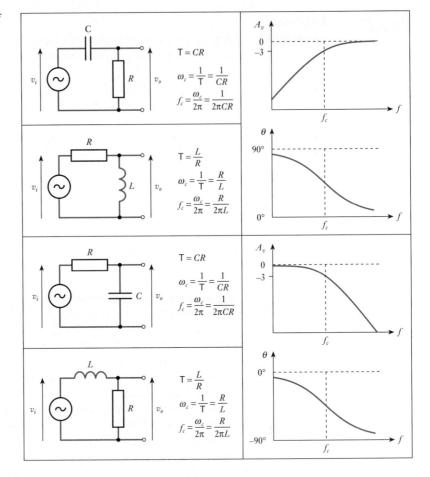

![File 17E icon]

File 17E

Computer Simulation Exercise 17.5

Calculate the time constants of the circuits of Figure 17.10 if $R = 1$ kΩ, $C = 1$ nF and $L = 1$ mH. Simulate the first of these circuits (using these component values) and use an AC sweep to plot the gain and phase responses of the circuit (as in the earlier simulation exercises in this chapter). Make a note of the cut-off frequency and confirm that this is a lower cut-off. Now interchange the capacitor and resistor and again plot the circuit's characteristics. Note the effect on the cut-off frequency and the nature of the cut-off (that is, whether it is now an upper or a lower cut-off).

Replace the capacitor by an inductor of 1 mH and again note the effect on the cut-off frequency and the nature of the cut-off. Finally, interchange the inductor and resistor and repeat the analysis. Hence confirm the form of the characteristics given in Figure 17.10.

17.7 Bode diagrams

Earlier we looked at Bode diagrams (also called Bode plots) as a means of describing the gain and phase response of a circuit (as in Figures 17.4 and 17.7). In the circuits we have considered, the gain at high and low frequencies has an asymptotic form, greatly simplifying the drawing of the diagram. The phase response is also straightforward, changing progressively between defined limits.

It is often sufficient to use a 'straight-line approximation' to the Bode diagram, simplifying its construction. For the circuits shown in Figure 17.10, we can construct the gain section of these diagrams simply by drawing the two asymptotes. One of these will be horizontal, representing the frequency range in which the gain is approximately constant. The other has a slope of +6 dB/octave (+20 dB/decade) or −6 dB/octave (−20 dB/decade), depending on whether this is a high-pass or low-pass circuit. These two asymptotes cross at the cut-off frequency of the circuit. The phase section of the response is often adequately represented by a straight-line transition between the two limiting values. The position of this line is defined by the phase shift at the cut-off frequency, which in these examples is 45°. A reasonable approximation to the response can be gained by drawing a straight line with a slope of −45°/decade through this point. Using this approach, the line starts one decade below the cut-off frequency and ends one decade above, making it very easy to construct. Straight-line approximations to the Bode diagrams for the circuits shown in Figure 17.10 are shown in Figure 17.11.

Once the straight-line Bode plots have been constructed, it is simple to convert these to a more accurate curved-line form if required. This can usually be done by eye by noting that the gain at the cut-off frequency is −3 dB, and that the phase response is slightly steeper at the cut-off frequency and slightly less steep near each end than the straight-line approximation. This is illustrated in Figure 17.12.

Figure 17.11 Simple straight-line Bode diagrams

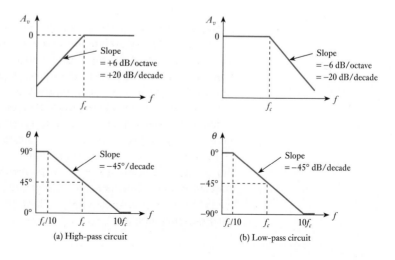

(a) High-pass circuit

(b) Low-pass circuit

Figure 17.12 Drawing Bode diagrams from their straight-line approximations

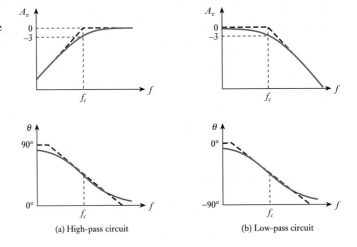

(a) High-pass circuit   (b) Low-pass circuit

| 17.8 | Combining the effects of several stages |

While simple circuits may produce a single cut-off frequency, more complex circuits often possess a number of elements that each have some form of frequency dependence. Thus a circuit might have both high-pass and low-pass characteristics, or might have several high- or low-pass elements.

One of the advantages of the use of Bode diagrams is that they make it very easy to see the effects of combining several different elements. We noted in Section 6.6 that, when several stages of amplification are connected in series, the overall gain is equal to the product of their individual gains, or the *sum* of their gains when these are expressed in decibels. Similarly, the phase shift produced by several amplifiers in series is equal to the sum of the phase shift produced by each amplifier separately. Therefore, the combined effects of a series of stages can be predicted by 'adding' the Bode diagrams of each stage. This is illustrated in Figure 17.13, which shows the effects of combining a high-pass and a low-pass element. In this case, the cut-off frequency of the high-pass element is lower than that of the low-pass element, resulting in a **band-pass filter** characteristic as shown in Figure 17.13(c). Such a circuit passes a given range of frequencies while rejecting lower- and high-frequency components.

Bode diagrams can also be used to investigate the effects of combining more than one high-pass or low-pass element. This is illustrated in Figure 17.14, which shows the effects of combining two elements that each contain a single high- and a single low-pass element. In this case, the cut-off frequencies of each element are different, resulting in four transitions in the characteristic. For obvious reasons, these frequencies are known as **break** or **corner frequencies**.

Figure 17.13 The combined effects of high- and low-pass elements

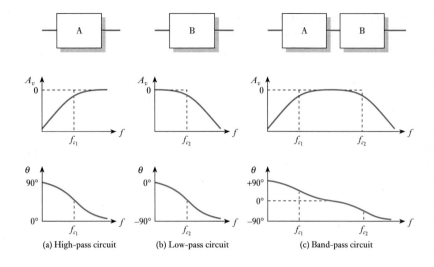

(a) High-pass circuit   (b) Low-pass circuit   (c) Band-pass circuit

In Figure 17.14, the first element is a band-pass amplifier that has a gain of A dB within its pass band, a lower cut-off frequency of f_1 and an upper cut-off frequency of f_3. The second element is also a band-pass amplifier, this time with a gain of B dB, a lower cut-off frequency of f_2 and an upper cut-off frequency of f_4. Within the frequency range from f_2 to f_3, the gains of both amplifiers are approximately constant, so the gain of the combination is also approximately constant, with a value of $(A + B)$ dB. In the range f_3 to f_4, the gain of the second amplifier is approximately constant, but the gain of the first falls at a rate of 6 dB/octave. Therefore, in this range the gain of the combination also falls at 6 dB/octave. At frequencies above f_4, the gain of both amplifiers is falling at a rate of 6 dB/octave, so the gain of the combination falls at 12 dB/octave. A similar combination of effects causes the gain to fall at first by 6 dB/octave, and then by 12 dB/octave as the frequency decreases below f_2. The result is a band-pass filter with a gain of $(A + B)$.

Within the pass band both amplifiers produce relatively little phase shift. However, as we move to frequencies above f_3 the first amplifier produces a phase shift that increases to $-90°$, and as we move above f_4 the second amplifier produces an additional shift, taking the total phase shift to $-180°$. This effect is mirrored at low frequencies, with the two amplifiers producing a total phase shift of $+180°$ at very low frequencies.

While the arrangement represented in Figure 17.14 includes a total of two lower and two upper cut-off frequencies, clearly more complex arrangements can include any number of cut-offs. As more are added, each introduces an additional 6 dB/octave to the maximum rate of increase or decrease of gain with frequency, and also increases the phase shift introduced at high or low frequencies by 90°.

Figure 17.14 Combinations of multiple high- and low-pass elements

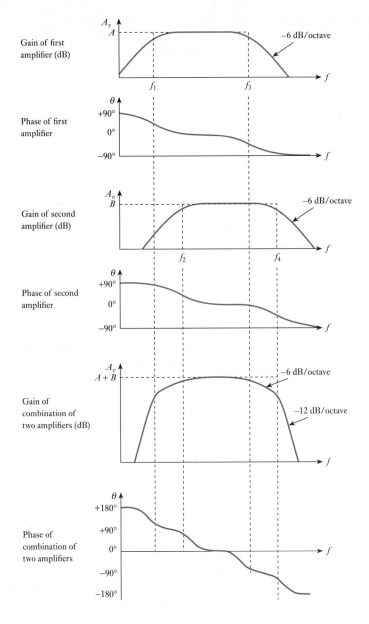

17.9 RLC circuits and resonance

17.9.1 Series RLC circuit

Having looked at *RC* and *RL* circuits, it is now time to look at circuits containing resistance, inductance and capacitance. Consider for example the series arrangement of Figure 17.15. This can be analysed in a similar manner to the circuits discussed above, by considering it as a potential divider.

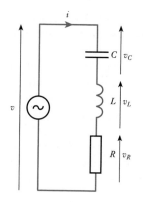

Figure 17.15 A series *RLC* arrangement

The voltages across each component can be found by dividing its complex impedance by the total impedance of the circuit and multiplying this by the applied voltage. For example, the voltage across the resistor is given by

$$v_R = v \times \frac{\mathbf{Z}_R}{\mathbf{Z}_R + \mathbf{Z}_L + \mathbf{Z}_C} = v \times \frac{R}{R + j\omega L + \dfrac{1}{j\omega C}} \qquad (17.14)$$

It is also interesting to consider the impedance of this arrangement, which is given by

$$\mathbf{Z} = R + j\omega L + \frac{1}{j\omega C} = R + j\left(\omega L - \frac{1}{\omega C}\right) \qquad (17.15)$$

It can be seen that, if the magnitude of the reactance of the inductor and the capacitor are equal (that is, if $\omega L = 1/\omega C$), the imaginary part of the impedance is zero. Under these circumstances, the impedance of the arrangement is simply equal to R. This condition occurs when

$$\omega L = \frac{1}{\omega C} \qquad \omega^2 = \frac{1}{LC} \qquad \omega = \frac{1}{\sqrt{LC}}$$

This situation is referred to as **resonance**, and the frequency at which it occurs is called the **resonant frequency** of the circuit. An arrangement that exhibits such behaviour is known as a **resonant circuit**. The angular frequency at which resonance occurs is given the symbol ω_0, and the corresponding cyclic frequency is given the symbol f_0. Therefore

$$\omega_0 = \frac{1}{\sqrt{LC}} \qquad (17.16)$$

$$f_0 = \frac{1}{2\pi\sqrt{LC}} \qquad (17.17)$$

From Equation 17.15 it is clear that in the circuit of Figure 17.15 the impedance is at a *minimum* at resonance, and therefore the current will be at a *maximum* under these conditions. Figure 17.16 shows the current in the circuit as the frequency varies above and below resonance. Since the current is at a maximum at resonance, it follows that the voltages across the capacitor and the inductor are also large. Indeed, at resonance the voltages

Figure 17.16 Variation of current with frequency for a series *RLC* arrangement

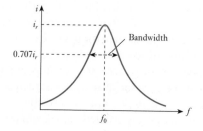

across these two components can be many times greater than the applied voltage. However, these two voltages are out of phase with each other and therefore cancel out, leaving only the voltage across the resistor.

We noted in Chapter 16 that power is not dissipated in capacitors or inductors but that these components simply store energy before returning it to the circuit. Therefore, the current flowing into and out of the inductor and capacitor at resonance results in energy being repeatedly stored and returned. This allows the resonant effect to be quantified by measuring the ratio of the energy stored to the energy dissipated during each cycle. This ratio is termed the **quality factor** or Q of the circuit. Since the energy stored in the inductor and the capacitor are equal, we can choose either of them to calculate Q. If we choose the inductor, we have

$$\text{quality factor } Q = \frac{I^2 X_L}{I^2 R} = \frac{X_L}{R} \tag{17.18}$$

and if we choose the capacitor, we have

$$\text{quality factor } Q = \frac{I^2 X_C}{I^2 R} = \frac{X_C}{R} \tag{17.19}$$

If we take either of these expressions and multiply top and bottom by I, we get the corresponding voltages across the associated component. Therefore, Q may also be defined as

$$\text{quality factor } Q = \frac{V_L}{V_R} = \frac{V_C}{V_R} \tag{17.20}$$

Since at resonance V_R is equal to the supply voltage, it follows that

$$\text{quality factor } Q = \frac{\text{voltage across } L \text{ or } C \text{ at resonance}}{\text{supply voltage}} \tag{17.21}$$

and thus Q represents the voltage magnification at resonance.

Combining Equations 17.16 and 17.21 gives us an expression for the Q of a series RLC circuit, which is

$$Q = \frac{1}{R}\sqrt{\left(\frac{L}{C}\right)} \tag{17.22}$$

The series RLC circuit is often referred to as an **acceptor circuit**, since it passes signals at frequencies close to its resonant frequency but rejects signals at other frequencies. From our earlier discussion of bandwidth, we can define the bandwidth B of a resonant circuit as the frequency range between the points where the gain (or in this case the current) falls to $1/\sqrt{2}$ (or 0.707) times its mid-band value. This is illustrated in Figure 17.16. An example of an application of an acceptor circuit is in a radio, where we wish to accept the frequencies associated with a particular station while rejecting others. In such situations, we need a resonant circuit with an appropriate bandwidth to accept the wanted signal while rejecting unwanted signals and interference. The 'narrowness' of the bandwidth is determined by the Q of the circuit, and it can be shown that the resonant frequency and the bandwidth are related by the expression

$$\text{quality factor } Q = \frac{\text{resonant frequency}}{\text{bandwidth}} = \frac{f_0}{B} \qquad (17.23)$$

Combining Equations 17.17, 17.22 and 17.23, we can obtain an expression for the bandwidth of the circuit in terms of its component values. This is

$$B = \frac{R}{2\pi L} \text{ Hz} \qquad (17.24)$$

It can be seen that reducing the value of R increases the Q of the circuit and reduces its bandwidth. In some situations it is desirable to have very high values of Q, and Equation 17.22 would suggest that if the resistor were omitted (effectively making $R = 0$) this would produce a resonant circuit with infinite Q. However, in practice all real components exhibit resistance (and inductors are particularly 'non-ideal' in this context), so the Q of such circuits is limited to a few hundred.

Example 17.5 | For the following arrangement, calculate the resonant frequency f_0, the impedance of the circuit at this frequency, the quality factor Q of the circuit and its bandwidth B.

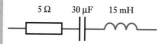

From Equation 17.17

$$f_0 = \frac{1}{2\pi\sqrt{LC}}$$

$$= \frac{1}{2\pi\sqrt{15 \times 10^{-3} \times 30 \times 10^{-6}}}$$

$$= 237 \text{ Hz}$$

At the resonant frequency the impedance is equal to R, so $\mathbf{Z} = 5 \ \Omega$.
From Equation 17.22

$$Q = \frac{1}{R}\sqrt{\left(\frac{L}{C}\right)} = \frac{1}{5}\sqrt{\left(\frac{15 \times 10^{-3}}{30 \times 10^{-6}}\right)} = 4.47$$

and from Equation 17.24

$$B = \frac{R}{2\pi L} = \frac{5}{2\pi \times 15 \times 10^{-3}} = 53 \text{ Hz}$$

File 17F

Computer Simulation Exercise 17.6

Simulate a circuit that applies a sinusoidal voltage to the arrangement of Example 17.5 and use an AC sweep to plot the variation of current with frequency. Measure the resonant frequency of the arrangement and its bandwidth, and hence calculate its Q. Measure the peak current in the circuit and, from a knowledge of the excitation voltage used, estimate the impedance of the circuit at resonance. Hence confirm the findings of Example 17.5 above.

17.9.2   Parallel *RLC* circuit

Consider now the parallel circuit of Figure 17.17. The impedance of this circuit is given by

$$Z = \cfrac{1}{\cfrac{1}{R} + j\omega C + \cfrac{1}{j\omega L}} = \cfrac{1}{\cfrac{1}{R} + j\left(\omega C - \cfrac{1}{\omega L}\right)} \qquad (17.25)$$

and it is clear that this circuit also has a resonant characteristic. When $\omega C = 1/\omega L$, the term within the brackets is equal to zero and the imaginary part of the impedance disappears. Under these circumstances, the impedance is purely resistive, and $Z = R$. The frequency at which this occurs is the resonant frequency f_0, which is given by

$$\omega L = \frac{1}{\omega C}$$

$$\omega^2 = \frac{1}{LC}$$

$$\omega = \frac{1}{\sqrt{LC}}$$

which is the same as for the series circuit. Therefore, as before, the resonant angular and cyclic frequencies are given by

$$\omega_0 = \frac{1}{\sqrt{LC}} \qquad (17.26)$$

Figure 17.17   A parallel *RLC* arrangement

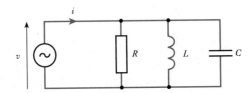

Figure 17.18 Variation of current with frequency for a parallel *RLC* arrangement

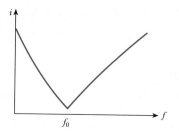

$$f_0 = \frac{1}{2\pi\sqrt{LC}} \tag{17.27}$$

From Equation 17.25, it is clear that the impedance of the parallel resonant circuit is a *maximum* at resonance and that it decreases at higher and lower frequencies. This arrangement is therefore a **rejector** circuit, and Figure 17.18 shows how the current varies with frequency.

As for the series resonant circuit, we can define both the bandwidth *B* and the quality factor *Q* for the parallel arrangement (although the definitions of these terms are a little different). The corresponding expressions for these quantities are

$$Q = R\sqrt{\left(\frac{C}{L}\right)} \tag{17.28}$$

and

$$B = \frac{1}{2\pi RC} \, \text{Hz} \tag{17.29}$$

A comparison between series and parallel resonant circuits is shown in Table 17.1. It should be noted that in a series resonant circuit *Q* is increased by *reducing* the value of *R*, while in a parallel resonant circuit *Q* is increased by *increasing* the value of *R*. In each case, *Q* is increased when the losses are reduced.

While the circuit of Figure 17.17 represents a generalised parallel *RLC* circuit, it is not the most common form. In practice, the objective is normally to maximise the *Q* of the arrangement, and this is achieved by removing the resistive element. However, in practice all inductors have appreciable resistance, so it is common to model this in the circuit as shown in Figure 17.19. Capacitors also exhibit resistance, but this is generally quite small and can often be ignored.

The resonant frequency of the circuit of Figure 17.19 is given by

$$f_0 = \frac{1}{2\pi}\sqrt{\frac{1}{LC} - \frac{R^2}{L^2}} \tag{17.30}$$

As the resistance of the coil tends to zero, this expression becomes equal to that of Equation 17.27. This circuit has similar characteristics to the earlier parallel arrangement and has a *Q* given by

$$Q = \sqrt{\frac{L}{R^2C} - 1} \qquad\qquad (17.31)$$

Table 17.1   Series and parallel resonant circuits

	Series resonant circuit	Parallel resonant circuit
Circuit	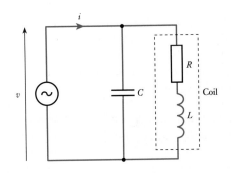	
Impedance, **Z**	$\mathbf{Z} = R + j\left(\omega L - \dfrac{1}{\omega C}\right)$	$\mathbf{Z} = \dfrac{1}{\dfrac{1}{R} + j\left(\omega C - \dfrac{1}{\omega L}\right)}$
Resonant frequency, f_0	$f_0 = \dfrac{1}{2\pi\sqrt{LC}}$	$f_0 = \dfrac{1}{2\pi\sqrt{LC}}$
Quality factor, Q	$Q = \dfrac{1}{R}\sqrt{\left(\dfrac{L}{C}\right)}$	$Q = R\sqrt{\left(\dfrac{C}{L}\right)}$
Bandwidth, B	$B = \dfrac{R}{2\pi L}$ Hz	$B = \dfrac{1}{2\pi RC}$ Hz

Figure 17.19   An *LC* resonant circuit

17.10   Filters

17.10.1   *RC* filters

Earlier in this chapter we looked at *RC* high-pass and low-pass networks and noted that these have the characteristics of filters since they pass signals of certain frequencies while attenuating others. These simple circuits, which contain only a single time constant, are called **first-order** or **single-pole filters**. Circuits of this type are often used in systems to select or remove components of a signal. However, for many applications the relatively slow roll-off of the gain (6 dB/octave) is inadequate to remove unwanted signals effectively. In such cases, filters with more than one time constant are used to provide a more rapid roll-off of gain. Combining two high-pass time constants produces a second order (two-pole) high-pass filter in which the gain will roll off at 12 dB/octave (as seen in Section 17.8). Similarly, the addition of three or four stages can produce a roll-off rate of 18 or 24 dB/octave.

In principle, any number of stages can be combined in this way to produce an nth order (n-pole) filter. This will have a cut-off slope of $6n$ dB/octave and produce up to $n \times 90°$ of phase shift. It is also possible to combine high-pass and low-pass characteristics into a single band-pass filter if required.

For many applications, an **ideal filter** would have a constant gain and zero phase shift within one range of frequencies (its **pass band**) and zero gain outside this range (its **stop band**). The transition from the pass band to the stop band occurs at the **corner frequency** f_0. This is illustrated for a low-pass filter in Figure 17.20(a).

Unfortunately, although adding more stages to the RC filter increases the *ultimate* rate of fall of gain within the stop band, the sharpness of the 'knee' of the response is not improved (see Figure 17.20(b)). To produce a circuit that more closely approximates an ideal filter, different techniques are required.

Figure 17.20 Gain responses of ideal and real low-pass filters

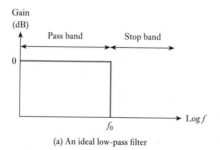

(a) An ideal low-pass filter

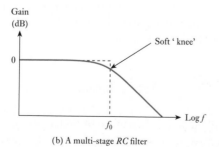

(b) A multi-stage RC filter

Figure 17.21   *LC* filters

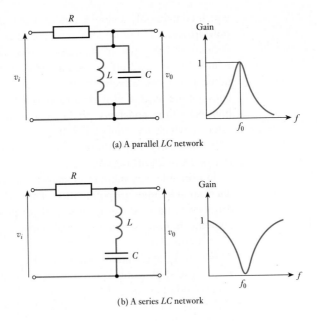

(a) A parallel *LC* network

(b) A series *LC* network

17.10.2   *LC* filters

The combination of inductors and capacitors allows the production of filters with a very sharp cut-off. Simple *LC* filters can be produced using the series and parallel resonant circuits discussed in the last section. These are also known as **tuned circuits** and are illustrated in Figure 17.21.

These combinations of inductors and capacitors produce narrow-band filters with centre frequencies corresponding to the resonant frequency of the tuned circuit, so

$$f_0 = \frac{1}{2\pi\sqrt{LC}} \qquad\qquad (17.32)$$

The bandwidth of the filters is determined by the **quality factor** Q as discussed in the last section.

Other configurations of inductors, capacitors and resistors can be used to form high-pass, low-pass, band-pass and band-stop filters and can achieve very high cut-off rates.

17.10.3   Active filters

Although combinations of inductors and capacitors can produce very high-performance filters, the use of inductors is inconvenient since they are expensive, bulky and suffer from greater losses than other passive

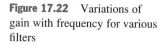

Figure 17.22 Variations of gain with frequency for various filters

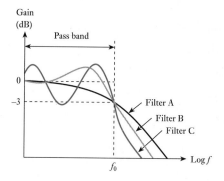

components. Fortunately, a range of very effective filters can be constructed using an operational amplifier and suitable arrangements of resistors and capacitors. Such filters are called **active filters**, since they include an active component (the operational amplifier) in contrast to the other filters we have discussed, which are purely passive (ignoring any buffering). A detailed study of the operation and analysis of active filters is beyond the scope of this text, but it is worth looking at the characteristics of these circuits and comparing them with those of the *RC* filters discussed earlier.

To construct multiple-pole filters, it is often necessary to cascade many stages. If the time constants and the gains of each stage are varied in a defined manner, it is possible to create filters with a wide range of characteristics. Using these techniques, it is possible to construct filters of a number of different types to suit particular applications.

In simple *RC* filters, the gain starts to fall towards the edge of the pass band and so is not constant throughout the band. This is also true of active filters, but here the gain may actually rise towards the edge of the pass band before it begins to fall. In some circuits the gain fluctuates by small amounts right across the band. These characteristics are illustrated in Figure 17.22.

The ultimate rate of fall of gain with frequency for any form of active filter is $6n$ dB/octave, where n is the number of poles in the filter, which is often equal to the number of capacitors in the circuit. Thus the performance of the filter in this respect is related directly to circuit complexity.

Although the ultimate rate of fall of gain of a filter is defined by the number of poles, the sharpness of the 'knee' of the filter varies from one design to another. Filters with a very sharp knee tend to produce more variation in the gain of the filter within the pass band. This is illustrated in Figure 17.22, where it is apparent that filters B and C have a more rapid roll-off of gain than filter A but also have greater variation in their gain within the pass band.

Of great importance in some applications is the **phase response** of the filter: that is, the variation of phase lag or lead with frequency as a signal passes through the filter. We have seen that *RC* filters produce considerable amounts of phase shift within the pass band. All filters produce a phase

Figure 17.23 A comparison of Butterworth, Chebyshev and Bessel filters

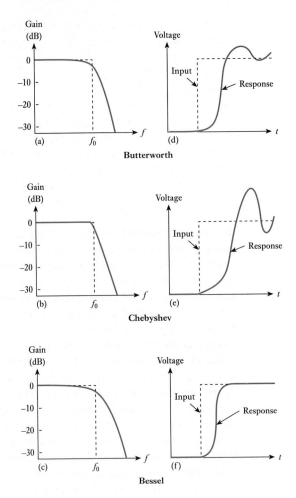

shift that varies with frequency. The way in which it is related to frequency varies from one type of filter to another. The phase response of a filter is of particular importance where pulses are to be used.

A wide range of filter designs are available, enabling one to be selected to favour any of the above characteristics. Unfortunately, the requirements of each are often mutually exclusive, so there is no universal optimum design, and an appropriate circuit must be chosen for a given application. From the myriad of filter designs, three basic types are discussed here, first because they are widely used and second because they are each optimised for a particular characteristic.

The **Butterworth filter** is optimised to produce a flat response within its pass band, which it does at the expense of a less sharp 'knee' and a less than ideal phase performance. This filter is sometimes called a **maximally flat filter** as it produces the flattest response of any filter type.

The **Chebyshev filter** produces a sharp transition from the pass band to the stop band but does this by allowing variations in gain throughout the pass band. The gain ripples within specified limits, which can be selected according to the application. The phase response of the Chebyshev filter is poor, and it creates serious distortion of pulse waveforms.

The **Bessel filter** is optimised for a linear phase response and is sometimes called a **linear phase filter**. The 'knee' is much less sharp than for the Chebyshev or the Butterworth types (though slightly better than a simple *RC* filter), but its superior phase characteristics make it preferable in many applications, particularly where pulse waveforms are being used. The phase shift produced by the filter is approximately linearly related to the input frequency. The resultant phase shift therefore has the appearance of a fixed time delay, with all frequencies being delayed by the same time interval. The result is that complex waveforms that consist of many frequency components (such as pulse waveforms) are filtered without distorting the phase relationships between the various components of the signal. Each component is simply delayed by an equal time interval.

Figure 17.23 compares the characteristics of these three types of filter. Parts (a), (b) and (c) show the frequency responses for Butterworth, Chebyshev and Bessel filters, each with six poles (the Chebyshev is designed for 0.5 dB ripple), while (d), (e) and (f) show the responses of the same filters to a step input.

Over the years a number of designs have emerged to implement various forms of filter. The designs have different characteristics, and each has advantages and disadvantages. Here we will look briefly at a single family of circuits that can be used to implement a range of filter types.

Figure 17.24 shows four filters, each constructed around a non-inverting amplifier. The circuits shown are **two-pole filters**, but several such stages may be cascaded to form higher-order filters. Circuits of this type are referred to as **Sallen–Key filters**. By appropriate choice of the component values, these circuits can be designed to produce the characteristics of various forms of filter, such as Bessel, Butterworth or Chebyshev. In general, the cascaded stages will not be identical but are designed such that the combination has the required characteristics. The component values shown produce Butterworth filters with $f_0 = 1/2\pi RC$ in each case. Other combinations of components will produce other types of filter, and the cut-off frequency may be slightly above or slightly below this value. The resistors R_1 and R_2 define the overall gain of each circuit, as in the non-inverting amplifier circuit described in Chapter 8. The gain in turn determines the Q of the circuit.

Files 17H
17J
17K

Computer Simulation Exercise 17.7

Simulate the filters of Figure 17.24 using $R = 16\ \text{k}\Omega$, $C = 10\ \text{nF}$, $R_1 = 5.9\ \text{k}\Omega$ and $R_2 = 10\ \text{k}\Omega$. Plot the frequency response of each arrangement and note the general shape of the response and its cut-off frequency (in the case of the high-pass and low-pass filters) or centre frequency (in the case of the band-pass and band-stop filters).

Figure 17.24 Operational
amplifier filter circuits

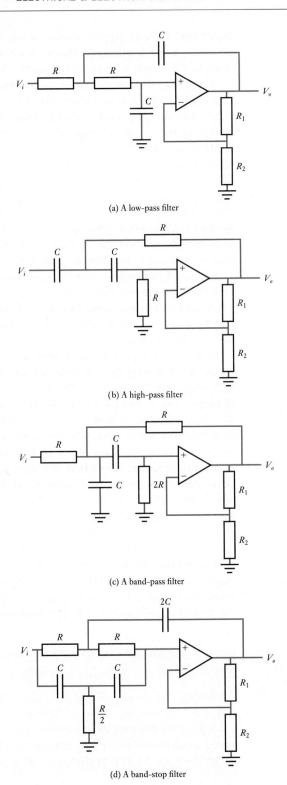

(a) A low-pass filter

(b) A high-pass filter

(c) A band-pass filter

(d) A band-stop filter

While active filters have several advantages over other forms of filter, it should be noted that they rely on the operational amplifier having sufficient gain at the frequencies being used. Active filters are widely used with audio signals (which are limited to a few tens of kilohertz) but are seldom used at very high frequencies. In contrast, *LC* filters can be used very successfully at frequencies up to several hundred megahertz. At very high frequencies, a range of other filter elements are available including SAW, ceramic and transmission line filters.

17.11 Stray capacitance and inductance

While many circuits will include a number of capacitors and inductors that have been intentionally introduced by the circuit designer, *all* circuits also include additional 'unintended' stray capacitances and stray inductances (as discussed in Chapters 13 and 14). Stray capacitance tends to introduce unintended low-pass filters in circuits, as illustrated in Figure 17.25(a). It also produces unwanted coupling of signals between circuits, resulting in a number of undesirable effects such as cross-talk. Stray inductance can also produce undesirable effects. For example, in Figure 17.25(b) a stray inductance L_s appears in series with a load resistor, producing an un-intended low-pass effect. Stray effects also have a dramatic effect on the stability of circuits. This is illustrated in Figure 17.25(c), where stray capacitance C_s across an inductor L results in an unintended resonant cir-cuit. We will return to look at stability in more detail in Chapter 24.

Stray capacitances and inductances are generally relatively small and therefore tend to be insignificant at low frequencies. However, at high frequencies they can have dramatic effects on the operation of circuits. In general, it is the presence of these unwanted circuit elements that limits the high-frequency performance of circuits.

Figure 17.25   The effects of stray capacitance and inductance

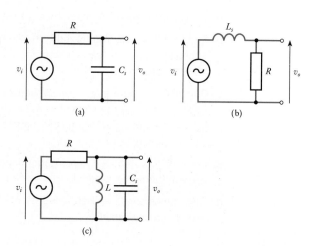

Key points

- The reactance of capacitors and inductors is dependent on frequency. Therefore, the behaviour of any circuit that contains these components will change with frequency.

- Since all real circuits include stray capacitance and stray inductance, all real circuits have characteristics that change with frequency.

- Combinations of a single resistor and a single capacitor, or a single resistor and a single inductor, can produce circuits with a single upper or lower cut-off frequency. In each case, the angular cut-off frequency ω_c is given by the reciprocal of the time constant T of the circuit.

- For an *RC* circuit T = CR, while in an *RL* circuit T = L/R.

- These single time constant circuits have certain similar characteristics:
  - their cut-off frequency $f_c = \omega_c/2\pi = 1/2\pi T$;
  - at frequencies well away from their cut-off frequency within their pass band, they have a gain of 0 dB and zero phase shift;
  - at their cut-off frequency, they have a gain of −3 dB and ±45° phase shift;
  - at frequencies well away from their cut-off frequency within their stop band, their gain changes by ±6 dB/octave (±20 dB/decade) and they have a phase shift of ±90°.

- Gain and phase responses are often given in the form of a Bode diagram, which plots gain (in dB) and phase against log frequency.

- When several stages are used in series, the gain of the combination at a given frequency is found by multiplying their individual gains, while the phase shift is found by adding their individual phase shifts.

- Combinations of resistors, inductors and capacitors can be analysed using the tools covered in earlier chapters. Of particular interest is the condition of resonance, when the reactance of the capacitive and inductive elements cancels. Under these conditions, the impedance of the circuit is simply resistive.

- The 'sharpness' of the resonance is measured by the quality factor *Q*.

- Simple *RC* and *RL* circuits represent first-order, or single-pole, filters. Although these are useful in certain applications, they have a limited 'roll-off' rate and a soft 'knee'.

- Combining several stages of *RC* filters increases the roll-off rate but does not improve the sharpness of the knee. Higher performance can be achieved using *LC* filters, but inductors are large, heavy and have high losses.

- Active filters produce high performance without using inductors. Several forms are available to suit a range of applications.

- Stray capacitance and stray inductance limit the performance of all high-frequency circuits.

Exercises

17.1 Calculate the reactance of a 1 μF capacitor at a frequency of 10 kHz, and the reactance of a 20 mH inductor at a frequency of 100 rad/s. In each case include the units in your answer.

17.2 Express an angular frequency of 250 rad/s as a cyclic frequency (in Hz).

17.3 Express a cyclic frequency of 250 Hz as an angular frequency (in rad/s).

17.4 Determine the transfer function of the following circuit.

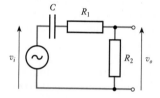

17.5 A series RC circuit is formed from a resistor of 33 kΩ and a capacitor or 15 nF. What is the time constant of this circuit?

17.6 Calculate the time constant T, the angular cut-off frequency ω_c and the cyclic cut-off frequency f_c of the following arrangement. Is this an upper- or a lower-frequency cut-off?

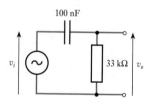

17.7 Simulate the arrangement of Exercise 17.6 and use an AC sweep to display the gain response. Measure the cut-off frequency of the circuit and hence confirm your results for the previous exercise.

17.8 Determine the frequencies that correspond to:

(a) an octave below 30 Hz;
(b) two octaves above 25 kHz;

(c) three octaves above 1 kHz;
(d) a decade above 1 MHz;
(e) two decades below 300 Hz;
(f) three decades above 50 Hz.

17.9 Calculate the time constant T, the angular cut-off frequency ω_c and the cyclic cut-off frequency f_c of the following arrangement. Is this an upper- or a lower-frequency cut-off?

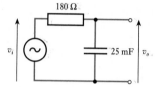

17.10 Simulate the arrangement of Exercise 17.9 and use an AC sweep to display the gain response. Measure the cut-off frequency of the circuit and hence confirm your results for the previous exercise.

17.11 A parallel RL circuit is formed from a resistor of 150 Ω and an inductor of 30 mH. What is the time constant of this circuit?

17.12 Calculate the time constant T, the angular cut-off frequency ω_c and the cyclic cut-off frequency f_c of the following arrangement. Is this an upper- or a lower-frequency cut-off?

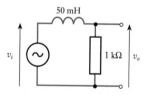

17.13 Simulate the arrangement of Exercise 17.12 and use an AC sweep to display the gain response. Measure the cut-off frequency of the circuit and hence confirm your results for the previous exercise.

17.14 Calculate the time constant T, the angular cut-off frequency ω_c and the cyclic cut-off frequency f_c

Exercises continued

of the following arrangement. Is this an upper- or a lower-frequency cut-off?

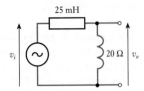

17.15 Simulate the arrangement of Exercise 17.14 and use an AC sweep to display the gain response. Measure the cut-off frequency of the circuit and hence confirm your results for the previous exercise.

17.16 Sketch a straight-line approximation to the Bode diagram of the circuit of Exercise 17.14. Use this approximation to produce a more realistic plot of the gain and phase responses of the circuit.

17.17 A circuit contains three high-frequency cut-offs and two low-frequency cut-offs. What are the rates of change of gain of this circuit at very high and very low frequencies?

17.18 Explain what is meant by the term 'resonance'.

17.19 Calculate the resonant frequency f_0, the quality factor Q and the bandwidth B of the following circuit.

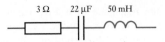

17.20 Simulate a circuit that applies a sinusoidal voltage to the arrangement of Exercise 17.19 and use an AC sweep to plot the variation of current with frequency. Measure the resonant frequency of the arrangement and its bandwidth, and hence calculate its Q. Hence confirm your results for the previous exercise.

17.21 Explain the difference between a passive and an active filter, giving examples of each.

17.22 Why are inductors often avoided in the construction of filters?

17.23 What form of active filter is optimised to produce a flat response within its pass band?

17.24 What form of active filter is optimised to produce a sharp transition from the pass band to the stop band?

17.25 What form of filter is optimised for a linear phase response?

17.26 Explain why stray capacitance and stray inductance affect the frequency response of electronic circuits.

Chapter 18

Transient Behaviour

Objectives

When you have studied the material in this chapter you should be able to:

- **explain concepts such as steady-state response, transient response and total response as they apply to electronic circuits;**
- **describe the transient behaviour of simple *RC* and *RL* circuits;**
- **predict the transient response of generalised first-order systems from a knowledge of its initial and final values;**
- **sketch increasing or decreasing exponential waveforms and identify key characteristics;**
- **describe the output of simple *RC* and *RL* circuits in response to a square-wave input;**
- **outline the transient behaviour of various forms of second-order system.**

18.1 Introduction

In earlier chapters, we looked at the behaviour of circuits in response to either fixed DC signals or constant AC signals. Such behaviour is often referred to as the **steady-state response** of the system. Now it is time to turn our attention to the performance of circuits before they reach this steady-state condition: for example, how the circuits react when a voltage or current source is initially turned on or off. This is referred to as the **transient response** of the circuit.

We will begin by looking at simple *RC* and *RL* circuits and then progress to more complex arrangements.

Figure 18.1   Capacitor charging

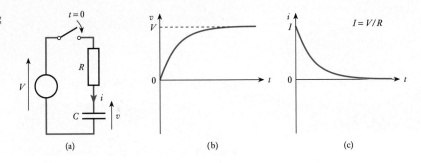

(a)   (b)   (c)

18.2.1   Capacitor charging

Figure 18.1(a) shows a circuit that charges a capacitor C from a voltage source V through a resistor R. The capacitor is assumed to be initially uncharged, and the switch in the circuit is closed at time $t = 0$.

When the switch is first closed the charge on the capacitor is zero, and therefore the voltage across it is also zero. Thus all the applied voltage is across the resistor, and the initial current is given by V/R. As this current flows into the capacitor the charge on it builds and the voltage across it increases. As the voltage across the capacitor increases, the voltage across the resistor decreases, causing the current in the circuit to fall. Gradually, the voltage across the capacitor increases until it is equal to the applied voltage, and the current goes to zero. We can understand this process more fully by deriving expressions for the voltage across the capacitor v and the current flowing into the capacitor i.

Applying Kirchhoff's voltage law to the circuit of Figure 18.1(a), we see that

$$iR + v = V$$

From Chapter 13, we know that the current in a capacitor is related to the voltage across it by the expression

$$i = C\frac{dv}{dt}$$

therefore, substituting

$$CR\frac{dv}{dt} + v = V$$

This is a first-order differential equation with constant coefficients and is relatively easy to solve. First we rearrange the expression to give

$$\frac{dv}{dt} = \frac{V - v}{CR}$$

and then again to give

$$\frac{\mathrm{d}t}{CR} = \frac{\mathrm{d}v}{V - v}$$

Integrating both sides then gives

$$\frac{t}{CR} = -\ln(V - v) + A$$

where A is the constant of integration.

In this case we know (from our assumption that the capacitor is initially uncharged) that when $t = 0$, $v = 0$. Substituting this into the previous equation gives

$$\frac{0}{CR} = -\ln(V - 0) + A$$

$$A = \ln V$$

Therefore

$$\frac{t}{CR} = -\ln(V - v) + \ln V = \ln\frac{V}{V - v}$$

and

$$e^{\frac{t}{CR}} = \frac{V}{V - v}$$

Finally, rearranging we have

$$v = V(1 - e^{-\frac{t}{CR}}) \tag{18.1}$$

From this expression, we can also derive an expression for the current I, since

$$i = C\frac{\mathrm{d}v}{\mathrm{d}t} = CV\frac{\mathrm{d}}{\mathrm{d}t}(1 - e^{-\frac{t}{CR}}) = \frac{V}{R}e^{-\frac{t}{CR}}$$

We noted earlier that at $t = 0$ the voltage across the capacitor is zero and the current is given by V/R. If we call this initial current I, then our expression for the current becomes

$$i = Ie^{-\frac{t}{CR}} \tag{18.2}$$

In Equations 18.1 and 18.2, you will note that the exponential component contains the term t/CR. You will recognise CR as the time constant T of the circuit, and thus t/CR is equal to t/T and represents time as a fraction of the time constant. For this reason, it is common to give these two equations in a more general form, replacing CR by T.

$$v = V(1 - e^{-\frac{t}{\mathsf{T}}}) \tag{18.3}$$

$$i = Ie^{-\frac{t}{\mathsf{T}}} \tag{18.4}$$

From Equations 18.3 and 18.4, it is clear that in the circuit of Figure 18.1(a) the voltage rises exponentially, while the current falls exponentially. These two waveforms are shown in Figure 18.1(b) and Figure 18.1(c).

Example 18.1 | The switch in the following circuit is closed at $t = 0$. Derive an expression for the output voltage v after this time and hence calculate the voltage on the capacitor at $t = 25$ s.

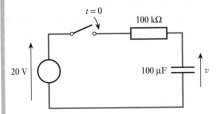

The time constant of the circuit $\mathsf{T} = CR = 100 \times 10^3 \times 100 \times 10^{-6} = 10$ s. From Equations 18.3

$$v = V(1 - e^{-\frac{t}{T}})$$

$$= 20(1 - e^{-\frac{t}{10}})$$

At $t = 25$ s

$$v = 20(1 - e^{-\frac{25}{10}})$$

$$= 18.36 \text{ V}$$

18.2.2   Inductor energising

Figure 18.2(a) shows a circuit that energises an inductor L using a voltage source V and a resistor R. The circuit is closed at time $t = 0$, and before that time no current flows in the inductor.

When the switch is first closed the current in the circuit is zero, since the nature of the inductor prevents the current from changing instantly. If the current is zero there is no voltage across the resistor, so all the applied voltage appears across the inductor. The applied voltage causes the current to

Figure 18.2 Inductor energising

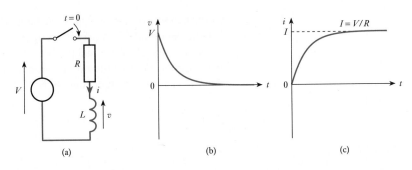

(a)   (b)   (c)

increase, producing a voltage drop across the resistor and reducing the voltage across the inductor. Eventually, the voltage across the inductor falls to zero and all the applied voltage appears across the resistor, producing a steady current of V/R. As before, it is interesting to look at expressions for v and i.

Applying Kirchhoff's voltage law to the circuit of Figure 18.2(a), we see that

$$iR + v = V$$

From Chapter 14, we know that the voltage across an inductor is related to the current through it by the expression

$$v = L\frac{di}{dt}$$

therefore, substituting

$$iR + L\frac{di}{dt} = V$$

This first-order differential equation can be solved in a similar manner to that derived for capacitors above. This produces the equations

$$v = Ve^{-\frac{Rt}{L}} \tag{18.5}$$

$$i = I(1 - e^{-\frac{Rt}{L}}) \tag{18.6}$$

where I represents the final (maximum) current in the circuit and is equal to V/R. In Equations 18.5 and 18.6, you will note that the exponential component contains the term Rt/L. Now L/R is the time constant T of the circuit, thus Rt/L is equal to t/T. We can therefore rewrite these two equations as

$$v = Ve^{-\frac{t}{\mathsf{T}}} \tag{18.7}$$

$$i = I(1 - e^{-\frac{t}{\mathsf{T}}}) \tag{18.8}$$

The forms of the v and i are shown in Figures 18.2(b) and 18.2(c). You might like to compare these figures with Figures 18.1(b) and 18.1(c) for a charging capacitor. You might also like to compare Equations 18.7 and 18.8, which describe the energising of an inductor, with Equations 18.3 and 18.4, which we derived earlier to describe the charging of a capacitor.

Example 18.2

An inductor is connected to a 15 V supply as shown below. How long after the switch is closed will the current in the coil reach 300 mA?

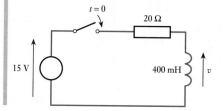

The time constant of the circuit $T = L/R = 0.4 \div 20 = 0.02$ s. The final current I is given by $V/R = 15/20 = 750$ mA.

From Equations 18.8

$$i = I(1 - e^{-\frac{t}{T}})$$

$$300 = 750(1 - e^{-\frac{t}{0.02}})$$

which can be evaluated to give

$$t = 10.2 \text{ ms}$$

18.3 Discharging of capacitors and de-energising of inductors

The charging of a capacitor or the energising of an inductor stores energy in that component that can be used at a later time to produce a current in a circuit. In this section, we look at the voltages and currents associated with this process.

18.3.1 Capacitor discharging

In order to look at the discharging of a capacitor, first we need to charge it up. Figure 18.3(a) shows a circuit in which a capacitor C is initially connected to a voltage source V and is then discharged though a resistor R. The discharge is initiated at $t = 0$ by opening one switch and closing another. In this diagram, the defining direction of the current i is *into* the capacitor, as in Figure 18.1(a), but clearly during the discharge process charge flows *out* of the capacitor, so i is negative.

The charged capacitor produces an electromotive force that drives a current around the circuit. Initially, the voltage across the capacitor is equal to the voltage of the source used to charge it (V), so the initial current is equal to V/R. However, as charge flows out of the capacitor its voltage deceases and the current falls. v and i can be determined in a similar manner to that used above for the charging arrangement. Applying Kirchhoff's voltage law to the circuit gives

Figure 18.3 Capacitor discharging

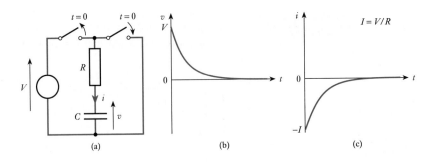

$$iR + v = 0$$

giving

$$CR\frac{dv}{dt} + v = 0$$

Solving this as before leads to the expressions

$$v = Ve^{-\frac{t}{CR}} = Ve^{-\frac{t}{\tau}} \tag{18.9}$$

$$i = -Ie^{-\frac{t}{CR}} = -Ie^{-\frac{t}{\tau}} \tag{18.10}$$

As before, the voltage and current have an exponential form, and these are shown in Figure 18.3(b) and Figure 18.3(c). Note that if i was defined in the opposite direction (as the current flowing out of the capacitor) then the polarity of the current in Figure 18.3(c) would be reversed. In this case, both the voltage and current would be represented by similar decaying exponential waveforms.

18.3.2 Inductor de-energising

In the circuit of Figure 18.4(a), a voltage source is used to energise an inductor by passing a constant current through it. At time $t = 0$ one switch is closed and the other is opened, so the energy stored in the inductor is now dissipated in the resistor. Since the current in an inductor cannot change instantly, initially the current flowing in the coil is maintained. To do this the inductor produces an electromotive force that is in the opposite direction to the potential created across it by the voltage source. With time, the energy stored in the inductor is dissipated and the e.m.f. decreases and the current falls.

As before, v and i can be determined by applying Kirchhoff's voltage law to the circuit. This gives

$$iR + v = 0$$

and thus

$$iR + L\frac{di}{dt} = 0$$

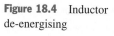

Figure 18.4 Inductor de-energising

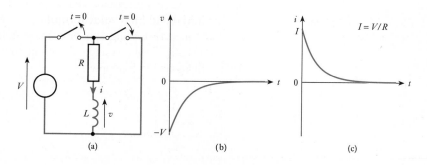

(a)   (b)   (c)

Solving this as before leads to the expressions

$$v = -Ve^{-\frac{Rt}{L}} = -Ve^{-\frac{t}{T}} \tag{18.11}$$

$$i = Ie^{-\frac{Rt}{L}} = Ie^{-\frac{t}{T}} \tag{18.12}$$

As before, the voltage and current have an exponential form, and these are shown in Figure 18.4(b) and Figure 18.4(c).

18.4 Generalised response of first-order systems

We have seen in Sections 18.2 and 18.3 that circuits containing resistance and either capacitance *or* inductance can be described by first-order differential equations. For this reason, such circuits are described as **first-order systems**. We have also seen that the transient behaviour of these circuits produces voltages and currents that change exponentially with time. However, although the various waveforms are often similar in form, they are not identical for different circuits. Fortunately, there is a simple method of determining the response of such systems to sudden changes in their environment.

18.4.1 Initial and final value formula

Increasing and decreasing exponential waveforms (for either voltage or current) can be found from the expressions:

$$v = V_f + (V_i - V_f)e^{-t/T} \tag{18.13}$$

$$i = I_f + (I_i - I_f)e^{-t/T} \tag{18.14}$$

where V_i and I_i are the *initial* values of the voltage and current, and V_f and I_f are the *final* values. The first element in these two expressions represents the steady-state response of the circuit, which lasts indefinitely. The second element represents the transient response of the circuit. This has a magnitude determined by the step change applied to the circuit, and it decays at a rate determined by the time constant of the arrangement. The combination of the steady state and the transient response gives the **total response** of the circuit. To see how these formulae can be used, Table 18.1 shows them applied to the circuits discussed in Sections 18.2 and 18.3.

The **initial and final value formula** is not restricted to situations where a voltage or current changes to, or from, zero. It can be used wherever there is a step change in the voltage or current applied to a first-order network. This is illustrated in Example 18.3.

Table 18.1 Transient response of first-order systems

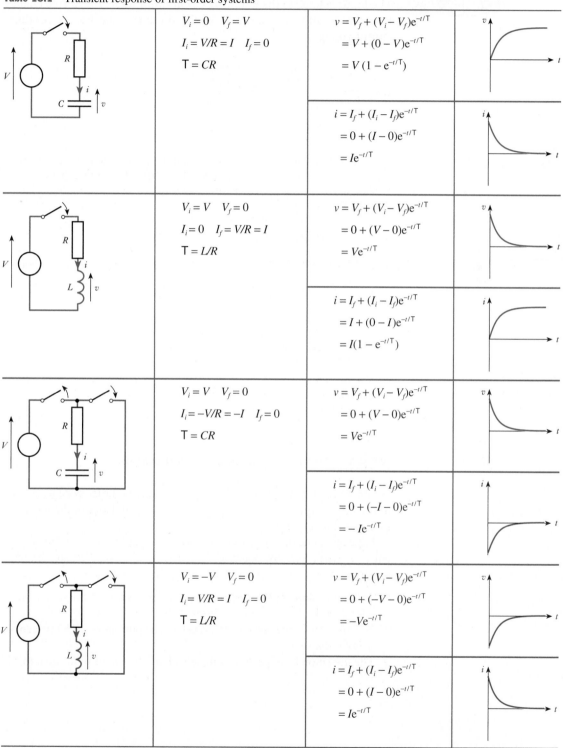

Example 18.3 The input voltage to the following *CR* network undergoes a step change from 5 V to 10 V at time $t = 0$. Derive an expression for the resulting output voltage.

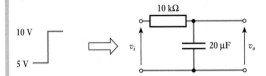

In this example the initial value is 5 V and the final value is 10 V. The time constant of the circuit is equal to $CR = 10 \times 10^3 \times 20 \times 10^{-6} = 0.2$ s.

Therefore, from Equation 18.13, for $t \geq 0$

$$v = V_f + (V_i - V_f)e^{-t/T}$$
$$= 10 + (5 - 10)e^{-t/0.2}$$
$$= 10 - 5e^{-t/0.2} \text{ V}$$

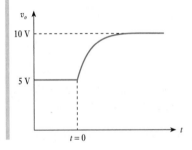

18.4.2 The nature of exponential curves

We have seen that the transients associated with first-order systems contain terms of the form $A(1 - e^{-t/T})$ or $Ae^{-t/T}$. The characteristics of these expressions are shown in Figure 18.5.

In general, one does not need to produce exact plots of such waveforms, but it is useful to know some of their basic properties. Perhaps the most important properties of exponential curves of this form are:

1. The initial slope of the curve crosses the final value of the waveform at a time $t = T$.
2. At a time $t = T$, the waveform has achieved approximately 63 percent of its total transition.
3. The transition is 99 percent complete after a period of time equal to $5T$.

Figure 18.5 Exponential waveforms

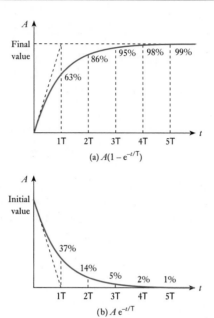

(a) $A(1 - e^{-t/T})$

(b) $A\,e^{-t/T}$

Response of first-order systems to pulse and square waveforms

Having looked at the transient response of first-order systems, we are now in a position to consider their response to pulse and square waveforms. Such signals can be viewed as combinations of positive-going and negative-going transitions and can therefore be treated in the same way as the transients discussed above. This is illustrated in Figure 18.6, which shows how a square waveform of fixed frequency is affected by RC and RL networks with different time constants.

Figure 18.6(a) shows the action of an RC network. We looked at the transient response of such an arrangement in Sections 18.2 and 18.3 and at typical waveforms in Figures 18.1 and 18.3. We noted that the response is exponential, with a rate of change that is determined by the time constant of the circuit. Figure 18.6(a) shows the effect of passing a square wave with a frequency of 1 kHz through RC networks with time constants of 0.01 ms, 0.1 ms and 1 ms. The first of these passes the signal with little distortion, since the wavelength of the signal is relatively long compared with the time constant of the circuit. As the time constant is increased to 0.1 ms and then to 1 ms, the distortion becomes more apparent as the network responds more slowly. When the time constant of the RC network is large compared with the period of the input waveform, the operation of the circuit resembles that of an **integrator**, and the output represents the integral of the input signal.

Figure 18.6 Response of
first-order systems to a square
wave

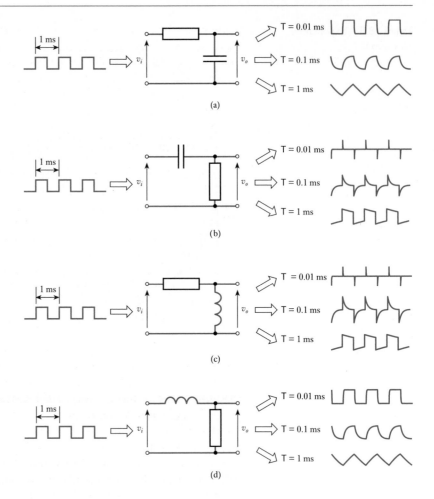

(a)

(b)

(c)

(d)

Transposing the positions of the resistor and the capacitor in the circuit
of Figure 18.6(a) produces the arrangement shown in Figure 18.6(b). The
output voltage is now the voltage across the resistor and is therefore proportional to the *current* in the circuit (and hence to the current in the capacitor). We would therefore expect the transients to be similar in shape to the
current waveforms shown in Figures 18.1 and 18.3. The steady-state value
of the output is zero in this circuit and, when a signal of 1 kHz is passed
though a network with a time constant of 0.01 ms, the signal is reduced to
a series of spikes. The circuit responds rapidly to the transient change in the
input, and the output then decays quickly to its steady-state output value of
zero. Here the time constant of the *RL* network is small compared with the
period of the input waveform, and the operation of the circuit resembles
that of a **differentiator**. As the time constant is increased, the output decays
more slowly and the output signal is closer to the input.

Figures 18.6(c) and 18.6(d) show first-order *RL* networks and again illustrate the effects of the time constant on the characteristics of the circuits. The pair of circuits produce similar signals to the *RC* circuits (when the configurations are reversed), and again one circuit approximates to an integrator while the other approximates to a differentiator.

File 18A

The shapes of the waveforms in Figure 18.6 are determined by the *relative* values of the time constant of the network and the period of the input waveform. Another way of visualising this relationship is to look at the effect of passing signals of different frequencies through the same network. This is shown in Figure 18.7. Note that the horizontal (time) axis is different in the various waveform plots.

The *RC* network of Figure 18.7(a) is a low-pass filter and therefore low-frequency signals are transmitted with little distortion. However, as the frequency increases the circuit has insufficient time to respond to changes in the input and becomes distorted. At high frequencies, the output resembles the **integral** of the input.

The *RC* network of Figure 18.7(b) is a high-pass filter and therefore high-frequency signals are transmitted with little distortion. At low frequencies, the circuit has plenty of time to respond to changes in the input signal and the output resembles that of a differentiator. As the frequency of the input increases, the network has progressively less time to respond and the output becomes more like the input waveform.

The *RL* network of Figure 18.7(c) represents a high-pass filter and therefore has similar characteristics to those of the *RC* network of Figure 18.7(b). Similarly, the circuit of Figure 18.7(d) is a low-pass filter and behaves in a similar manner to the circuit of Figure 18.7(a).

Figure 18.7 Response of first-order systems to square waves of different frequencies

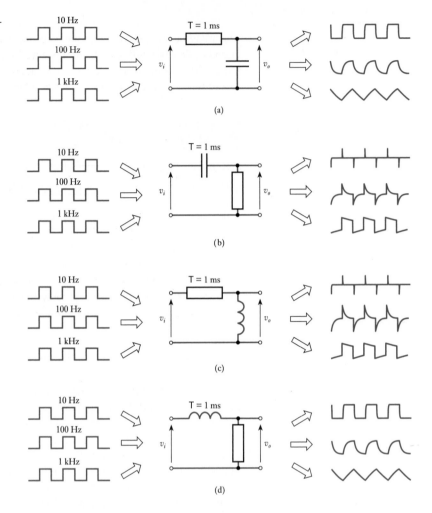

Computer Simulation Exercise 18.2

Simulate the circuit of Figure 18.7(a) choosing appropriate component values to produce a time constant of 1 ms. Use a digital clock generator to produce a square-wave input signal to this circuit, setting the frequency of the clock to 10 Hz. Observe the output of this circuit and compare this with that predicted in Figure 18.7. Change the frequency of the clock generator to 100 Hz and then to 1 kHz, observing the output in each case. Hence confirm the form of the waveforms shown in the figure. Experiment with both higher and lower frequencies and note the effect on the output.

Repeat this exercise for the remaining three circuits of Figure 18.7.

18.5 Second-order systems

Circuits that contain both capacitance and inductance are normally described by **second-order differential equations** (which may also describe some other circuit configurations). Arrangements described by these equations are termed **second-order systems**. Consider for example the *RLC* circuit of Figure 18.8. Applying Kirchhoff's voltage law to this circuit gives

$$L\frac{di}{dt} + Ri + v_C = V$$

Since *i* is equal to the current in the capacitor, this is equal to $C dv_C/dt$. Differentiating this with respect to *t* gives $di/dt = C d^2 v_C/dt^2$, and therefore

$$LC\frac{d^2 v_C}{dt^2} + RC\frac{dv_C}{dt} + v_C = V$$

which is a second-order differential equation with constant coefficients.

When a step input is applied to a second-order system, the form of the resultant transient depends on the relative magnitudes of the coefficients of its differential equation. The general form of the differential equation is

$$\frac{1}{\omega_n^2}\frac{d^2 y}{dt^2} + \frac{2\zeta}{\omega_n}\frac{dy}{dt} + y = x$$

where ω_n is the **undamped natural frequency** in rads/s and ζ (Greek letter *zeta*) is the **damping factor**.

The characteristics of second-order systems with different values of ζ are illustrated in Figure 18.9. This shows the response of such systems to a step change at the input. Small values of the damping factor ζ cause the system to respond more rapidly, but values less than unity cause the system to **overshoot** and oscillate about the final value. When $\zeta = 1$, the system is said to be **critically damped**. This is often the ideal situation for a control system, since this condition produces the fastest response in the absence of overshoot. Values of ζ greater than unity cause the system to be **overdamped**, while values less than unity produce an **underdamped** arrangement. As the damping is reduced, the amount of overshoot produced and the **settling time** both increase. When $\zeta = 0$, the system is said to be **undamped**. This produces a continuous oscillation of the output with a natural frequency of ω_n and a peak height equal to that of the input step.

Figure 18.8 A series *RLC* arrangement

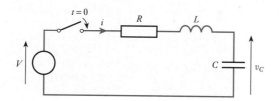

Figure 18.9   Response of
second-order systems

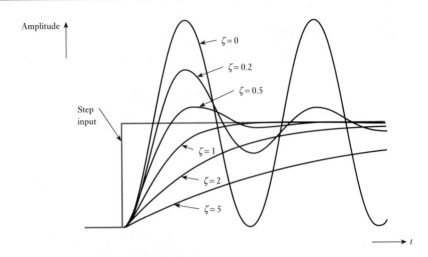

File 18C

Computer Simulation Exercise 18.3

Simulate the circuit of Figure 18.8, replacing the voltage source and the switch by a digital clock generator. Use values of 100 Ω, 10 mH and 100 μF for R, L and C, respectively, and set the frequency of the clock generator to 2.5 Hz. Use transient analysis to look at the output voltage over a period of 1 s.

Observe the output of the circuit and note the approximate time taken for the output to change. Increase the value of R to 200 Ω and note the effect on the output waveform. Progressively increase R up to 1 kΩ and observe the effect.

Now look at the effect of progressively reducing R below 100 Ω (down to 1 Ω or less). Estimate from your observations the value of R that corresponds to the circuit being critically damped.

18.6   Higher-order systems

Higher-order systems, that is, those that are described by third-order, fourth-order or higher-order equations, often have a transient response that is similar to that of the second-order systems described in the last section. Because of the complexity of the mathematics of such systems, they will not be discussed further here.

Key points

- The reaction of a circuit to instantaneous changes at its input is termed its transient response.

- The charging or discharging of a capacitor, and the energising or de-energising of an inductor, are each associated with exponential voltage and current waveforms.

- Circuits that contain resistance, and either capacitance *or* inductance, may be described by first-order differential equations and are therefore called first-order systems.

- The increasing or decreasing exponential waveforms associated with first-order systems can be found using the initial and final value formula.

- The transient response of first-order systems can be used to determine their response to both pulse and square waveforms.

- At high frequencies, low-pass networks approximate to integrators.

- At low frequencies, high-pass networks approximate to differentiators.

- Circuits that contain both capacitance and inductance are normally described by second-order differential equations and are termed second-order systems.

- Such systems are characterised by their undamped natural frequency ω_n and their damping factor ζ. The latter determines how rapidly a system responds, while the former dictates the frequency of undamped oscillation.

Exercises

18.1 Explain the meanings of the terms 'steady-state response' and 'transient response'.

18.2 When a voltage is suddenly applied across a series combination of a resistor and an uncharged capacitor, what is the initial current in the circuit? What is the final, or steady-state, current in the circuit?

18.3 The switch in the following circuit is closed at $t = 0$. Derive an expression for the current in the circuit after this time and hence calculate the current in the circuit at $t = 4$ s.

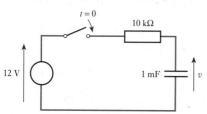

18.4 Simulate the arrangement of Exercise 18.3 and use transient analysis to investigate the current in the circuit. Use a switch element that *closes* at

$t = 0$ to start the charging process, and use a second switch that *opens* at $t = 0$ to ensure that the capacitor is initially discharged (this second switch should be connected directly across the capacitor). Use your simulation to verify your answer to Exercise 18.3.

18.5 When a voltage is suddenly applied across a series combination of a resistor and an inductor, what is the initial current in the circuit? What is the final, or steady-state, current in the circuit?

18.6 The switch in the following circuit is closed at $t = 0$. Deduce an expression for the output voltage of the circuit and hence calculate the time at which the output voltage will be equal to 8 V.

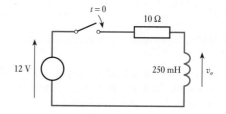

Exercises continued

18.7 Simulate the arrangement of Exercise 18.6 and use transient analysis to investigate the output voltage of the circuit. Use a switch element that closes at $t = 0$ to start the energising process, and use your simulation to verify your answer to Exercise 18.6.

18.8 A capacitor of 25 μF is initially charged to a voltage of 50 V. At time $t = 0$, a resistance of 1 kΩ is connected directly across its terminals. Derive an expression for the voltage across the capacitor as it is discharged and hence determine the time taken for its voltage to drop to 10 V.

18.9 An inductor of 25 mH is passing a current of 1 A. At $t = 0$, the circuit supplying the current is instantly replaced by a resistor of 100 Ω connected directly across the inductor. Derive an expression for the current in the inductor as a function of time and hence determine the time taken for the current to drop to 100 mA.

18.10 What is meant by a 'first-order system', and what kind of circuits fall within this category?

18.11 Explain how the equation for an increasing or decreasing exponential waveform may be found using the initial and final values of the waveform.

18.12 The input voltage to the following *CR* network undergoes a step change from 20 V to 10 V at time $t = 0$. Derive an expression for the resulting output voltage.

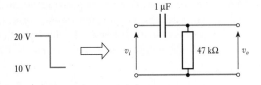

18.13 Sketch the exponential waveform $v = 5\,e^{-t/10}$.

18.14 For each of the following circuit arrangements, sketch the form of the output voltage when the period of the square-wave input voltage is:

(a) much greater than the time constant of the circuit;
(b) equal to the time constant of the circuit;
(c) much less than the time constant of the circuit.

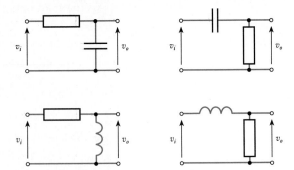

18.15 Simulate each of the circuit arrangements of Exercise 18.14, selecting component values to give a time constant of 1 ms in each case. Use a digital clock generator to apply a square-wave input voltage to the circuit and use transient analysis to observe the form of the output for input frequencies of 200 Hz, 1 kHz and 5 kHz. Compare these observations with your results for Exercise 18.14.

18.16 Under what circumstances does the behaviour of a first-order high-pass filter resemble that of a differentiator?

18.17 Under what circumstances does the behaviour of a first-order low-pass filter resemble that of an integrator?

18.18 What is meant by a 'second-order system', and what kind of circuits fall within this category?

18.19 Derive an expression for the *current* in the circuit of Figure 18.8.

18.20 Explain what is meant by the terms 'undamped natural frequency' and 'damping factor' as they apply to second-order systems.

18.21 What is meant by 'critical damping' and what value of the damping factor corresponds to this situation?

Chapter 19

Semiconductor Diodes

Objectives

When you have studied the material in this chapter you should be able to:

- **explain the basic function of diodes in electrical circuits and describe the characteristics of an ideal diode;**
- **describe the electrical characteristics of conductors, insulators and semiconductors;**
- **discuss the doping of semiconductor materials and the construction of semiconductor diodes;**
- **describe the characteristics of a typical diode and sketch its current–voltage characteristics;**
- **outline the use of several forms of special-purpose semiconductor devices, including Zener, tunnel and varactor diodes;**
- **design a range of circuits that exploit the characteristics of semiconductor diodes.**

19.1 Introduction

Throughout this text, we have adopted a 'top-down' approach to electrical and electronic systems and have looked at the overall behaviour of circuit elements before analysing their characteristics in detail. Following this doctrine, we have so far taken a 'black-box' view of components such as operational amplifiers and have considered only their external properties. It is now time to take a look 'inside the box' by considering the operation of the various components that are used to construct them. We have already looked at the characteristics of components such as resistors, capacitors and inductors but have yet to consider the operation of the various semi-conductor components, such as diodes and transistors. We will begin this process in this chapter by considering semiconductor diodes and will then move on to transistors in the following chapters.

Continuing our top-down philosophy, we will begin this chapter by looking at the general (and idealised) characteristics of diodes. We will then consider the basic properties of semiconductor materials and see how this allows us to produce a semiconductor diode. We will then look at the behaviour of such devices and at a range of circuits that make use of them.

While the treatment of the physics of semiconductor materials is fairly superficial, some readers may wish to ignore this part of the chapter (Sections 19.3 to 19.5) and concentrate on only the external characteristics of the devices. These sections can be omitted without affecting the comprehension of the remaining material, but they are required if readers wish to understand the physical operation of transistors in later chapters.

19.2 Diodes

Simplistically, a **diode** is an electrical component that conducts electricity in one direction but not the other. One could characterise an **ideal diode** as a component that conducts no current when a voltage is applied across it in one direction but appears as a short circuit when a voltage is applied in the opposite direction. One could picture such a device as an electrical equivalent of a hydraulic non-return valve, which allows water to flow in one direction but not the other.

The characteristics of an ideal diode are shown in Figure 19.1(a), while Figure 19.1(b) shows the circuit symbol for a diode. A diode has two electrodes, called the **anode** and the **cathode**, and the latter diagram also shows the polarity of the voltage that must be applied across the diode in order for it to conduct. It can be seen that the symbol for a diode resembles an arrow that points in the direction of current flow.

Diodes have a wide range of applications, including the **rectification** of alternating voltages. This process is illustrated in Figure 19.2. Here the diode conducts for the positive half of the input waveform but opposes the flow of current during the negative half-cycle. When diodes are used in such circuits they are often referred to as **rectifiers**, and the arrangement of Figure 19.2 would be described as a **half-wave rectifier**. We will return to look at this circuit in Section 19.8 when we consider some applications of semiconductor diodes.

In practice, no real component has the properties of an ideal diode, but semiconductor diodes, in the form of *pn* junctions, can produce a good

Figure 19.1 An ideal diode

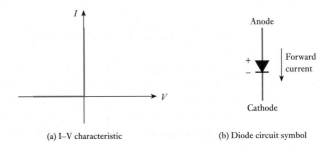

(a) I–V characteristic   (b) Diode circuit symbol

Figure 19.2 A diode as a rectifier

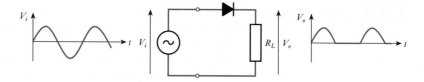

approximation to these characteristics. To understand how we can construct such devices, we need to know a little about the electrical properties of materials.

19.3 Electrical properties of solids

Solid materials can be divided with respect to their electrical properties into three categories: conductors, insulators and semiconductors. The different characteristics of these groups are produced by the atomic structure of the materials and in particular by the distribution of electrons in the outer orbits of the atoms. These outermost electrons are termed **valence electrons**, and they play a major part in determining many of the properties of the material.

19.3.1 Conductors

Conductors such as copper or aluminium have a cloud of 'free' electrons at all temperatures above absolute zero. This is formed by the weakly bound 'valence' electrons in the outermost orbits of the atoms. If an electric field is applied across such a material, electrons will flow causing an electric current.

19.3.2 Insulators

In insulating materials, such as polythene, the valence electrons are tightly bound to the nuclei of the atoms, and very few are able to break free to conduct electricity. The application of an electric field does not cause a current to flow, since there are no free electrons.

19.3.3 Semiconductors

At very low temperatures, semiconductors have the properties of an insulator. However, at higher temperatures some electrons are free to move and the materials take on the properties of a conductor – albeit a poor one. Nevertheless, semiconductors have some useful characteristics that make them distinct from both insulators and conductors.

19.4 Semiconductors

Semiconductor materials have very interesting electrical properties that make them extremely useful in the production of electronic devices. The most commonly used semiconductor material for such applications is **silicon**, but **germanium** is also used, as are several more exotic materials such as **gallium arsenide** and **gallium nitride**. Many metal oxides have semiconducting properties (for example, the oxides of manganese, nickel and cobalt).

19.4.1 Pure semiconductors

At temperatures near absolute zero, the valence electrons in a pure semiconductor are tightly bound to their nuclei, leaving no electrons free to conduct electricity. This gives the material the properties of an insulator. However, as the temperature rises, thermal vibration of the crystal lattice results in some of the bonds being broken, generating a few **free electrons**, which are able to move throughout the crystal. This also leaves behind **holes**, which accept electrons from adjacent atoms and therefore also move about. Electrons are negative charge carriers and will move *against* an applied electric field, generating an electrical current. Holes, being the absence of an electron, act like positive charge carriers and will move *in the direction of* an applied electric field and will also contribute to current flow.

At normal room temperatures, the number of charge carriers present in pure silicon is small and consequently it is a poor conductor. This form of conduction is called **intrinsic conduction**.

19.4.2 Doping

The addition of small amounts of impurities to a semiconductor can drastically affect its properties. This process is known as **doping**. Adding appropriate materials produces an excess of charge carriers, which can then be used for conduction. Materials that produce an excess of *negative* charge carriers (electrons) are called **donor impurities**, and semiconductors containing such impurities are called **n-type semiconductors**. Phosphorus is a commonly used donor impurity in silicon. Materials that produce an excess of *positive* charge carriers (holes) are called **acceptor impurities**, and semiconductors containing such impurities are called **p-type semiconductors**. Boron can be used to produce this effect in silicon.

It is important to remember that a piece of doped semiconductor in isolation will be electrically neutral. Therefore, the presence of *mobile* charge carriers of a particular polarity must be matched by an equal number of *fixed* (or *bound*) charge carriers of the opposite polarity. Thus, in an *n*-type semiconductor the free electrons produced by the doping will be matched by an equal number of positive charges bound within the atoms in the

Figure 19.3 Charges in doped semiconductors

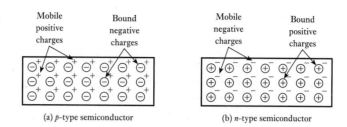

(a) *p*-type semiconductor

(b) *n*-type semiconductor

lattice. Similarly, in a *p*-type semiconductor, free holes are matched by an equal number of bound negative charges. This process is illustrated in Figure 19.3.

Both *n*-type and *p*-type semiconductors have much greater conductivities than that of the intrinsic material, the magnitude depending on the doping level. This is called **extrinsic conductivity**. The dominant charge carriers in a doped semiconductor (that is, electrons in an *n*-type material and holes in a *p*-type material) are called **majority charge carriers**. The other charge carriers are called **minority charge carriers**.

19.5 *pn* junctions

Although *p*-type and *n*-type semiconductor materials have some useful characteristics individually, they are of greater interest when they are used together.

When *p*-type and *n*-type materials are joined, the charge carriers in each interact in the region of the junction. Although each material is electrically neutral, each has a much higher concentration of majority charge carriers than of minority charge carriers. Thus on the *n*-type side of the junction there are far more free electrons than on the *p*-type side. Consequently, electrons diffuse across the junction from the *n*-type side to the *p*-type side, where they are absorbed by recombination with free holes, which are plentiful in the *p*-type region. Similarly, holes diffuse from the *p*-type side to the *n*-type side and combine with free electrons.

This process of diffusion and recombination of charge carriers produces a region close to the junction that has very few mobile charge carriers. This region is referred to as a **depletion layer** or sometimes as a **space–charge layer**. The diffusion of negative charge carriers in one direction and positive charge carriers in the other generates a net charge imbalance across the junction. The existence of positive and negative charges on each side of the junction produces an electric field across it. This produces a **potential barrier**, which charge carriers must overcome to cross the junction. This process is illustrated in Figure 19.4.

Only a small number of *majority* charge carriers have sufficient energy to surmount this barrier, and these generate a small **diffusion current** across the junction. However, the field produced by the space–charge region does not oppose the movement of *minority* charge carriers across the

Figure 19.4   A *pn* junction

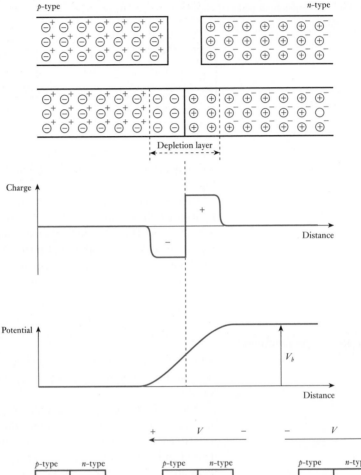

Figure 19.5   Currents in a *pn* junction

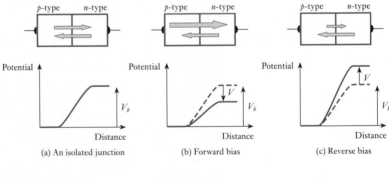

junction; rather, it assists it. Any such charge carriers that stray into the depletion layer, or that are formed there by thermal vibration, are accelerated across the junction forming a small **drift current**. In an isolated junction, a state of dynamic equilibrium exists in which the diffusion current exactly matches the drift current. This situation in shown in Figure 19.5(a). The application of an external potential across the device will affect the height of the potential barrier and change the state of dynamic equilibrium.

19.5.1 Forward bias

If the p-type side of the device is made positive with respect to the n-type side, the applied potential neutralises some of the space–charge and the width of the depletion layer decreases. The height of the barrier is reduced and a larger proportion of the majority carriers in the region of the junction now have sufficient energy to surmount it. The diffusion current produced is therefore much larger than the drift current and a net current flows across the junction. This situation is shown in Figure 19.5(b).

19.5.2 Reverse bias

If the p-type side of the device is made negative with respect to the n-type side, the space–charge increases and the width of the depletion layer is increased. This produces a larger potential barrier and reduces the number of majority carriers that have sufficient energy to surmount it, reducing the diffusion current across the junction. This situation is shown in Figure 19.5(c).

Even a small negative bias, of perhaps 0.1 V, is sufficient to reduce the diffusion current to a negligible value. This leaves a net imbalance in the currents flowing across the junction, which are now dominated by the drift current. Since the magnitude of this current is determined by the rate of thermal generation of minority carriers in the region of the junction, it is not related to the applied voltage. At normal room temperatures this reverse current is very small, typically a few nanoamps for silicon devices and a few microamps for germanium devices. However, it is exponentially related to temperature and doubles for a temperature rise of about 10 °C. Reverse current is proportional to the junction area and so is much greater in large power semiconductors than in small, low-power devices.

19.5.3 Forward and reverse currents

The current flowing through a pn junction can be approximately related to the applied voltage by the expression

$$I = I_s\left(\exp\frac{eV}{\eta kT} - 1\right)$$

where I is the current through the junction, e is the electronic charge, V is the applied voltage, k is Boltzmann's constant, T is the absolute temperature and η (Greek letter *eta*) is a constant in the range 1 to 2 determined by the junction material. Here a positive applied voltage represents a forward-bias voltage and a positive current a forward current.

The constant η is approximately 1 for germanium and about 1.3 for silicon. However, for our purposes it is reasonable to use the approximation that

Figure 19.6 Current–voltage characteristic of a *pn* junction

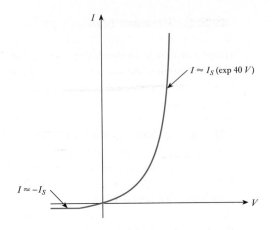

$$I \approx I_s \left(\exp \frac{eV}{kT} - 1 \right) \tag{19.1}$$

and we will make this assumption for the remainder of this text.

At normal room temperatures e/kT has a value of about 40 V^{-1}. If V is less than about -0.1 V, the exponential term within the brackets in Equation 19.1 is small compared with 1, and I is given by

$$I \approx I_s(0 - 1) = -I_s \tag{19.2}$$

Similarly, if V is greater than about $+0.1$ V, the exponential term is much greater than 1, and I is given by

$$I \approx I_s \left(\exp \frac{eV}{kT} \right) = I_s(\exp 40V) \tag{19.3}$$

We therefore have a characteristic for which the reverse-bias current is approximately constant at $-I_s$ (the **reverse saturation current**), and the forward-bias current rises exponentially with the applied voltage.

In fact, the expressions of Equations 19.1 to 19.3 are only approximations of the junction current in a real device, as effects such as **junction resistance** and **minority carrier injection** tend to reduce the current flowing. However, this analysis gives values that indicate the form of the relationship and are adequate for our purposes. Figure 19.6 shows the current–voltage characteristic of a *pn* junction.

19.6   Semiconductor diodes

A *pn* junction is not an ideal diode, but it does have a characteristic that approximates to such a device. When viewed on a large scale, the relationship between the current and the applied voltage is as shown in Figure 19.7. When forward-biased, a *pn* junction exhibits an exponential current–voltage characteristic. A small forward voltage is required to make the

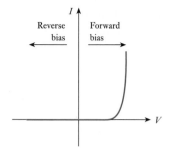

Figure 19.7 Forward and reverse currents in a semiconductor diode

device conduct, but then the current increases rapidly as this voltage is increased. When reverse-biased, the junction passes only a small reverse current, which is almost always negligible. The *pn* junction therefore represents a reasonable approximation to an ideal diode and is widely used in diode applications.

19.6.1 Diode characteristics

While the graph shown in Figure 19.7 provides an overview of diode behaviour, we often need a more detailed view of the component's characteristics. As we have seen, when reverse-biased a semiconductor diode passes only a very small current – the reverse saturation current. For silicon devices, this saturation current is typically 1 nA and is negligible in almost all applications. The reverse current is approximately constant as the reverse voltage is increased to a critical voltage, called the **reverse breakdown voltage** V_{br}. If the negative voltage is increased beyond this point, the junction breaks down and begins to conduct. This limits the useful voltage range of the diode. The reverse characteristics of a typical silicon diode are shown in Figure 19.8(a). The value of the reverse breakdown voltage will depend on the type of diode and may have a value from a few volts to a few hundred volts.

When a semiconductor diode is forward-biased a negligible current will flow for a small applied voltage, but this increases exponentially as the voltage is increased. *When viewed on a large scale*, it appears that the current is zero until the voltage reaches a so-called **turn-on voltage**, and that as the voltage is increased beyond this point the junction begins to conduct and the current increases rapidly. This turn-on voltage is about 0.5 V for a silicon junction. A further increase in the applied voltage causes the junction current to increase rapidly. This results in the current–voltage characteristic being almost vertical, showing that the voltage across the diode is approximately constant, irrespective of the junction current. The forward characteristic of a silicon diode is shown in Figure 19.8(a). In many applications it is reasonable to approximate the characteristic by a straight-line

Figure 19.8 Semiconductor diode characteristics

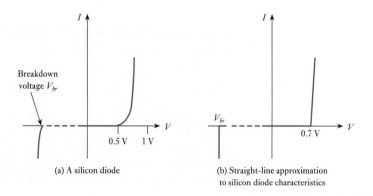

response, as shown in Figure 19.8(b). This simplified form represents the diode by a forward voltage drop (of about 0.7 V for silicon devices) combined with a forward resistance. The latter results in the slope of the characteristic above the turn-on voltage. In many cases, the forward resistance of the diode may be ignored and the diode considered simply as a near ideal diode with a small forward voltage drop. This voltage drop is termed the **conduction voltage** of the diode. As the current through the diode increases, the voltage across the junction also increases. At 1 A the conduction voltage might be about 1 V for a silicon diode, rising to perhaps 2 V at 100 A. In practice, most diodes would be destroyed long before the current reached such large values.

So far we have considered only diodes constructed from silicon and have seen that such devices have a turn-on voltage of about 0.5 V and a conduction voltage of about 0.7 V. While silicon is the most common material for the manufacture of semiconductor diodes, many other materials are also used. Examples include germanium (which has a turn-on voltage of about 0.2 V and a conduction voltage of about 0.25 V) and gallium arsenide (which has a turn-on voltage of about 1.3 V and a conduction voltage of about 1.4 V).

Diodes are used for a number of purposes in electronic circuits. In many cases, relatively low voltages and currents are present, and devices for such applications are usually called **signal diodes**. A typical device might have a maximum forward current of 100 mA and a reverse breakdown voltage of 75 V. Other common applications for diodes include their use in power supplies to convert alternating currents into direct currents. Such diodes will usually have a greater current-handling capacity (usually measured in amperes or tens of amperes) and are generally called rectifiers rather than diodes. Reverse breakdown voltages for such devices will vary with the application but are typically hundreds of volts.

Diodes and rectifiers can be made using a variety of semiconductor materials and may use other techniques in place of simple *pn* junctions. This allows devices to be constructed with a wide range of characteristics in terms of current-handling capability, breakdown voltage and speed of operation.

File 19A

Computer Simulation Exercise 19.1

Use simulation to investigate the relationship between the current and the applied voltage in a small signal diode (such as a 1N4002). Measure the current while the applied voltage is swept from 0 to 0.8 V and plot the resulting curve.

Look at the behaviour of the device over different voltage ranges, including both forward- and reverse-bias conditions. Estimate from these experiments the reverse breakdown voltage of the diode.

19.7 Special-purpose diodes

We have already encountered several forms of semiconductor diode in this text. These include the signal diode and rectifier discussed in the last section, the *pn* junction temperature sensor and photodiode discussed in Chapter 3, and the light-emitting diode (LED) described in Chapter 4. Several other forms of diode are widely used, each having its own unique characteristics and applications. We will look briefly at some of the more popular forms.

19.7.1 Zener diodes

When the reverse breakdown voltage of a diode is exceeded, the current that flows is generally limited only by external circuitry. If steps are not taken to limit this current, the power dissipated in the diode may destroy it. However, if the current is limited by the circuitry connected to the diode, the breakdown of the junction need not cause any damage to the device. This effect is utilised in special-purpose devices called **Zener diodes**. From Figure 19.8, it is clear that when the junction is in the breakdown region the junction voltage is approximately constant irrespective of the reverse current flowing. This allows the device to be used as a **voltage reference**. In such devices, the breakdown voltage is often given the symbol V_Z. Zener diodes are available with a variety of breakdown voltages to allow a wide range of reference voltages to be produced.

A typical circuit using a Zener diode is shown in Figure 19.9, which also shows the special symbol given to this component. Here a poorly regulated voltage V is applied to a series combination of a resistor and a Zener diode. The diode is connected so that it is reserved-biased by the positive applied voltage V. If V is greater than V_Z the diode junction will break down and will conduct, drawing current from the resistance R. The diode prevents the output voltage going above its breakdown voltage V_Z and thus generates an approximately constant output voltage irrespective of the value of the input voltage, provided that it remains greater than V_Z. If V is less that V_Z the reverse-biased diode will conduct negligible current, and the output V_o will be approximately equal to V. In this situation, the Zener diode has no effect on the circuit.

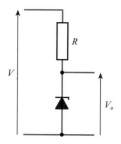

Figure 19.9 A simple voltage reference using a Zener diode

File 19B

Computer Simulation Exercise 19.2

The D1N750 is a 4.7 V Zener diode. Simulate the circuit of Figure 19.9 using this diode (or a similar component) and a suitable resistor.

Apply a swept DC input voltage to the circuit and plot the output voltage against the input voltage for a range of values of R. Investigate the effect of connecting a load resistor to the circuit.

19.7.2 Schottky diodes

Unlike conventional *pn* junction diodes, which are formed at the junction of two layers of doped semiconductor material, **Schottky diodes** are formed by a junction between a layer of metal (such as aluminium) and a semiconductor. The rectifying contact formed relies only on majority charge carriers and is consequently much faster in operation than *pn* junction devices, which are limited in speed by the relatively slow recombination of minority charge carriers.

Schottky diodes also have a low forward voltage drop of about 0.25 V. This characteristic is used to great effect in the design of high-speed logic gates.

19.7.3 Tunnel diodes

The **tunnel diode** uses high doping levels to produce a device with a very narrow depletion region. This region is so thin that a quantum mechanical effect known as *tunnelling* can take place. This results in charge carriers being able to cross the depletion layer, even though they do not have sufficient energy to surmount it. The combination of the tunnelling effect and conventional diode action produces a characteristic as shown in Figure 19.10.

This rather strange characteristic finds application in a number of areas. Of particular interest is the fact that for part of its operating range the voltage across the device falls for an increasing current. This corresponds to a region where the incremental resistance of the device is *negative*. This property is utilised in high-frequency oscillator circuits, in which the negative resistance of the tunnel diode is used to cancel losses in passive components.

Figure 19.10 Characteristic of a tunnel diode

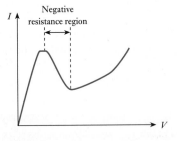

19.7.4 Varactor diodes

A reverse-biased diode has two conducting regions of *p*- and *n*-type semiconductor separated by a depletion region. This structure resembles a capacitor, with the depletion region forming the insulating dielectric. Small silicon signal diodes have a capacitance of a few picofarads, which changes with the reverse-bias voltage since this varies the thickness of the depletion region.

This effect is used by **varactor diodes**, which act as voltage-dependent capacitors. A typical device might have a capacitance of 160 pF at 1 V, falling to about 9 pF at 10 V. Such devices are used at the heart of many automatic tuning arrangements, where the varactor is used in an *LC* or *RC* tuned circuit. The capacitance of the device, and therefore the frequency characteristics of the circuit, may then be varied by the applied reverse-bias voltage.

19.8   Diode circuits

In this section we will look at just a few of the many circuits that make use of diodes of one form or another.

19.8.1   A half-wave rectifier

One of the most common uses of diodes is as a rectifier in a power supply to generate a direct voltage from an alternating supply. A simple arrangement to achieve this is the half-wave rectifier (discussed in Section 19.2), which is shown in Figure 19.11. While the input voltage is greater than the turn-on voltage of the diode, the diode conducts and the input voltage (minus the small voltage drop across the diode) appears across the load. During the part of the cycle in which the diode is reverse-biased, no current flows in the load.

To produce a steadier output voltage, a **reservoir capacitor** is normally added to the circuit as shown. This is charged while the diode is conducting and maintains the output voltage when the diode is turned off by

Figure 19.11   A half-wave rectifier

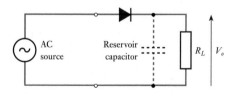

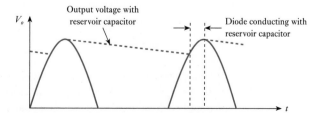

supplying current to the load. This current gradually discharges the capacitor, causing the output voltage to decay. One effect of adding a reservoir capacitor is that the diode conducts for only short periods of time. During these periods the diode current is thus very high. The magnitude of the *ripple* in the output voltage is affected by the current taken by the load, the size of the capacitor and the frequency of the incoming signal. Clearly, as the supply frequency is increased the time for which the capacitor must maintain the output is reduced.

| Example 19.1 | A half-wave rectifier connected to a 50 Hz supply generates a peak voltage of 10 V across a 10 mF reservoir capacitor. Estimate the peak ripple voltage produced if this arrangement is connected to a load that takes a constant current of 200 mA. |

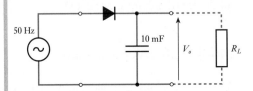

The voltage V across a capacitor is related to its charge Q and its capacitance C by the expression

$$V = \frac{Q}{C}$$

Differentiating with respect to time gives

$$\frac{dV}{dt} = \frac{1}{C}\frac{dQ}{dt} = \frac{i}{C}$$

where i is the current into, or out of, the capacitor. In this example the current is constant at 200 mA, and the capacitance is 10 mF, so

$$\frac{dV}{dt} = \frac{i}{C} = \frac{0.2}{0.01} = 20 \text{ V/s}$$

Therefore the output voltage will fall at a rate of 20 V/s.

During each cycle, the capacitor is discharged for a time almost equal to the period of the input, which is 20 ms. Therefore, during this time the voltage on the capacitor (the output voltage) will fall by 20 ms × 20 V/s = 0.4 V.

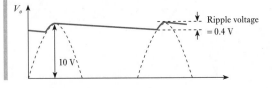

Computer Simulation Exercise 19.3

Simulate the circuit of Figure 19.11 and investigate the behaviour of the circuit.

While typical half-wave rectifier arrangements might have input voltages of several hundred volts, the operation of the circuit is more apparent if smaller voltages are used so that the turn-on voltage of the diode is more easily observed.

Simulate the circuit with and without a reservoir capacitor and use transient analysis to study the circuit's behaviour. Plot the output voltage and the current through the diode, and see how these relate to the input voltage. Vary the input frequency and see how this affects the output.

Note the peak voltage at the output and the ripple voltage for a given set of circuit parameters (for use in the next simulation exercise).

19.8.2 A full-wave rectifier

One simple method of effectively increasing the frequency of the waveform applied to the capacitor in the previous circuit is to use a full-wave rectifier arrangement as shown in Figure 19.12. When terminal A of the supply is positive with respect to B, diodes D2 and D3 are forward-biased and diodes D1 and D4 are reverse-biased. Current therefore passes from terminal A, through D2, through the load R_L, and returns to terminal B through D3. This makes the output voltage V_o positive. When terminal B is positive with respect to terminal A, diodes D1 and D4 are forward-biased and D2 and D3 are reverse-biased. Current now flows from terminal B through D4, through the load R_L and returns to terminal A through D1. Since the direction of the

Figure 19.12 A full-wave rectifier

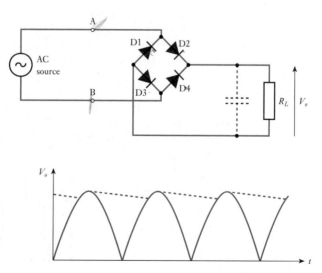

current in the output resistor is the same, the polarity of the output voltage is unchanged. Thus both positive and negative half-cycles of the supply produce positive output peaks, and the time during which the capacitor must maintain the output voltage is reduced.

Example 19.2

Determine the effect on the ripple voltage of replacing the half-wave rectifier of Example 19.1 with a full-wave arrangement, assuming that the reservoir capacitor and load remain the same.

Since the capacitor and load are unchanged, the rate of change of the voltage on the capacitor is also unchanged at 20 V/s. However, in this case the time between successive peaks in the output voltage is equal to half the period of the input, which is 10 ms. Hence the ripple voltage is now 10 ms × 20 V/s = 0.2 V. Thus the ripple voltage is halved.

File 19D

Computer Simulation Exercise 19.4

Repeat the investigations of Computer Simulation Exercise 19.3 for the full-wave rectifier circuit of Figure 19.11. Compare the peak voltage at the output and the ripple voltage with those obtained using the earlier circuit for similar circuit parameters.

19.8.3   A signal rectifier

In Chapter 5, we looked at the use of modulation and demodulation in the transmission of signals and discussed **full amplitude modulation (full AM)** in radio broadcasting. Such signals consist of a high-frequency carrier component, the amplitude of which is modulated by a lower-frequency signal. It is this low-frequency signal that conveys the useful information and that must be recovered by demodulating the signal. A simple circuit to perform this task is shown in Figure 19.13.

Figure 19.13   A signal rectifier

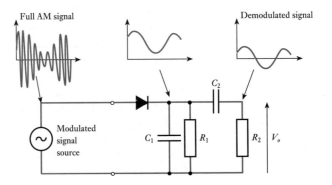

The demodulator works in a similar manner to the half-wave rectifier described earlier. The modulated signal is passed through a diode, which removes the negative half-cycles of the waveform and applies the positive half of each cycle to a parallel RC network formed by R_1 and C_1. This behaves as a low-pass filter, and the values of R_1 and C_1 are chosen such that they produce a high-frequency cut-off that is above the signal frequency but below that of the radio frequency carrier. Thus the carrier is removed, leaving only the required signal plus a DC component. This direct component is removed by a second capacitor C_2, which applies the demodulated signal to R_2. This second RC network is effectively a high-pass filter that removes the DC component but has a cut-off frequency sufficiently low to pass the signal frequency.

The output voltage developed across R_2 represents the envelope of the original signal. For this reason, the circuit is often called an **envelope detector**. This arrangement forms the basis of most AM radio receivers from simple **crystal sets**, which consist largely of the envelope detector and a simple frequency selective network, to complex **superheterodyne receivers**, which use sophisticated circuitry to select and amplify the required signal.

19.8.4   Signal clamping

Diodes may be used in a number of ways to change the form of a signal. Such arrangements come under the general heading of **wave-shaping circuits**, and Figure 19.14 shows a few examples.

Figure 19.14(a) shows a simple arrangement for limiting the negative excursion of a signal. When the input signal is positive, the diode is reverse-biased and so has no effect. However, when the input is negative and larger than the turn-on voltage of the diode, the diode conducts, clamping the output signal. This prevents the output from going more negative than the turn-on voltage of the diode (about 0.7 V for a silicon device). If a second diode is added in parallel with the first but connected in the opposite sense, the output will be clamped to ±0.7 V.

If the diode of Figure 19.14(a) is replaced with a Zener diode, as shown in Figure 19.14(b), the waveform is clamped for both positive and negative excursions of the input. If the input goes more positive than the breakdown voltage of the Zener diode V_Z, breakdown will occur, preventing the output from rising further. If the input goes negative by more than the forward turn-on voltage of the Zener, it will conduct, again clamping the output. The output will therefore be restricted to the range $+V_Z > V_o > -0.7$ V.

Two Zener diodes may be used, as shown in Figure 19.14(c), to clamp the output voltage to any chosen positive and negative voltages. Note that the voltages at which the output signal is clamped are the sums of the breakdown voltage of one of the Zener diodes V_Z and the turn-on voltage of the other Zener diode.

Figure 19.14   Signal-clamping circuits

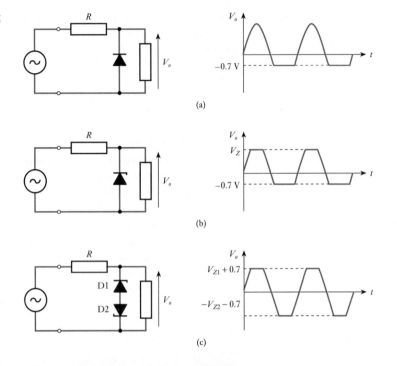

(a)

(b)

(c)

Files 19E
19F
19G

Computer Simulation Exercise 19.5

Use simulation to investigate the behaviour of the various circuits of Figure 19.14. Apply a sinusoidal input voltage of 10 V peak and use various combinations of simple diodes and Zener diodes. Suitable components might include 1N4002 signal diodes and D1N750 4.7 V Zener diodes.

Use transient analysis to look at the relationship between the input and the output waveforms.

19.8.5   Catch diodes

We saw in Chapter 4 that many actuators are inductive in nature. Examples include relays and solenoids. One problem with such actuators is that a large back e.m.f. is produced if they are turned off rapidly. This effect is used to advantage in some automotive ignition systems, in which a circuit breaker (the 'points') is used to interrupt the current in a high-voltage coil. The large back e.m.f. produced is used to generate the spark required to ignite the fuel in the engine. In other electronic systems, these reverse voltages can do serious damage to delicate equipment if they are not removed.

Figure 19.15 Use of a catch diode

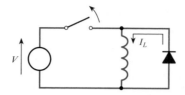

Fortunately, in many cases the solution is very simple. This involves placing a **catch diode** across the inductive component, as shown in Figure 19.15, to reduce the magnitude of this reverse voltage.

The diode is connected so that it is reverse-biased by the applied voltage and is therefore normally non-conducting. However, when the supply voltage is removed any back e.m.f. produced by the inductor will forward bias the diode, which therefore conducts and dissipates the stored energy. The diode must be able to handle a current equal to the forward current flowing before the supply was removed.

Key points

- Diodes allow current to flow in one direction but not the other.

- The electrical properties of materials are brought about by their atomic structure.

- At very low temperatures, semiconductors have the properties of an insulator. At higher temperatures, thermal vibration of the atomic lattice leads to the generation of mobile charge carriers.

- Pure semiconductors are poor conductors even at high temperatures. However, the introduction of small amounts of impurities dramatically changes their properties.

- Doping of semiconductors with appropriate materials can lead to the production of *n*-type or *p*-type materials.

- A junction between *n*-type and *p*-type semiconductors (a *pn* junction) has the properties of a diode.

- Semiconductor diodes approximate ideal diodes but have a conduction voltage. Silicon diodes have a conduction voltage of about 0.7 V.

- In addition to conventional *pn* junction diodes, there is a wide variety of more specialised diodes, such as Zener diodes, Schottky diodes, tunnel diodes and varactor diodes.

- Diodes are used in a range of applications in both analogue and digital systems, including rectification, demodulation and signal clamping.

Exercises

19.1 Explain the basic function of a diode and sketch the current–voltage characteristic of an ideal device.

19.2 What is the difference between a diode and a rectifier?

19.3 Describe briefly the electrical properties of conductors, insulators and semiconductors.

19.4 Name three materials commonly used for semiconductor devices. Which material is most widely used for this purpose?

19.5 Outline the effect of an applied electric field on free electrons and on holes.

19.6 What are meant by the terms 'intrinsic conduction' and 'extrinsic conduction'? What form of charge carriers are primarily responsible for conduction in doped semiconductors?

19.7 Explain what is meant by a 'depletion layer' and why this results in a potential barrier.

19.8 Explain the diode action of a *pn* junction in terms of the effects of an external voltage on the drift and diffusion currents.

19.9 Sketch the current–voltage characteristic of a silicon diode for both forward- and reverse-bias conditions.

19.10 Explain what is meant by the 'turn-on voltage' and the 'conduction voltage' of a diode. What are typical values of these quantities for a silicon diode?

19.11 What are typical values for the turn-on voltage and the conduction voltage of diodes formed from germanium and gallium arsenide?

19.12 Sketch a simple circuit that uses a Zener diode to produce a constant output voltage of 5.6 V from an input voltage that may vary from 10 to 12 V. Select appropriate component values such that the circuit will deliver a current of at least 100 mA to an external load, and estimate the maximum power dissipation in the diode.

19.13 A half-wave rectifier is connected to a 50 Hz supply and generates a peak output voltage of 100 V across a 220 µF reservoir capacitor. Estimate the peak ripple voltage produced if this arrangement is connected to a load that takes a constant current of 100 mA.

19.14 What would be the effect on the ripple voltage calculated in the last exercise of replacing the half-wave rectifier with a full-wave rectifier of similar peak output voltage?

19.15 Sketch the output waveforms of the following circuits. In each case the input signal is a sine wave of ±5 V peak.

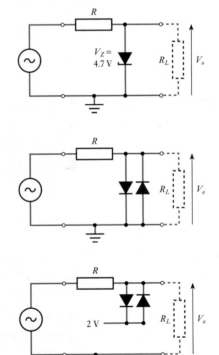

19.16 Use circuit simulation to verify your answers to the last exercise. How does the value of R_L affect the operation of the circuits?

19.17 Design a circuit that will pass a signal unaffected, except that it limits its excursion to the range $+10.4 \text{ V} > V > -0.4 \text{ V}$.

19.18 Use circuit simulation to verify your solution to the last exercise.

Chapter 20

Field-effect Transistors

Objectives

When you have studied the material in this chapter you should be able to:

- **list the various major forms of field-effect transistor (FET);**
- **describe the characteristics common to all forms of FET and explain how these characteristics make them suitable for use in amplifiers;**
- **explain the notation and symbols used in FET circuits;**
- **describe the physical operation of both MOSFETs and JFETs and explain how this influences the characteristics of these devices;**
- **outline the behaviour of FETs;**
- **discuss the use of FETs in a range of amplifier circuits;**
- **suggest a number of other uses of FETs in electronic circuits.**

20.1 Introduction

Field-effect transistors or FETs are probably the simplest form of transistor to understand and are widely used in both analogue and digital applications. They are characterised by a very high input resistance and small physical dimensions, and they can be used to create circuits with low power consumption, making them ideal for use in **very large-scale integration (VLSI)** circuits. There are two main forms of field-effect transistor, namely the *insulated-gate FET* and the *junction-gate FET*. We will look at both forms in this chapter.

Continuing our 'top-down' approach, we will begin by looking at the general behaviour of FETs and then turn out attention to their physical construction and operation. We will then consider the characteristics of these devices and look briefly at some simple circuits. The chapter ends by looking at the use of FETs in a range of analogue and digital applications.

Some readers may wish to omit the sections dealing with the physical construction and operation of the devices and concentrate on their external characteristics. Such readers may skip Sections 20.3 to 20.5 without compromising their understanding of the remainder of the chapter.

20.2   An overview of field-effect transistors

While there are many forms of field-effect transistor, the general operation and characteristics of these devices are essentially the same. In each case, a voltage applied to a control input produces an electric field, which affects the current that flows between two of the terminals of the device.

In Chapter 6, when considering simple amplifiers, we looked at an arrangement using an unspecified 'control device' to produce amplification. This arrangement is repeated in Figure 20.1(a). Here the control device varies the current flowing through a resistor in response to some input voltage. The output voltage V_o of this arrangement is equal to the supply voltage V minus the voltage across the resistor R. Therefore $V_o = V - IR$, where I is the current flowing through the resistor. The resistor current I is equal to the current flowing into the control device (ignoring any current flowing through the output terminal) and hence the output voltage is directly affected by the control device. If the current flowing in the control device is in turn determined by its input voltage V_i, then the output voltage of the circuit is controlled by this input voltage. Given an appropriate 'gain' in the control device this arrangement can be used to create voltage amplification.

In simple electronic circuits the control device of Figure 20.1(a) will often be some form of **transistor**, and it will generally be either a *field-effect transistor*, as discussed in this chapter, or a *bipolar transistor*, as discussed in the next chapter. Figure 20.1(b) shows a simple amplifier based on a FET. This diagram shows the circuit symbol for a junction-gate FET, although other forms of device could also be used. The input voltage to the FET controls the current that flows through the device and hence determines the output voltage as described above.

FETs have three terminals, which are called the **drain**, the **source** and the **gate**, labelled d, s and g, respectively, in Figure 20.1(b). It can be seen that the gate represents the **control input** of the device, and a voltage applied to this terminal will affect the current flowing from the drain to the source.

Figure 20.1   A FET as a control device

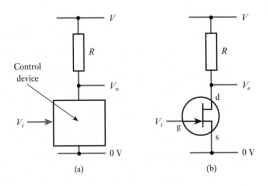

(a)   (b)

20.2.1   Notation

When describing FET circuits, we are often interested in the voltages between its various terminals and the current flowing into these terminals. We normally adopt a notation whereby voltages are given symbols of the form V_{XY}, where X and Y correspond to the symbols for two of the device's terminals. This symbol then represents the voltage on X with respect to Y. For example, V_{GS} would be used to represent the voltage on the *gate* with respect to the *source*. Device currents are given labels to represent the associated terminal. For example, the current into the drain would be labelled I_D. As discussed in earlier chapters, we normally use capital letters for steady voltages and currents and small letters for varying quantities. For example, V_{GS} and I_D represent steady quantities, while v_{gs} and i_d represent varying quantities. A special notation is used to represent the power supply voltages and currents in FET circuits. The voltage and current associated with the supply line that is connected (directly or indirectly) to the drain of the FET are normally given the labels V_{DD} and I_{DD}. Similarly, the corresponding labels for the supply connected to the source are V_{SS} and I_{SS}. In many cases, V_{SS} is taken as the zero volts reference (or ground) of the circuit. Passive components connected to the various terminals are often given corresponding labels, so that a resistor connected to the gate of a FET might be labelled R_G. This notation is illustrated in Figure 20.2.

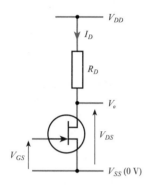

Figure 20.2   Labelling of voltages and currents in FET circuits

20.3   Insulated-gate field-effect transistors

In field-effect transistors, conduction between the drain and the source electrodes takes place through a **channel** of semiconductor material. Insulated-gate field-effect transistors are so-called because their metal-gate electrode is *insulated* from the conducting channel by a layer of insulating oxide. Such devices are often referred to as **IGFETs** (insulated-gate field-effect transistors) but are more commonly described as **MOSFETs** (metal oxide semiconductor field-effect transistors). Digital circuits constructed using such techniques are usually described as using **MOS technology**. Here we will refer to insulated-gate devices as MOSFETs.

The channel in a MOSFET can be made of either *n*-type or *p*-type semiconductor material, leading to two polarities of transistor, which are termed *n*-channel and *p*-channel devices. Figure 20.3 illustrates the construction of each form of device. The behaviour of these two forms is similar, except that the polarities of the various currents and voltages are reversed. To avoid duplication, we will concentrate on *n*-channel devices in this section.

An *n*-channel device is formed by taking a piece of *p*-type semiconductor material (the **substrate**) and forming *n*-type regions within it to represent the drain and the source. Electrical connections are made to these regions, forming the drain and source electrodes. A thin *n*-type channel

Figure 20.3 Insulated-gate field-effect transistors – MOSFETs

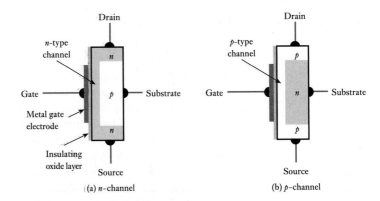

(a) *n*–channel

(b) *p*–channel

is then formed to join these two regions. This channel is covered by an insulating oxide layer and then by a metal-gate electrode. Electrical connections are made to the gate and to the substrate, although the latter is often internally connected to the source to form a device with three external connections.

20.3.1 MOSFET operation

The channel between the drain and the source represents a conduction path between these two electrodes and therefore permits a flow of current. With zero volts applied to the gate, a voltage applied between the drain and the source will produce an electric field that will cause the mobile charge carriers in the channel to flow, thus creating a current. The magnitude of this current will be determined by the applied voltage and the number of charge carriers available in the channel.

The application of a voltage to the gate of the device will affect the number of charge carriers in the channel and hence the flow of electricity between the drain and source. The metal-gate electrode and the semiconductor channel represent two conductors separated by an insulating layer. The construction therefore resembles a capacitor, and the application of a voltage to the gate will induce the build-up of charge on each side. If the gate of an *n*-channel MOSFET is made *positive* with respect to the channel this will attract electrons to the channel region, increasing the number of mobile charge carriers and increasing the apparent thickness of the channel. Under these circumstances, the channel is said to be **enhanced**. If the gate is made *negative* this will repel electrons from the channel, reducing its thickness. Here the channel is said to be **depleted**. Thus the voltage on the gate directly controls the effective thickness of the channel and the resistance between the drain and the source. This process is illustrated in Figure 20.4.

Note that the junction between the *p*-type substrate and the various *n*-type regions represents a *pn* junction, which will have the normal properties of a semiconductor diode. However, in normal operation the voltages

Figure 20.4 The effect of gate voltage on a MOSFET

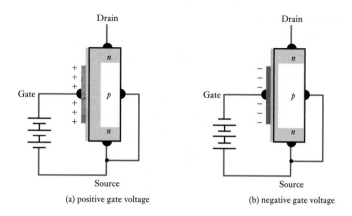

(a) positive gate voltage   (b) negative gate voltage

applied ensure that this junction is always reverse-biased, so that no current flows. For this reason, the substrate can be largely ignored when considering the operation of the device.

20.3.2 Forms of MOSFET

The MOSFETs shown in Figure 20.3 and described above can be used with both positive and negative gate voltages, which can be used to enhance or deplete the channel. Such a device is called a **depletion–enhancement MOSFET** or **DE MOSFET** (or sometimes simply a **depletion MOSFET**).

Other forms of MOSFET are constructed in a similar manner, except that no channel is formed during manufacture. In the absence of a gate voltage there is no conduction path between the drain and the source, and no current will flow. However, the application of a positive voltage to the gate (of an *n*-channel device) will attract electrons and repel holes from the region around the gate to form a conducting *n*-type channel. This region is called an inversion layer, since it represents an *n*-type layer in a *p*-type material. A device of this type can be used in an enhancement mode, as with the depletion–enhancement MOSFET, but it cannot be used in a depletion mode. For this reason, such devices are called **enhancement MOSFETs**.

To avoid confusion, different circuit symbols are used for the various types of MOSFET, and these are shown in Figure 20.5. The thick vertical line within the symbols represents the channel and is shown solid in a DE MOSFET (since it is present even in the absence of a gate voltage) and dotted in an enhancement MOSFET (since it is present only when an appropriate gate voltage is applied). The arrow on the substrate indicates the polarity of the device. This represents the *pn* junction between the substrate and the channel and points in the same direction as the symbol of an equivalent diode. That is, it points towards an *n*-type channel and away from a *p*-type channel. Note that within the circuit symbols the substrate is given the label 'b' (which stands for *bulk*) to avoid confusion with the source. As

Figure 20.5 MOSFET circuit symbols

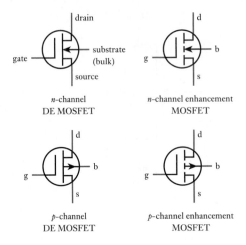

noted earlier, the substrate is often joined to the source internally to give a device with only three terminals. If the significance of the various elements of the symbols is considered, then it is relatively easy to remember the symbols for the various types of device.

20.4 Junction-gate field-effect transistors

As in a MOSFET, conduction in a junction-gate transistor takes place through a channel of semiconductor material. However, in this case the conduction within the channel is controlled not by an insulated gate but by a gate formed by a reverse-biased *pn* junction. Junction-gate FETs are sometimes referred to as **JUGFETs**, but here we will use another widely used abbreviation – **JFET**.

The form of a JFET is illustrated in Figure 20.6, which shows both *n*-channel and *p*-channel versions. As with MOSFETs, the operation of *n*-channel and *p*-channel devices is similar except for the polarity of the voltages and currents involved. Here we will concentrate on *n*-channel devices.

Figure 20.6(a) shows an *n*-channel JFET. Here a substrate of *n*-type material is used, with electrical connections at each end to form the drain and the source. A region of *p*-type material is now added between the drain and the source to form the gate. The fusion of the *n*-type and *p*-type materials forms a *pn* junction, which has the electrical properties of a semiconductor diode. If this junction is forward-biased (by making the gate *positive* with respect to the other terminals of the device) current will flow across this junction. However, in normal operation the gate is kept *negative* with respect to the rest of the device, reverse-biasing the junction and preventing any current from flowing across it. This situation is shown in Figure 20.7.

Figure 20.6 Junction field-effect transistors – JFETs

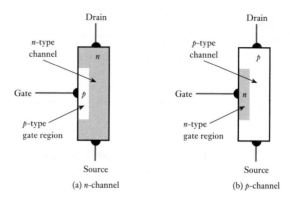

(a) *n*-channel (b) *p*-channel

Figure 20.7 An *n*-channel JFET with negative gate bias

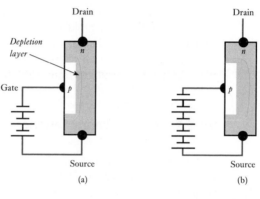

(a) (b)

Figure 20.8 JFET circuit symbols

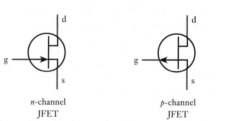

n-channel
JFET

p-channel
JFET

In the last chapter we noted that reverse biasing a *pn* junction creates a **depletion layer** about the junction in which there are very few mobile charge carriers. Such a region is effectively an insulator, and the formation of such a region about the gate in a JFET reduces the cross-sectional area of the channel and increases its effective resistance. This is shown in Figure 20.7(a). As the magnitude of the reverse-bias voltage is increased, the thickness of the depletion layer increases and the channel is further reduced, as shown in Figure 20.7(b). Thus the resistance of the channel is controlled by the voltage applied to the gate.

The circuit symbols used for *n*-channel and *p*-channel JFETs are shown in Figure 20.8. As in the MOSFET the arrow shows the polarity of the device, and as before it points *towards* an *n*-type channel and *away from* a *p*-type channel.

While MOSFETs and JFETs operate in somewhat different ways, their characteristics are in many ways quite similar. When considering amplifiers in Chapter 6, we characterised them in terms of the nature of their input, their output and the relationship between them. We can consider the characteristics of field effect transistors in a similar manner.

20.5.1   Input characteristics

Both MOSFETs and JFETs have a very high input resistance. In a MOSFET the gate electrode is insulated from the rest of the device by an oxide layer, which prevents any current flowing into this terminal. In a JFET the gate takes the form of a *pn* junction that is kept in a reverse-biased state. In Chapter 19, we noted that the current across such a reverse-biased junction is negligible in almost all cases. Thus in both MOSFETs and JFETs the gate is effectively insulated from the remainder of the device.

20.5.2   Output characteristics

The output characteristics of a device describe how the output voltage affects the output current. When considering amplifiers in Chapter 6, we saw that in many cases this relationship can be adequately represented by a simple, fixed output resistance. However, in transistors the situation is slightly more complicated, and we need to look in more detail at how the device operates when connected to an external power supply.

In most circuits that use an *n*-channel FET a voltage is applied across the device such that the drain is positive with respect to the source, so that V_{DS} is positive. The applied voltage produces a **drain current** I_D through the channel between the drain and the source. As the current flows through the channel, its resistance produces a potential drop such that the potential gradually falls along the length of the channel. As a result, the voltage between the gate and the channel varies along the length of the channel.

The effect of this variation in gate-to-channel voltage on a MOSFET is shown in Figure 20.9(a). Here the voltage on the gate is positive with respect to the source, and a correspondingly larger positive voltage is applied to the drain. In this arrangement, the potential in the channel at the end adjacent to the drain terminal makes it more positive than the gate terminal, and the channel in this area is *depleted*. At the other end of the channel, the potential is approximately equal to that of the source, which is negative with respect to the gate, and the channel is *enhanced*. Thus the apparent thickness of the channel changes along its length, being narrower near the drain and wider nearer the source. In this example, the gate voltage is positive with respect to the source, but in a DE MOSFET the gate voltage could alternatively be zero or negative with respect to the source.

Figure 20.9 Typical FET circuit configuration

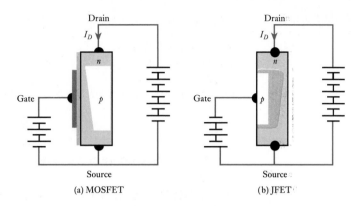

(a) MOSFET   (b) JFET

This would reduce the average thickness of the channel, but the tapering effect of the variation in channel potential would still be present.

Figure 20.9(b) shows the corresponding situation in a JFET. In this case, the gate-to-source voltage is negative and the drain-to-source voltage is positive. Here the gate-to-channel voltage is negative along the length of the channel (ensuring that the gate junction is reverse-biased throughout its length). However, at the drain end of the channel this reverse-bias voltage is much greater than at the source end, so the depletion layer is much thicker near the drain than near the source. This results in a tapered channel, as shown in the figure, where the channel is much narrower at the drain end than at the source end.

It can be seen that in both the MOSFET and the JFET the thickness of the channel is controlled by the voltage applied to the gate, but it is also influenced by the drain-to-source voltage. For small values of V_{DS}, as the gate-to-source voltage V_{GS} is made more positive (for n-channel FETs) the channel thickness increases and the effective resistance of the channel is decreased. The behaviour of the channel resembles that of a resistor, with the drain current I_D being proportional to the drain voltage V_{DS}. The value of this effective resistance is controlled by the gate voltage V_{GS}, the resistance getting smaller as V_{GS} is made more positive. This is referred to as the **ohmic region** of the device's operation.

As V_{DS} is increased the channel becomes more tapered, and eventually the channel thickness is reduced to approximately zero at the end near the drain. The channel is now said to be **pinched-off**, and the drain-to-source voltage at which this occurs is called the **pinch-off voltage**. This does not stop the flow of current through the channel, but it does prevent any further increase in current. Thus, as the drain voltage is increased above the pinch-off voltage the current remains essentially constant. This is referred to as the **saturation region** of the characteristic.

If the gate voltage is held constant and the drain voltage is gradually increased, the current initially rises linearly with the applied voltage (in the ohmic region) and then, above the pinch-off voltage, becomes essentially constant (in the saturation region). This behaviour is shown in Figure 20.10(a). Varying the voltage on the gate changes the effective resistance of the

Figure 20.10 FET output
characteristics

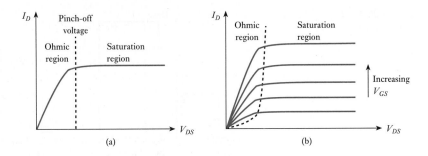

channel in the ohmic region and also changes the value of the steady current produced in the saturation region. This can be illustrated by drawing the output characteristic of Figure 20.10(a) for a range of different gate voltages. This is shown in Figure 20.10(b), which represents the **output characteristics** of the device. This figure shows a generic set of curves that could represent any form of FET. The range of values of V_{GS} will depend on the type of device concerned, as discussed in earlier sections.

From the output characteristics, it is clear that FETs have two distinct regions of operation, namely the ohmic region and the saturation region. In the first of these the device resembles a **voltage-controlled resistance**, and there are several applications that make use of this characteristic. In the second region (saturation), the output current (I_D) is largely independent of the applied voltage and is controlled by the input voltage (V_{GS}). It is this second region that is normally used in the creation of amplifiers. When used in this region, it is the slight slope of the output characteristic that indicates how the output current changes with the output voltage. This slope therefore represents the **output resistance** of the device in that operating region.

20.5.3   Transfer characteristic

Having considered the input and the output characteristics of FETs, we can now turn our attention to the relationship between the input and the output. This is often termed the **transfer characteristic** of the device.

In Chapter 6, when we looked at voltage amplifiers, we represented the relationship between the input and the output by the gain of the circuit, this being equal to V_o/V_i. However, in the case of a FET the input quantity is the gate voltage, while the output quantity is the drain current. It is also clear from Figure 20.10(b) that there is not a linear relationship between these two quantities. If we arrange that the device remains within the saturation region we can plot the relationship between the input voltage V_{GS} and the output current I_D, and this is shown in Figure 20.11 for various forms of FET.

It can be seen that the basic form of the transfer function is similar for each device, although the characteristics are offset with respect to each

Figure 20.11 FET transfer characteristics

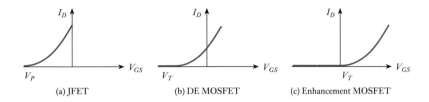

other because of their different gate voltage ranges. For most FETs, the relationship between I_D and V_{GS} is approximately parabolic and is described by an expression of the form

$$I_D = K(V_{GS} - V')^2 \tag{20.1}$$

where K is a constant that depends on the physical parameters of the device, and V' is the voltage at which the FET begins to conduct. In MOSFETs V' represents the **threshold voltage** V_T of the device, while in JFETs it is the pinch-off voltage V_P. The voltages corresponding to V_T and V_P are shown in Figure 20.11.

While the characteristics of Figure 20.11 are clearly not linear, over a small range of values of V_{GS} they may be said to approximate to a linear relationship. Thus, if we restrict V_{GS} to small fluctuations about a particular mean value (termed the **operating point**), the relationship between variations in V_{GS} and the resultant variations in I_D is *approximately* linear. Figure 20.12 illustrates the operating points and normal operating ranges for the three forms of FET discussed earlier.

When the device is constrained to operation about its operating point, then the transfer characteristics of a FET are described by the *change* in the output that is produced by a corresponding *change* in the input. This corresponds to the slope of the curve at the operating point in the graphs of Figure 20.12. This quantity has the units of current/voltage, which is the reciprocal of resistance. In Chapter 2 we defined this as **conductance**. Since this quantity describes the transfer properties of the FET it is given the name **transconductance**, and like conductance it has the units of siemens. The symbol given to the transconductance is g_m.

It should be noted that g_m represents the slope of the transfer characteristic at the operating point, *not* the ratio of the drain current to the gate

Figure 20.12 Normal operating ranges for FETs

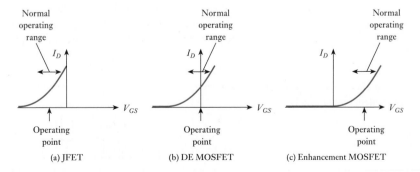

voltage at this point. Thus, if a small change in the gate voltage ΔV_{GS} produces a small change in the drain current ΔI_D, then

$$g_m = \frac{\Delta I_D}{\Delta V_{GS}} \tag{20.2}$$

$$g_m \neq \frac{I_D}{V_{GS}}$$

From Figure 20.12, it is clear that the slope of the transfer function varies along the curve and hence g_m is not constant for a given device. In fact, g_m is proportional to the square root of the drain current, which is determined by the design of the circuit.

20.5.4 Equivalent circuit of a FET

In Chapter 6, we noted the usefulness of equivalent circuits in describing the characteristics of amplifiers. We are now in a position to construct an equivalent circuit for a FET, and this is shown in Figure 20.13. The circuit shows no input resistor, since the input resistance is so high that it can normally be considered to be infinite. Since the output of a FET is normally considered to be the drain *current*, the equivalent circuit models the output using a Norton equivalent circuit rather than the Thévenin arrangement used in Chapter 6. The equivalent circuit describes the behaviour of the device when its input fluctuates by a small amount about its normal operating point, rather than describing its behaviour in response to constant (DC) voltages. For this reason, it is referred to as a **small-signal equivalent circuit**.

The equivalent circuit represents the transfer characteristics of the FET using the dependent current source, which produces a current of $g_m v_{gs}$, where v_{gs} is the fluctuating, or small-signal, input voltage. This flows *downwards*, since the drain current i_d is assumed to flow *into* rather than *out of* the device. The resistor r_d models the way in which the output current is affected by the output voltage. If the lines in the output characteristic of Figure 20.10(b) were completely horizontal then the output current would be independent of the output voltage. In practice these have a slight slope, giving rise to r_d, which is termed the small-signal **drain resistance**. This resistance is given by the slope of the output characteristic, and for this reason it is also known as the output **slope resistance**.

Figure 20.13 A small-signal equivalent circuit of a FET

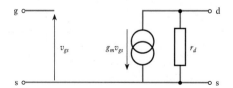

20.6 Summary of FET characteristics

At this stage it may be useful to summarise the characteristics of FETs, if only for the benefit of those readers who have decided to omit the earlier sections, which cover FET operation and characteristics. To a large extent we can understand the use of FETs (at least at a superficial level) simply by knowing that:

- FETs have three terminals called the drain, the source and the gate (some FETs have a fourth terminal called the substrate, but this is usually connected internally, or externally, to the source).
- The control input of a FET is the gate, and the voltage applied to the gate controls the current that flows from the drain to the source.
- Two polarities of FET are available, namely **n-channel FETs** and **p-channel FETs**. The characteristics of these two forms of device are similar, except that the polarities of the voltages and currents used are reversed. In this chapter, we are concentrating on n-channel devices to avoid duplication.
- In n-channel FETs, current consists of a flow of *negative* charge carriers (electrons) in the channel. Similarly, in p-channel devices current takes the form of a flow of *positive* charge carriers (holes). For this reason, FETs are referred to as **unipolar transistors**, since the current consists of only one polarity of charge carriers in each case.
- Two main forms of FET are used, namely MOSFETs and JFETs. MOSFETs can then be subdivided into DE MOSFETs and enhancement MOSFETs. Although the operation of these different forms is often quite different, their characteristics are very similar.
- All FETs have a very high input resistance. This is usually so high that input currents can almost always be ignored.
- In circuits using n-channel FETs the drain is normally more positive than the source, and the drain current represents the controlled (output) quantity.
- The drain current is controlled by the voltage applied to the gate, and in n-channel FETs making the gate voltage more positive increases the drain current. The way in which variations in gate voltage affect the drain current is described by the transconductance g_m of the device. If a small change in the gate voltage ΔV_{GS} produces a small change in the drain current ΔI_D, then $g_m = \Delta I_D / \Delta V_{GS}$.
- The operation of FETs is essentially non-linear. However, approximately linear control can be achieved for small variations of the gate voltage about an appropriate **operating point**. This is termed small-signal operation.
- The operating point required depends on the form and type of FET used. When using n-channel devices the operating point will normally be negative in JFET circuits, zero in DE MOSFET circuits and positive in enhancement MOSFET circuits.
- The circuit symbols for the various forms of FET are summarised in Figure 20.14.

Figure 20.14   FET circuit
symbols

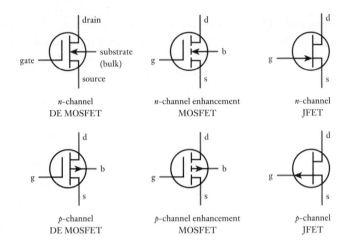

20.7   FET amplifiers

The basic amplifier of Figure 20.1(b) can be adapted to suit any form of
FET by adding additional circuitry to 'bias' the gate to the appropriate
operating point for the FET used. For example, when using a DE MOSFET
it is normal to bias the circuit so that the operating point is at zero volts.
In other words, the circuit is designed so that, in the absence of any input
signal, the voltage on the gate (with respect to the source) is zero. This can
be achieved simply by connecting a resistor from the gate to ground. A
coupling capacitor (also called a **blocking capacitor**) is then used to *couple*
the input signal to the amplifier while *blocking* any DC component from
upsetting the biasing of the FET. Such a circuit is shown in Figure 20.15.

When using an enhancement MOSFET, it is normal to bias the gate to
an appropriate positive voltage. This can be done by replacing the single-
gate resistor in Figure 20.15 with a pair of resistors forming a potential
divider between V_{DD} and V_{EE}. The values of these resistors are chosen to
produce a bias voltage appropriate to the FET used. The normal operating
point for a JFET requires a negative-bias voltage. This can be produced by
connecting the gate resistor to a negative voltage supply (if this is available)

Figure 20.15   A simple DE
MOSFET amplifier

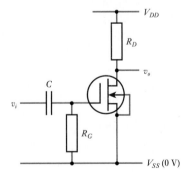

or by connecting the gate to zero volts and adding a resistor between the source and the V_{SS}. Current flowing through this source resistor will make the source positive with respect to V_{SS} and hence with respect to the gate. The gate will therefore be negative with respect to the source as required.

The circuit of Figure 20.15 cannot amplify DC signals because of the presence of the capacitor C. However, AC signals applied to the input will be coupled through the capacitor and change the gate voltage. The circuit is therefore an **AC-coupled amplifier** or simply an **AC amplifier**. In the absence of any input voltage, the circuit is said to be in its *quiescent* state. The drain current flowing though the FET under these conditions is termed the **quiescent drain current**. This current will flow through the drain resistor, and the resulting voltage drop will determine the **quiescent output voltage**. When an input signal causes the gate to become more *positive*, this will increase the current through the FET (and hence through R_D), increasing the voltage drop across the resistor and hence making the output more *negative*. Similarly, when the input makes the gate more *negative*, this will reduce the current in the FET and the resistor and make the output more *positive*. The arrangement is therefore an inverting amplifier.

Since the input resistance of the FET is extremely large, the resistance seen looking into the input of the circuit of Figure 20.15 is determined by C and R_G. Normally, C will be chosen so that it looks like a short circuit at the frequencies to be amplified, so the capacitor can usually be ignored. Thus in most cases the input resistance is equal to R_G, which can be selected to suit the application. Further analysis of the circuit shows that the output resistance of the circuit is approximately equal to R_D, while the gain of the circuit is approximately $-g_m R_D$ (the minus sign indicating that this is an inverting amplifier). As mentioned earlier, the value of C is normally chosen to minimise its effects, but its presence produces a low-frequency cut-off as discussed in Chapter 17, with a cut-off frequency f_c equal to $1/2\pi CR$, where R is the input resistance of the amplifier.

Example 20.1

Determine the input resistance, the output resistance, the small-signal voltage gain and the low-frequency cut-off of the following circuit, given that $g_m = 2$ **mS**.

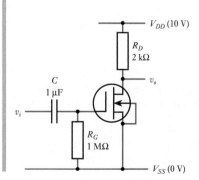

Since the input resistance of the FET is extremely high, the input resistance of the amplifier is simply equal to R_G, which is 1 MΩ.

From the above, the output resistance is approximately equal to R_D, which is 2 kΩ.

The small-signal voltage gain is given by

$$G = -g_m R_D = -2 \times 10^{-3} \times 2 \times 10^3 = -4$$

where the minus sign indicates that this is an inverting amplifier.

The low-frequency cut-off is given by

$$f_c = \frac{1}{2\pi CR} = \frac{1}{2\pi \times 10^{-6} \times 10^6} = 0.16 \text{ Hz}$$

20.7.1   Biasing considerations

Having considered the small-signal (or AC) behaviour of the circuit of Figure 20.15, we now need to consider the **biasing** of the circuit. This refers to those aspects of the design that determine its quiescent (or DC) operation. Of central importance is the quiescent drain current $I_{D(\text{quies})}$, which in turn determines the quiescent output voltage $V_{o(\text{quies})}$. Since the output voltage is equal to the supply voltage minus the voltage across R_D, it follows that

$$V_{o(\text{quies})} = V_{DD} - I_{D(\text{quies})} R_D$$

In many cases, V_{DD} will be fixed and the design will require a particular value for the quiescent output voltage. This will require that an appropriate combination of $I_{D(\text{quies})}$ and R_D be used to produce the required output voltage.

The quiescent drain current in the circuit will be determined partly by the FET's characteristics and partly by the circuit's biasing arrangement. The device's data sheet will give information on its characteristics and will indicate how the drain current relates to the gate-to-source voltage. The information given will vary depending on the type of FET being used. For example, for a DE MOSFET this might include the threshold voltage V_T and the magnitude of the drain current when the gate-to-source voltage is zero. This latter parameter is termed the gate-to-source saturation current, which is given the symbol I_{DSS}. From the information in the datasheet the designer must pick appropriate circuit components to set the required quiescent conditions, in addition to satisfying the small-signal requirements.

Example 20.2

The circuit of Figure 20.15 is constructed using a supply voltage V_{DD} of 12 V and a DE MOSFET that has an I_{DSS} of 4 mA. Select a value for R_D to give a quiescent output voltage of 8 V.

In this circuit the gate is joined to ground through R_G, so $V_{GS} = 0$. Since I_{DSS} is defined as the drain current when $V_{GS} = 0$, it follows that the quiescent drain-to-source current is given by

$$I_{D(\text{quies})} = I_{DSS} = 4 \text{ mA}$$

Now

$$V_{o(\text{quies})} = V_{DD} - I_{D(\text{quies})} \times R_D$$

Therefore, if $V_o = 8$ V

$$8 = 12 - 4 \times 10^{-3} \times R_D$$

$$R_D = 1 \text{ k}\Omega$$

20.7.2   Negative feedback amplifier

In Chapter 6, we noted that all active components suffer from considerable variability in their characteristics, and this leads to major problems in the circuit of Figure 20.15. Variations in the transconductance of the device affect both the quiescent output voltage and the gain of the circuit. In Chapter 7, we saw how negative feedback could be used to tackle variability, and in Chapter 8 we looked at the application of such techniques to operational amplifier circuits. We can also use negative feedback to tackle variability in FET amplifiers, and an example of such a circuit is shown in Figure 20.16.

While the circuit of Figure 20.16 is based on an enhancement MOSFET, similar circuits can be constructed using other forms of FET, provided that appropriate adjustments are made to the biasing arrangement. Since the normal operating point of an enhancement MOSFET requires a positive gate voltage, the single-gate resistor used in the circuit of Figure 20.15 is replaced by a potential divider formed by R_1 and R_2. A source resistor R_S has also been added between the source and ground. The voltage across this resistor is directly proportional to the source current which is equal to the drain current. When an input voltage is applied to the circuit this appears between the gate and ground. The voltage across the source resistor is then subtracted from this input voltage to produce the voltage that is applied between the gate and the source of the FET. Thus a voltage proportional to the output current is subtracted from the input, and we have negative feedback. The use of negative feedback reduces the gain of the circuit but produces a circuit with much more stable characteristics.

Figure 20.16   A negative feedback amplifier

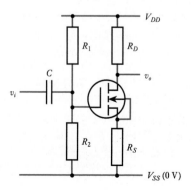

Computer Simulation Exercise 20.1

Simulate the feedback amplifier of Figure 20.16 using an enhancement MOSFET such as the IRF150. Suitable component values when using the IRF150 are $V_{DD} = 12$ V, $R_1 = 2$ MΩ, $R_2 = 1$ MΩ, $R_D = 3.3$ kΩ, $R_S = 1$ kΩ and $C = 1$ µF. If your simulation package does not provide this particular MOSFET, then you may have to experiment with the component values.

Apply a 1 kHz sinusoidal input voltage of 1 V peak and use transient analysis to investigate the relationship between the input and output waveforms. What happens to the output of your amplifier if the magnitude of the input is progressively increased?

20.7.3   The source follower

Another form of FET amplifier is shown in Figure 20.17. This is similar to the circuit of Figure 20.15, except that a source resistor has been added and the output is now taken from the source rather than the drain. This circuit again uses negative feedback, but here the feedback drives the output to be equal to the input such that the output *follows* the input. For this reason, the circuit is often called a source follower. It has the advantage of having a very high input resistance and a very low output resistance and is therefore useful as a unity-gain buffer amplifier.

Figure 20.17   Source follower
amplifier

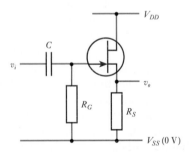

Computer Simulation Exercise 20.2

Simulate the source follower circuit of Figure 20.17 using a JFET such as the 2N3819. Suitable component values when using this JFET are $V_{DD} = 12$ V, $R_G = 100$ kΩ, $R_S = 3.3$ kΩ and $C = 1$ µF. If your simulation package does not provide this particular JFET then you may have to experiment with the component values.

Apply a 1 kHz sinusoidal input voltage of 1 V peak and use transient analysis to investigate the relationship between the input and output waveforms. What happens to the output of your amplifier if the magnitude of the input is progressively increased?

20.8 Other FET applications

In addition to their uses in amplifiers, FETs can also be used to produce a range of other functions in both analogue and digital circuits. Here we will look at just a few examples, but we will see other uses of FETs in later chapters. In each example only a single form of FET is shown, although in many cases other forms could be used with minor modifications to the circuits.

20.8.1 A voltage-controlled attenuator

For small values of drain-to-source voltage, FETs resemble voltage-controlled resistors, where the apparent resistance between the drain and the source is controlled by the gate voltages. This can be used to form a **voltage-controlled attenuator** as shown in Figure 20.18. The FET is used in a potential divider together with a fixed resistor. The gate voltage V_G determines the effective resistance of the FET and hence the effective gain of the circuit. An arrangement of this form is often used in the feedback path of an amplifier to achieve **automatic gain control**. Here the output voltage of an amplifier is measured and used to control the gain, such that the output amplitude is kept constant. This technique is used, for example, to keep the volume of a radio receiver constant, even if the strength of the radio signal changes.

Figure 20.18 A voltage-controlled attenuator

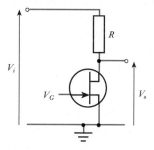

20.8.2 A FET as an analogue switch

By applying a suitable voltage to the gate of a FET, the effective drain-to-source resistance can be varied from a resistance that approximates a short circuit (or a 'closed' switch) to a resistance that approximates an open circuit (or an 'open' switch). This allows the device to be used as an **electrically controlled switch** for analogue signals. Both series and shunt arrangements are possible, as shown in Figure 20.19. In Figure 20.19(a), the FET is placed in series with the signal path, so that when the switch is 'closed' the signal is transmitted, and when it is 'open' the signal is blocked. In Figure 20.19(b), the switch is connected across the signal path, so that when the switch is 'closed' the signal is effectively shorted to ground, and when the switch is 'open' the signal is transmitted.

Figure 20.19 A FET as an analogue switch

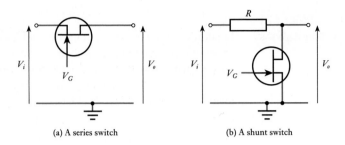

(a) A series switch   (b) A shunt switch

20.8.3   A FET as a logical switch

In addition to their use in analogue circuits, FETs (particularly MOSFETs) are widely used in digital applications. For example, Figure 20.20(a) shows a logical inverter (a NOT gate) implemented using a MOSFET. In this circuit, logical 0 is represented by 0 V, and logical 1 is represented by a voltage equal to V_{DD}. An input voltage of 0 V will turn the enhancement FET off, so that no current flows through the device. This will produce no current through the resistor R, so the voltage across it will be zero and the output voltage will be V_{DD} (which corresponds to logical 1). If the input voltage is equal to V_{DD} the FET will be turned on, drawing current through the resistor and pulling the output down towards zero volts (logical 0). Thus the circuit has the characteristics of a logical inverter.

The circuit of Figure 20.20(a) is quite acceptable when discrete components are used but is unattractive when an integrated circuit is to be produced. One of the reasons why MOSFETs are so widely used in digital integrated circuits is that each transistor requires a very small area of silicon, allowing a large number of devices to be fabricated on a single chip. Resistors, on the other hand, occupy a proportionately larger area, making them components to be avoided wherever possible. When producing logical inverters using MOSFETs it is more efficient to use the circuit shown in Figure 20.20(b), where a second MOSFET is used as an **active load**. However, even better performance can be achieved using combinations of *n*-channel and *p*-channel devices to form complementary MOS (or CMOS) circuits. These will be discussed when we return to look at logic circuits in more detail in Chapter 25.

Figure 20.20 Logical inverters using MOSFETs

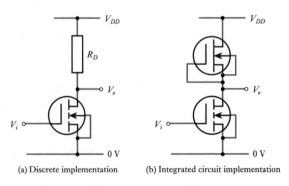

(a) Discrete implementation   (b) Integrated circuit implementation

Key points

- Field-effect transistors are widely used in both analogue and digital applications.

- They are characterised by a very high input resistance and a small physical size, and they can be used to produce circuits with very low power consumption.

- There are two basic forms of field-effect transistor, namely the MOSFET and the JFET. Although these operate in rather different ways, many of their characteristics are similar.

- All FETs are voltage-controlled devices in which the voltage on the gate controls the current that flows between the drain and the source.

- The characteristics of DE MOSFETs, enhancement MOSFETs and JFETs are similar, except that they require different bias voltages.

- The use of coupling capacitors prevents the amplification of DC and produces AC-coupled amplifiers.

- Like other active devices, FETs suffer from variability in their characteristics. This can be overcome through the use of negative feedback.

- FETs can be used not only to produce amplifiers but also in a range of other applications such as voltage-controlled attenuators, analogue switches and logic gates.

Exercises

20.1 Why are field-effect transistors so called?

20.2 What characteristics of FETs make them ideal for use in integrated circuits?

20.3 Name the three terminals common to all FETs. Which of these constitutes the control input?

20.4 What is meant by the symbols V_{DS}, I_D, V_{GS}, v_{gs}, R_S, V_{DD} and V_{SS}?

20.5 What name is given to the conductive path between the drain and the source in a FET?

20.6 Explain the difference between an IGFET and a MOSFET.

20.7 What is the difference between an n-channel FET and a p-channel FET? How do their characteristics compare?

20.8 What is meant by the 'substrate' of a MOSFET? What character is used to represent this terminal in its circuit symbol?

20.9 Explain what is meant by 'enhancement' of a channel. In an n-channel MOSFET, what polarity of gate-to-source voltage will cause the channel to be enhanced?

20.10 Explain what is meant by 'depletion' of a channel. In an n-channel MOSFET, what polarity of gate-to-source voltage will cause the channel to be depleted?

20.11 Explain the difference between a DE MOSFET and an enhancement MOSFET.

20.12 What is used in a JFET to replace the insulated gate found in a MOSFET?

Exercises continued

20.13 In a circuit based on an *n*-channel JFET, what would be the normal polarity of the gate-to-source voltage? Why?

20.14 Explain the significance of the depletion layer in a JFET.

20.15 How is the polarity of a FET (that is, *n*-channel or *p*-channel) indicated in its circuit symbol?

20.16 Describe the input characteristics of a MOSFET and a JFET.

20.17 Sketch typical output characteristics for a FET, indicating the ohmic region, the saturation region and the pinch-off voltage.

20.18 In which region of its characteristic does a FET resemble a voltage-controlled resistance?

20.19 Explain the concept of an operating point.

20.20 Define the term 'transconductance' and give its units.

20.21 Sketch the small-signal equivalent circuit of a FET.

20.22 Explain the functions of R_G and C in the circuit of Figure 20.15.

20.23 What is meant by the quiescent output voltage of a circuit?

20.24 Determine the input resistance, the output resistance, the small-signal voltage gain and the low-frequency cut-off of the following circuit, given that $g_m = 3$ mS.

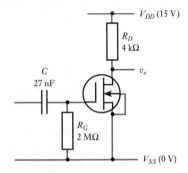

20.25 Repeat Computer Simulation Exercise 20.1 and note the effect of increasing the input voltage until the output becomes grossly distorted. What are the maximum and minimum values of the output voltage obtainable from the circuit? What causes these limitations?

20.26 What is the voltage gain of a source follower amplifier?

20.27 Repeat Computer Simulation Exercise 20.2 and note the effect of increasing the input voltage until the output becomes grossly distorted. What are the maximum and minimum values of the output voltage obtainable from the circuit? What causes these limitations?

20.28 Explain the use of a FET in an automatic gain control circuit.

20.29 Discuss the use of a FET as an analogue switch.

20.30 Explain why the circuit of Figure 20.20(b) might be more attractive than that of Figure 20.20(a) in some situations.

Chapter 21

Bipolar Transistors

Objectives

When you have studied the material in this chapter you should be able to:

- **explain the importance of bipolar transistors in modern electronic circuits;**
- **describe the construction, operation and characteristics of bipolar transistors;**
- **analyse simple amplifier circuits based on transistors and determine their operating conditions and voltage gain;**
- **discuss the importance of negative feedback in overcoming variability in such circuits;**
- **describe the low-frequency behaviour of amplifiers that use coupling capacitors;**
- **list a range of applications of bipolar transistors in addition to their uses in amplifiers.**

21.1 Introduction

Bipolar transistors are one of the main 'building blocks' in electronic systems and are used in both analogue and digital applications. The devices incorporate two *pn* junctions and are also known as **bipolar junction transistors** or **BJTs**. It is common to refer to bipolar transistors as simple 'transistors', the term FET being used to identify field-effect transistors.

Bipolar transistors get their name from the fact that current is carried by both polarities of charge carriers (that is, by electrons and by holes), unlike FETs, which are *unipolar*. Bipolar transistors generally have a higher gain than FETs and can often supply more current. However, they have a lower input resistance than FETs, are more complex in operation and often consume more power.

We will start by looking at the general behaviour of bipolar transistors before turning our attention to their physical construction, operation and characteristics. We will then see how such devices can be used in amplifier circuits and a range of other applications. Readers who wish to omit the sections dealing with the physical operation of the devices can skip Sections 21.3 and 21.4 without compromising their understanding of the remainder of the chapter.

21.2   An overview of bipolar transistors

In Chapter 6, we considered the use of a 'control device' in the construction of an amplifier, and in Chapter 20 we saw how a FET could be used in such an arrangement. Bipolar transistors may be used in a similar manner, and this is shown in Figure 21.1. The devices have three terminals called the **collector**, the **base** and the **emitter**, which are given the symbols c, b and e. The base is the control input, and signals applied to this terminal affect the flow of current between the collector and the emitter. As in the corresponding FET circuit, variations in the input to the transistor alter the current flowing through the resistor R, and hence determine the output voltage V_o. However, the behaviour of bipolar transistors differs from that of FETs, producing circuits with somewhat different characteristics.

While FETs are 'voltage-controlled' devices, bipolar transistors are often considered to be **current-controlled** components. When a control *current* is supplied to the base of a transistor, this causes a larger current to flow from the collector to the emitter (provided that external circuitry is able to supply this current). When used in this way the transistor acts as an almost linear **current amplifier**, where the output current (the current flowing into the collector) is directly proportional to the input current (the current flowing into the base). This relationship is illustrated in Figure 21.2. The **current gain** produced by a transistor might be a hundred or more and is relatively constant for a given device (although it will vary with temperature).

While it is common to view the transistor as a current-controlled device, an alternative view is to consider the input to be the *voltage* applied to its base. This voltage produces an input current, which is then amplified to produce an output current. The behaviour of the device is then described by the relationship between the output current and the input voltage (the

Figure 21.1   A bipolar transistor as a control device

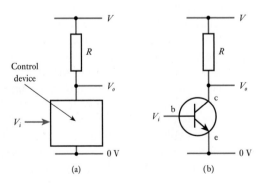

(a)   (b)

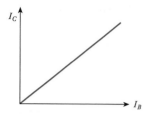

Figure 21.2 The relationship between the collector current and the base current in a bipolar transistor

transconductance), as in the FET. Unfortunately, the relationship between the input voltage and the input current is not linear, so, unlike the current gain, the transconductance is not constant but varies with the magnitude of the output current.

Regardless of which 'model' is adopted, an input signal applied to the base controls the current flowing into the collector of the transistor. Therefore, when used in a circuit of the form shown in Figure 21.1(b), variations in the input will cause variations in the current through the resistor and will control the output voltage.

21.2.1 Construction

Bipolar transistors are formed from three layers of semiconductor material. Two device polarities are possible. The first is formed by placing a thin layer of *p*-type semiconductor between two layers of *n*-type material to form an **npn transistor**. The second is formed by placing a thin layer of *n*-type material within two layers of *p*-type material to give a **pnp transistor**. Both types of device are widely used, and circuits often combine components of these two polarities. The operation of the two forms is similar, differing mainly in the polarities of the voltages and currents (and in the polarities of the charge carriers involved). Figure 21.3 shows the form of each kind of transistor together with its circuit symbol. It can be seen that in each case the sandwich construction produces two *pn* junctions (diodes). However, the operation of the transistors is very different from that of two connected diodes.

Since the operation of *npn* and *pnp* transistors is similar, in this chapter we will concentrate on the former to avoid duplication. In general, the operation of *pnp* devices is similar to that of *npn* devices if the polarities of the various voltages and currents are reversed.

Figure 21.3 *npn* and *pnp* transistors

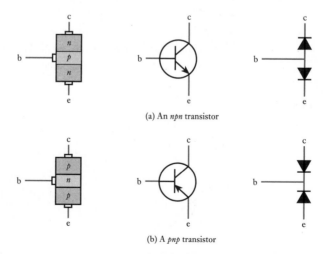

(a) An *npn* transistor

(b) A *pnp* transistor

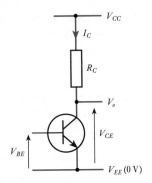

Figure 21.4 Labelling of voltages and currents in bipolar transistor circuits

21.2.2 Notation

The notation used to represent the various voltages and currents in bipolar transistor circuits is similar to that used in FET circuits. For example, V_{CE} would be used to represent the voltage on the *collector* with respect to the *emitter*, and the current into the base would be labelled I_B. Again we use capital letters for steady voltages and currents and small letters for varying quantities. The voltage and current associated with the supply line connected (directly or indirectly) to the collector of the transistor are normally given the labels V_{CC} and I_{CC}, while the corresponding labels for the supply connected to the emitter are V_{EE} and I_{EE}. In circuits using *npn* transistors, V_{CC} is normally positive and V_{EE} is taken as the zero volts reference (or ground). This notation is illustrated in Figure 21.4.

21.3 Bipolar transistor operation

The *npn* or *pnp* structure produces two *pn* junctions connected 'back-to-back', as shown in Figure 21.3. If a voltage is connected across the device between the collector and the emitter, with the base open circuit, one or other of these junctions is reverse-biased, so negligible current will flow. If a transistor was nothing more than two 'back-to-back diodes' it would have little practical use. However, the construction of the device, in particular the fact that the base region is very thin, allows the base to act as a control input. Signals applied to this electrode can be used to produce, and control, currents between the other two terminals. To see why this is so, consider the circuit configuration shown in Figure 21.5(a).

The normal circuit configuration for an *npn* transistor is to make the collector more positive than the emitter. Typical voltages between the collector and the emitter (V_{CE}) might be a few volts. With the base open circuit, the only current flowing from the collector to the emitter will be a small leakage current I_{CEO}, the subscript specifying that it is the current from the Collector to the Emitter with the base Open circuit. This leakage current is small and can normally be neglected. If the base is made positive with respect to the emitter, this will forward-bias the base–emitter junction, which will behave in a manner similar to a diode (as described in

Figure 21.5 Transistor operation

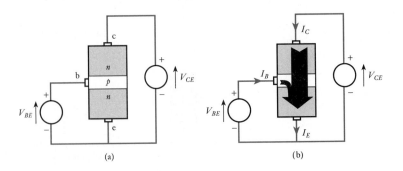

Chapter 19). For small values of the base-to-emitter voltage (V_{BE}) very little current will flow, but as V_{BE} is increased beyond about 0.5 V (for a silicon device) the base current begins to rise rapidly.

The fabrication of the device defines that the emitter region is heavily doped, while the base is lightly doped. The heavy doping in the emitter region results in a large number of majority charge carriers, which are electrons in an *npn* transistor. The light doping in the base region generates a smaller number of holes, which are the majority carriers in the *p*-type base region. Thus in an *npn* transistor the base current is dominated by electrons flowing from the emitter to the base. In addition to being lightly doped, the base region is very thin. Electrons that pass into the base from the emitter as a result of the base–emitter voltage become minority charge carriers in the *p*-type base region. Since the base is very thin, electrons entering the base find themselves close to the space–charge region formed by the reverse bias of the base–collector junction. While the reverse-bias voltage acts as a barrier to majority charge carriers near the junction, it actively propels minority charge carriers across it. Thus any electrons entering the junction area are swept across into the collector and give rise to a collector current. Careful design of the device ensures that the majority of the electrons entering the base get swept across the junction into the collector. Thus the flow of electrons from the emitter to the collector is many times greater than the flow from the emitter to the base. This allows the transistor to function as a current-amplifying device, with a small base current generating a larger collector current. Since conventional current flow is in the opposite direction to the flow of the negatively charged electrons, a flow of electrons from the emitter to the collector represents a flow of conventional current in the opposite direction, as shown in Figure 21.5(b). This phenomenon of current amplification is referred to as **transistor action**.

The relationship between the collector current I_C and the base current I_B for a typical silicon bipolar transistor is shown in Figure 21.6(a).

Because of a slight non-linearity in the relationship between I_B and I_C, there are two ways of specifying the current gain of the device. The first is the **DC current gain**, h_{FE} or β, which is found simply by dividing the collector current by the base current. This is usually given at a particular value of I_C. Because of the slight non-linearity of the characteristic, it is slightly different at different values of I_C. The DC current gain is used in large-signal calculations, and therefore

Figure 21.6 Characteristics of a typical silicon bipolar transistor

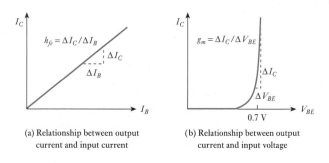

(a) Relationship between output current and input current

(b) Relationship between output current and input voltage

$$I_C = h_{FE}I_B \qquad (21.1)$$

When considering small signals, we need to know the relationship between a small change in I_B (ΔI_B) and the corresponding change in I_C (ΔI_C). The ratio $\Delta I_C / \Delta I_B$ is called the small-signal current gain and is given the symbol h_{fe}. It is also called the **AC current gain** of the device. The value of h_{fe} may be obtained from the slope of the characteristic given in Figure 21.6(a), and

$$i_c = h_{fe}i_b \qquad (21.2)$$

For most practical purposes, h_{FE} and h_{fe} can be considered to be equal. A typical value for a general-purpose silicon transistor would be in the range 100 to 300, but the current gain of bipolar transistors varies considerably with temperature and operating conditions. There is also a considerable spread of characteristics between devices of the same nominal type, and even within the same batch.

The characteristics of a bipolar transistor may also be described by the relationship between the output current I_C and the input voltage V_{BE}, as shown in Figure 21.6(b). Since the base–emitter junction resembles a simple *pn* junction, the input current I_B is exponentially related to the input voltage V_{BE}, as illustrated in Figure 19.7. Since the output current I_C is approximately linearly related to the input current (by the current gain h_{FE}), the relationship between I_C and V_{BE} has the same shape (although with correspondingly larger values of current). The slope of this curve at any point is given by the ratio $\Delta I_C / \Delta V_{BE}$, which represents the transconductance of the device g_m (you may like to compare this with the similar discussion of the FET given in Section 20.5). In the limit

$$g_m = \frac{\mathrm{d}I_C}{\mathrm{d}V_{BE}} \qquad (21.3)$$

Unlike h_{FE}, which is approximately constant for a given device, g_m varies with the collector current (and hence the emitter current) at which the circuit is operated.

21.4	Bipolar transistor characteristics

21.4.1   Transistor configurations

Bipolar transistors are very versatile components and can be used in a number of circuit configurations. These arrangements differ in the way that signals are applied to the device and the way in which the output is produced. In each configuration, control signals are applied to the transistor by an 'input circuit', and a controlled quantity is sensed by an 'output circuit'. In Figure 21.5, we considered a circuit where the input takes the form of a voltage applied to the base (with respect to the emitter) and the output is represented by the voltage on the collector (with respect to the emitter). This arrangement is represented in Figure 21.7. It can be seen that the

Figure 21.7 A common-emitter arrangement

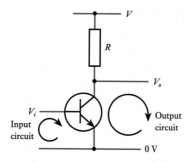

emitter terminal is common to both the input and the output circuits. For this reason, this arrangement is known as a **common-emitter** circuit. It should perhaps be noted at this point that the 'E' or 'e' in h_{FE} and h_{fe} each stand for 'emitter' as these are the current gains of the transistor when used in a common-emitter configuration. Common-collector and common-base circuits are also used, and these have different current gains and different characteristics. This allows us to select from a range of characteristics to suit our needs. In this section, we will concentrate on the characteristics associated with common-emitter circuits.

In the previous chapter, when looking at FETs, we noted that the behaviour of such a transistor may be understood by considering its input characteristics, its output characteristics and the relationship between the input and the output (the transfer characteristics). We will now consider these three aspects of the behaviour of a bipolar transistor.

21.4.2 Input characteristics

From Figure 21.5(a), it is clear that the input of the transistor takes the form of a forward-biased *pn* junction, and the input characteristics are therefore similar to the characteristics of a semiconductor diode. The input characteristics of a typical silicon device are shown in Figure 21.8.

Figure 21.8 Input characteristics of a bipolar transistor in the common-emitter configuration

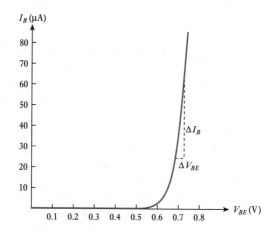

In Chapter 19, we deduced that the current in a semiconductor diode is given by

$$I \approx I_S \exp(40V)$$

where the '40' represents q/kT and therefore varies with temperature.

In this case the current I is the base current of the transistor I_B, and the junction voltage V is the base-to-emitter voltage V_{BE}. We therefore have

$$I_B \approx I_{BS} \exp(40V_{BE}) \tag{21.4}$$

where I_{BS} is a constant determined by the base characteristics. This equation represents the **input characteristic** of the device.

The slope, at any point, of the line in Figure 21.8 represents the relationship between a small change in the base-to-emitter voltage ΔV_{BE} and the corresponding change in the base current ΔI_B. The slope therefore indicates the **small-signal input resistance** of the arrangement, which is given the symbol h_{ie}. Clearly, the magnitude of the input resistance varies with position along the characteristic. The value at any point may be found by differentiating Equation 21.4 with respect to I_B, which gives the simple result that

$$h_{ie} = \frac{dV_{BE}}{dI_B} \approx \frac{1}{40I_B} \Omega \tag{21.5}$$

Since a typical value for I_B might be a few tens of microamps, a typical value for h_{ie} might be a few kilohms.

The input characteristics of a bipolar transistor in the common-emitter configuration may therefore be described by very simple expressions for the base current and the small-signal input resistance. However, it should be noted that the 'e' subscript in h_{ie} stands for 'emitter' and that h_{ie} is the small-signal input resistance in the *common-emitter configuration*. Other circuit configurations will have a different input resistance.

21.4.3　Output characteristics

Figure 21.9 shows the relationship between the collector current I_C and the collector voltage V_{CE} for a typical device for various values of the base current I_B. These two quantities represent the output current and the output voltage of the transistor in the common-emitter configuration. The relationship between them is often referred to as the common-emitter **output characteristic**. You might like to compare this characteristic with that obtained for FETs in Figure 20.10.

For a given base current, the collector current initially rises rapidly with the collector voltage as this increases from zero. However, it soon reaches a steady value, and any further increase in the collector voltage has little effect on the collector current. The value of collector current at which the characteristic stabilises is determined by the base current. The ratio

Figure 21.9 A typical common-emitter output characteristic

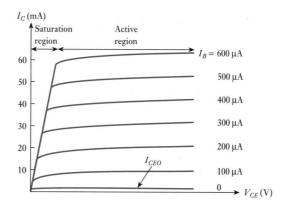

between this steady value of collector current and the value of the base current represents the DC current gain of the device h_{FE}.

The section of the characteristic over which the collector current is approximately linearly related to the base current is referred to as the **active region**. Most linear amplifier circuits operate in this region. The section of the characteristic close to the origin where this linear relationship does not hold is called the **saturation region**. The saturation region is generally avoided in linear circuits but is widely used in non-linear arrangements, including digital circuitry. It is important to note that the term 'saturation' has a different meaning when applied to bipolar transistors than when discussing FETs. Saturation occurs in bipolar transistors when V_{CE} is very low, because the efficiency of the transistor action is reduced and many charge carriers pass from the emitter to the base without being swept into the collector region.

In an ideal bipolar transistor, the various lines in the output characteristic would be horizontal, indicating that the output current was completely independent of the collector voltage. In practice this is not the case, and all real devices have a slight gradient, as shown in Figure 21.9. The slope of these lines indicates the change in output current with output voltage and is therefore a measure of the **output resistance** of the arrangement. A typical value for this resistance might be of the order of 100 kΩ.

Notice in Figure 21.9 that the collector current is not zero when the base current is zero, because of the presence of the **leakage current** I_{CEO}. The effect of I_{CEO} is magnified in the figure to allow it to be visible. In silicon devices its effects are usually negligible.

21.4.4 Transfer characteristics

Figure 21.6 represents the transfer characteristics of a bipolar transistor in a common-emitter configuration. From this figure, it is clear that the characteristics can be described in two ways: the first, in terms of the current gain of the device; and the second, in terms of its transconductance. For a given device, the current gain tends to be relatively constant regardless of how the device is used (although the gain *will* vary with temperature and between devices). In contrast, the transconductance varies with the operating conditions of the circuit, as shown in Figure 21.6.

From Figure 21.5(b), it is clear that the emitter current I_E must be given by the sum of the collector current I_C and the base current I_B. Thus

$$I_E = I_C + I_B$$

and since

$$I_C = h_{FE} I_B$$

it follows that

$$I_E = h_{FE} I_B + I_B = (h_{FE} + 1) I_B$$

Since h_{FE} is usually much greater than unity, we may make the approximation that

$$I_E \approx h_{FE} I_B = I_C \tag{21.6}$$

Combining the results of Equations 21.3, 21.4 and 21.6 it can be shown that

$$g_m \approx 40 I_C \approx 40 I_E \text{ siemens} \tag{21.7}$$

21.4.5 Equivalent circuits for a bipolar transistor

We are now in a position to create equivalent circuits for a bipolar transistor, and Figure 21.10 shows two such circuits. You might wish to compare these models with that shown in Figure 20.13 for a field-effect transistor. As in the earlier chapter, the models in Figure 21.10 are small-signal equivalent circuits.

It can be seen that the two models in Figure 21.10 are very similar, differing only in the magnitude of the current produced by their current generator. In Figure 21.10(a), the behaviour of the device is modelled by its

Figure 21.10 Small-signal equivalent circuits for a bipolar transistor

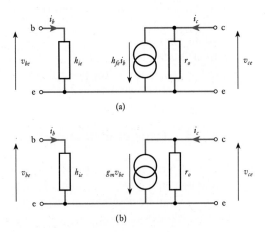

(a)

(b)

current gain h_{fe}, which for general-purpose small-signal transistors would normally have a value of between 100 and 300 (but could be much lower in power transistors). In Figure 21.10(b), the transistor is modelled by its transconductance g_m. The value of g_m depends on the circuit configuration, and an estimate of its value can be obtained from Equation 21.7. The input and output resistance of each model are the same (as one would expect since these are alternative representations of the same device). The magnitude of the input resistance h_{ie} depends on the circuit arrangement, and an estimate of its value can be obtained from Equation 21.5. The magnitude of the output resistance corresponds to the slope of the output characteristic and would typically be between 10 kΩ and 1 MΩ.

The equivalent circuit used in a particular application will depend on the circuit and the nature of the analysis required. More sophisticated models are also available to describe the frequency-dependent characteristics of the device.

21.5 Summary of bipolar transistor characteristics

To a large extent, one can understand the operation and use of bipolar transistors by knowing that:

- Bipolar transistors have three terminals, called the collector, the emitter and the base.
- The control input is the base, and the signal applied to the base controls the current that flows from the collector to the emitter.
- Bipolar transistors are three-layer devices and may be of either an *npn* or a *pnp* form.
- In circuits using *npn* transistors, the collector is normally more positive than the emitter, and the collector current represents the controlled (output) quantity.
- A current I_B injected into the base will cause a larger current to flow from the collector to the emitter. The ratio of these two currents is called the current gain of the device and is given the symbol h_{FE}. Therefore, the collector current I_C is given by $I_C = h_{FE} I_B$.
- The relationship between I_B and I_C is approximately linear (as shown in Figure 21.2 but does vary slightly with current. This leads to the concept of a small signal voltage gain h_{fe}, which represents the relationship between small changes in I_B and corresponding changes in I_C at a particular current. Thus $i_c = h_{fe} i_b$.
- It is normally reasonable to assume that $h_{fe} = h_{FE}$.
- Since the current flowing into the terminals of the transistor must sum to zero, the emitter current is equal to the sum of the base current and the collector current. Thus $I_E = I_B + I_C$. Since, in general, $I_C \gg I_B$, it is usually reasonable to assume that $I_E \approx I_C$.
- In a transistor, a *pn* junction exists between the base and the emitter terminals. Therefore, when a voltage is applied to the base (with respect to

the emitter), the relationship between this voltage and the resulting base current is similar to that of a semiconductor diode (as described in Chapter 19).

- Consequently, when a current is flowing into the base of a (silicon) transistor, the base-to-emitter voltage (V_{BE}) is generally about 0.7 V (since this is the typical conduction voltage for a silicon semiconductor diode).

21.6 Bipolar transistor amplifiers

21.6.1 A simple amplifier

In Figure 21.1(b), we looked at a basic amplifier based on a bipolar transistor. Varying the voltage applied to the input will vary the current flowing into the base of the transistor. This base current will be amplified by the transistor to produce a correspondingly larger collector current. This collector current, flowing through the collector resistor, will determine the output voltage.

While the arrangement of Figure 21.1 has some interesting properties, it does not represent a useful amplifier. The relationship between the base current and the collector current is relatively linear (as shown in Figure 21.2), but the device can only be used with positive base currents. If we wish to operate with a bipolar signal (that is, one that goes negative as well as positive), we must offset the input from zero in order to amplify the complete signal. This is done by **biasing** the input of the amplifier.

A simple amplifier with a biasing arrangement is shown in Figure 21.11. The base resistor R_B applies a positive voltage to the base of the transistor, forward-biasing the base–emitter junction and producing a base current. This in turn produces a collector current, which flows through the collector load resistor R_C, producing a voltage drop and making the output voltage V_o less than the collector supply voltage V_{CC}. When no input is applied to the circuit it is said to be in its quiescent state. The value of the base resistor R_B will determine the **quiescent base current**, which in turn will determine the **quiescent collector current** and the quiescent output voltage. The values of R_B and R_C must be chosen carefully to ensure the correct operating point for the circuit.

Figure 21.11   A simple amplifier

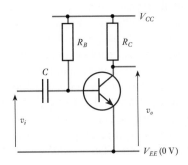

If a *positive* voltage is applied to the input of the amplifier, this will tend to increase the voltage on the base of the transistor and thus increase the base current. This in turn will raise the collector current, increasing the voltage drop across the collector resistor R_C and *decreasing* the output voltage V_o. If a *negative* voltage is applied to the input, this will decrease the current through the transistor and thus *increase* the output voltage. We therefore have an inverting amplifier. As with the MOSFET amplifiers described in the previous chapter, a **coupling capacitor** is used to prevent input voltages from affecting the mean voltage applied to the base. The circuit therefore cannot be used to amplify DC signals and is an AC-coupled amplifier. You will notice that in this arrangement the input is applied between the base and the emitter, while the output is measured between the collector and the emitter. The emitter is thus common to the input circuit and the output circuit. For this reason, such an arrangement is called a **common-emitter amplifier**.

| Example 21.1 | Determine the quiescent collector current and the quiescent output voltage of the following circuit, given that the h_{FE} of the transistor is 200. |

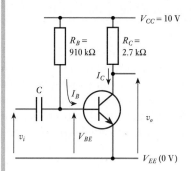

The base-to-emitter junction of the transistor resembles a forward-biased *pn* junction, therefore we will assume that the base-to-emitter voltage V_{BE} is 0.7 V.

From a knowledge of V_{BE} we also know the voltage across R_B since this is simply $V_{CC} - V_{BE}$, which in turn enables us to calculate the base current I_B. In this case

$$I_B = \frac{V_{CC} - V_{BE}}{R_B} = \frac{10 - 0.7 \text{ V}}{910 \text{ k}\Omega} = 10.2 \text{ }\mu\text{A}$$

The collector current I_C is now given by

$$I_C = h_{FE}I_B = 200 \times 10.2 \text{ }\mu\text{A} = 2.04 \text{ mA}$$

The quiescent output voltage is simply the supply voltage minus the voltage drop across R_C and is therefore

$$V_o = V_{CC} - I_C R_C = 10 - 2.04 \times 10^{-3} \times 2.7 \times 10^3 \approx 4.5 \text{ V}$$

Thus the circuit has a quiescent collector current of about 2 mA and a quiescent output voltage of approximately 4.5 V.

File 21A

Computer Simulation Exercise 21.1

Simulate the circuit of Example 21.1 using an appropriate bipolar transistor. Most simulation programs do not provide a standard transistor with a gain of 200, so you will need to choose from the parts available in your particular simulation package. If you are using PSpice, then you might choose the 2N2222 (part Q2N2222 in PSpice). If you are using another simulation package, you may need to experiment with other parts, perhaps adjusting the resistor values to suit.

Apply no input signal to the circuit and measure the quiescent collector current and quiescent output voltage. From these measurements, and a knowledge of the base current, calculate the current gain h_{FE} of the transistor in your simulation. Experiment with different values for the various resistors and note their effects on the quiescent voltages and currents within the circuit.

Now apply a small alternating voltage to the input and observe the resultant variations at the output.

From the analysis of Example 21.1, it is clear that the quiescent collector current and the quiescent output voltage are both determined by the value of h_{FE}, which varies considerably between devices. For this reason, this simple circuit arrangement is rarely used. In earlier chapters, we have seen how feedback can be used to overcome such problems, and this approach may also be applied to bipolar transistors.

21.6.2 A negative feedback amplifier

Consider the circuit of Figure 21.12(a). You will notice that an emitter resistor has been added. The voltage across this resistor is clearly

Figure 21.12 Amplifiers with negative feedback

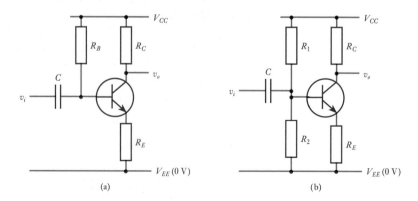

(a)   (b)

proportional to the emitter current, which is (almost exactly) equal to the collector current. Since the input voltage is applied to the base, the voltage across the emitter resistor is effectively subtracted from the input to produce the voltage that is applied to the transistor. Therefore, we have negative feedback, with a voltage proportional to the output current being subtracted from the input. This will stabilise the voltage gain, making it less affected by variations in the current gain of the device. From the discussion of the effects of negative feedback in Chapter 7, it is clear that this form of feedback will also increase both the input resistance and the output resistance of the circuit.

A further development of this circuit is shown in Figure 21.12(b). Here two resistors are used to provide base bias rather than the single resistor used in the earlier circuit. This arrangement produces good stabilisation of the DC operating conditions of the circuit and also gives an arrangement that is very easy to analyse. We will begin by looking at the quiescent or DC analysis of the circuit and then consider its AC behaviour.

Example 21.2 | Determine the quiescent output voltage of the following circuit.

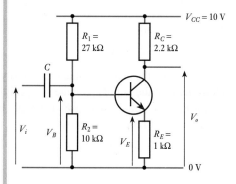

We are not told the current gain of the transistor in this circuit, but it is safe to assume that this will be relatively high. Therefore, the emitter current will be approximately equal to the collector current and the base current will be negligible.

If we assume that the base current is negligible, since no constant current can flow through the input capacitor, the **quiescent base voltage** is determined simply by the supply voltage V_{CC} and by the potential divider formed by R_1 and R_2. Hence

$$V_B \approx V_{CC} \frac{R_2}{R_1 + R_2}$$

Therefore, in our example

$$V_B \approx 10 \frac{10 \text{ k}\Omega}{27 \text{ k}\Omega + 10 \text{ k}\Omega} \approx 2.7 \text{ V}$$

It is clear from Examples 21.2 and 21.3 that both the DC and the AC behaviour of such circuits can be determined quickly and easily. It is interesting to note that at no time in the analysis did we need to know the value of the current gain of the transistor. This indicates that the performance of the circuit is largely independent of the gain of the device.

File 21B

Computer Simulation Exercise 21.2

Simulate the circuit of Examples 21.2 and 21.3 using an appropriate bipolar transistor (for example a 2N2222). Measure the quiescent voltages and currents in the circuit and compare these with the values calculated in Example 21.2.

Use a coupling capacitor of 1 µF and apply an input voltage of 50 mV peak at 1 kHz. Use transient analysis to measure the voltage gain of the circuit and compare this with the value calculated in Example 21.3.

21.6.3　Input and output resistance of the feedback amplifier

Since the supply lines are at a constant voltage, they have no small-signal voltages associated with them. Therefore, as far as small-signal (AC) signals are concerned, the various supply rails (such as V_{CC} and V_{EE}) are effectively joined together. They each represent a virtual 'ground' as far as AC signals are concerned.

If we now consider the circuit of Figure 21.12(b), it can be seen that (as far as small signals are concerned) the input signal 'sees' three parallel conduction paths to ground. These are through R_2 to ground, through R_1 to V_{CC} (which is effectively ground as far as small signals are concerned) and through the base–emitter junction of the transistor and R_E to ground. The effective resistance of the base–emitter junction is generally of the order of a few kilohms (see Section 21.4 for a discussion of the input characteristics of the transistor), but the effects of R_E are magnified by the current gain of the transistor, making it appear much larger. For this reason, the conduction path through the transistor can almost always be ignored, and the input resistance of the circuit is approximately equal to the parallel combination of R_1 and R_2.

$$R_i \approx R_1 // R_2 \tag{21.10}$$

The output resistance of the circuit is similarly given by the parallel combination of the collector resistor R_C and the resistance seen looking into the collector. In practice, the latter is almost always much larger than R_C, so the output resistance is approximately equal to R_C.

$$R_o \approx R_C \tag{21.11}$$

21.6.4 The effects of the coupling capacitor

We noted earlier that the value of the coupling capacitor is normally chosen so that its effects are negligible at the frequencies of interest. The coupling capacitor is used to prevent the DC component of any input signal from upsetting the biasing of the transistor. However, its presence will inevitably introduce a low-frequency cut-off as discussed in Section 17.2. The frequency of this cut-off is given by

$$f_{co} = \frac{1}{2\pi CR_i} \tag{21.12}$$

where C is the value of the coupling capacitor and R_i is the input resistance of the amplifier. The input resistance of the feedback amplifier is discussed above.

| Example 21.4 | Determine the low-frequency cut-off of the following circuit. |

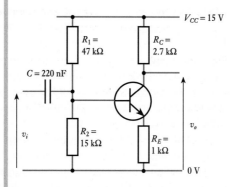

From Equation 21.10, the input resistance of the circuit is given by

$$R_i \approx R_1 /\!/ R_2$$

$$= 15 \text{ k}\Omega / 47 \text{ k}\Omega$$

$$= 11.37 \text{ k}\Omega$$

Therefore, from Equation 21.12, the low-frequency cut-off is

$$f_{co} = \frac{1}{2\pi CR_i}$$

$$= \frac{1}{2 \times \pi \times 220 \times 10^{-9} \times 11.37 \times 10^3}$$

$$\approx 64 \text{ Hz}$$

Computer Simulation Exercise 21.3

Use your circuit from computer Simulation Exercise 21.2 to investigate the effects of the coupling capacitor. Estimate the input resistance of the amplifier as described above and calculate the low-frequency cut-off produced by the capacitor. Use an AC sweep to measure the frequency response of the amplifier and compare the actual cut-off frequency with your calculated value.

21.6.5   A common-collector amplifier

We noted earlier that transistors can be used in a number of circuit configurations, and we have already looked at several aspects of the common-emitter configuration. Another widely used technique is the common-collector configuration, and an example of such a circuit is shown in Figure 21.13.

The circuit of Figure 21.13 is similar to the feedback circuit used earlier except that the output is taken from the emitter rather than the collector, eliminating the need for a collector resistor. The collector is connected directly to the positive supply, which is at earth potential for AC signals since there are no AC voltages between the supply and ground. Input signals are applied between the base and ground, while output signals are measured between the emitter and ground. Since the collector is at ground potential, for small signals the collector is common to both the input and output circuits. Hence the arrangement is a **common-collector amplifier**.

As with the feedback amplifier considered earlier, this circuit uses negative feedback. However, in this case, the voltage that is subtracted from the input is related to the output *voltage* rather than the output current, since the emitter voltage *is* the output voltage of the circuit. From the discussion of Chapter 7, it is clear that this will *increase* the input resistance, as before, but *decrease* the output resistance.

Analysis of the circuit is similar, but much simpler, than that of the earlier feedback amplifier.

Figure 21.13   A common-collector amplifier

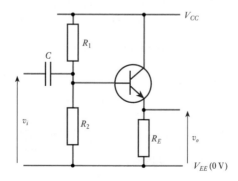

Example 21.5

Determine the quiescent output voltage and the small-signal voltage gain of the following circuit.

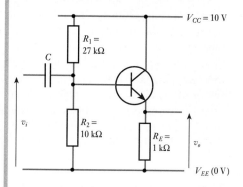

The component values in this circuit are identical to the corresponding components in the circuit of Example 21.2, and thus the first part of the analysis is identical. If we adopt similar notations for the various currents and voltages as in the earlier examples, then (as before)

$$V_B \approx V_{CC} \frac{R_2}{R_1 + R_2} = 10 \frac{10 \text{ k}\Omega}{27 \text{ k}\Omega + 10 \text{ k}\Omega} \approx 2.7 \text{ V}$$

and

$$V_E = V_B - V_{BE} = 2.7 - 0.7 = 2.0 \text{ V}$$

However, in this circuit V_E represents our output voltage, so our quiescent output voltage is 2.0 V.

As before the input signal is applied to the base of the transistor through the coupling capacitor C, and therefore as before

$$v_e \approx v_b \approx v_i$$

However, in this circuit v_e is the output voltage, and since the output is equal to the input, the voltage gain is unity.

Thus the small-signal voltage gain of the common-collector amplifier is approximately unity, and the emitter simply follows the input signal. In fact, the emitter tracks the base with a constant offset voltage (typically about 0.7 V) caused by the conduction voltage of the base–emitter junction. For this reason, this form of amplifier is often called an **emitter follower amplifier**.

The fact that the emitter follower has a voltage gain of approximately unity might at first sight appear to make it of little use. However, its relatively high input impedance and very low output impedance make it of interest as a unity-gain buffer amplifier. Note also that this is a non-inverting amplifier.

Computer Simulation Exercise 21.4

Simulate the circuit of Examples 21.5 using an appropriate bipolar transistor (for example a 2N2222). Measure the quiescent voltages and currents in the circuit and compare these with the values calculated above.

Use a coupling capacitor of 1 μF and apply an input voltage of 500 mV peak at 1 kHz. Use transient analysis to investigate the behaviour of the circuit and display the input and output signals simultaneously. Observe the relationship between these two waveforms and hence confirm that the circuit behaves as expected.

21.6.6   Common-base amplifiers

The common-base configuration is the least widely used transistor arrangement. Input signals are applied to the emitter, and the output is taken from the collector. This results in a non-inverting amplifier with a low input resistance and a high output resistance. These characteristics are not generally associated with good voltage amplifiers but make them useful as **trans-impedance amplifiers** (amplifiers that take an input *current* and produce a related output *voltage*). Common-base amplifiers are regularly used in wide-bandwidth radio frequency amplifiers, but these specialist circuits will not be discussed further here.

21.7   Other bipolar transistor applications

21.7.1   A phase splitter

Consider the circuit of Figure 21.14. This produces two output signals that are inverted with respect to each other. By taking outputs from both the collector and the emitter, we have combined the inverting and non-inverting

Figure 21.14   A phase splitter

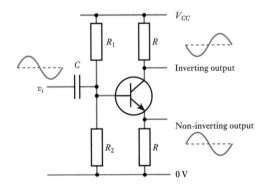

amplifiers discussed above. From Equation 21.9, we know that the magnitude of the voltage gain of the feedback amplifier is $-R_C/R_E$. In this circuit, the two resistors are equal, giving a gain of -1. The emitter follower has a gain of $+1$, so the two signals are identical in size but are of opposite polarity.

It should be remembered that the output resistances of the two outputs are very different. Therefore, the signals should be fed into high input resistance circuits if the correct relative signal magnitudes are to be maintained. It is also worth noting that the DC quiescent output voltages are different on the two outputs.

File 21D

> ### Computer Simulation Exercise 21.5
>
> Simulate the circuit of Figure 21.14 using a suitable *npn* transistor (for example a 2N2222). A suitable choice of components would be $R = 1\ \text{k}\Omega$, $R_1 = 6.7\ \text{k}\Omega$, $R_2 = 3.3\ \text{k}\Omega$ and $C = 1\ \mu\text{F}$. Use a supply voltage $V_{CC} = 10\ \text{V}$ and apply a 1 kHz sinusoidal input of 500 mV peak.
>
> Display the two output signals and compare their relative magnitude and phase. Then connect load resistors of 1 kΩ between each output and ground and again compare the outputs.

21.7.2 A bipolar transistor as a voltage regulator

The emitter follower configuration discussed earlier produces a voltage that is determined by its input voltage (with an offset of about 0.7 V). It also has a very low output resistance, meaning that its output voltage is not greatly affected by the load connected to it. These two characteristics allow this circuit to form the basis of a **voltage regulator**, as shown in Figure 21.15(a). The resistor and Zener diode form a constant voltage reference V_Z (as discussed in Section 19.7), which is applied to the base of the transistor. The output voltage will be equal to this voltage minus the approximately constant base-to-emitter voltage of the transistor.

The circuit of Figure 21.15(a) is shown redrawn as Figure 21.15(b), which illustrates a more common way of representing the circuit. The

Figure 21.15 A simple voltage regulator

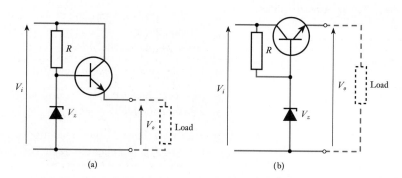

(a)   (b)

arrangement as shown provides a considerable amount of regulation but for large fluctuations in output current suffers from the variation of V_{BE} with current. More effective regulator circuits will be discussed in Chapter 22.

21.7.3 A bipolar transistor as a switch

We saw in Chapter 20 that FETs can be used as both linear and logical switches. Bipolar transistors are not usually used in linear switching applications but can be used as logical switching elements.

Figure 21.16 shows a simple form of logical inverter based on a bipolar transistor. The circuit resembles the common-emitter amplifier considered earlier and indeed its operation is similar. The main difference between this circuit and a linear amplifier is that the inputs are restricted to two distinct ranges. Input voltages close to zero (representing logical '0') are insufficient to forward-bias the base of the transistor, and it is therefore turned OFF. Negligible collector current flows, and the output voltage is therefore close to the supply voltage. Input voltages close to the supply voltage (representing a logical '1') forward-bias the base junction, turning ON the transistor. The resistor R_B is chosen such that the base current is sufficient to saturate the transistor, producing an output voltage equal to the transistor's collector saturation voltage (normally about 0.2 V). Thus an input of logical '0' produces an output of logical '1', and vice versa. The circuit is therefore a logical inverter.

If the transistor were an ideal switch, it would have an infinite resistance when turned OFF and zero resistance when turned ON, and it would operate in zero time. In fact, the bipolar transistor is a good, but not an ideal, switch. When turned OFF, the only currents that are passed are small leakage currents, which are normally negligible. When turned ON, the device has a low ON resistance but a small saturation voltage, as described above. The speed of operation of bipolar transistors is very high, with devices able to switch from one state to another in a few nanoseconds (or considerably faster for high-speed devices). We will return to look in more detail at the characteristics of transistors as switches in Chapter 24.

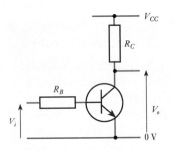

Figure 21.16 A logical inverter based on a bipolar transistor

Key points

- Bipolar transistors are one of the most important forms of electronic component, and an understanding of their operation and use is essential for anyone working in this area.

- They are used in a wide variety of both analogue and digital circuits.

- Bipolar transistors can be considered as either current-controlled or voltage-controlled devices.

- If we view them as current-controlled devices, we describe their performance by their current gain. Two forms of current gain are used:

the DC current gain h_{FE} and the AC current gain h_{fe}. For most purposes, these two current gains may be considered to be equal.

■ If we choose to view them as voltage-controlled devices, we describe their behaviour using their transconductance g_m.

■ The current gain of a transistor varies with temperature and between components. Wherever possible, we attempt to design circuits in which the actual value of the gain is unimportant.

■ Feedback can be used to stabilise the characteristics of the circuits so that they are less affected by the devices used.

■ A major advantage of the use of feedback is that, by making the circuit less dependent on the characteristics of the transistor, the analysis of the circuit is made much simpler.

■ Although the majority of transistor circuits apply the input signal to the base and take the output signal from the collector, other configurations are also used.

■ The common-collector (emitter follower) mode produces a unity gain amplifier with a high input resistance and a low output resistance. Such circuits make good buffer amplifiers.

■ Bipolar transistors are used in a wide range of applications in addition to their use as simple amplifiers.

Exercises

21.1 Why are bipolar transistors so called?

21.2 Is a bipolar transistor a voltage-controlled or a current-controlled device?

21.3 Sketch the relationship between the base current and the collector current in a bipolar transistor (when the device is in its normal operating environment).

21.4 Sketch the construction of the two polarities of bipolar transistor.

21.5 What is meant by the symbols V_{CC}, V_{CE}, V_{BE}, v_{be} and i_c.

21.6 Explain what is meant by 'transistor action'.

21.7 Explain the terms h_{FE} and h_{fe} and describe their relative magnitudes.

21.8 Sketch the relationship between the base voltage and the collector current in a bipolar transistor (when the device is in its normal operating environment).

21.9 What is meant by the transconductance of a transistor? How does this quantity relate to the characteristic described in the previous exercise?

21.10 Determine the quiescent collector current and the quiescent output voltage of the following circuit, given that the h_{FE} of the transistor is 100.

Exercises continued

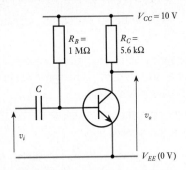

21.11 Repeat Exercise 21.10 with the transistor replaced by one with a current gain of 200. Is this circuit useful as an amplifier?

21.12 Calculate the quiescent collector current, the quiescent output voltage and the small-signal voltage gain of the following circuit.

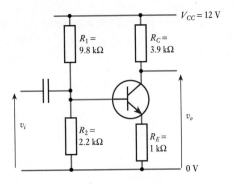

21.13 For the circuit of Exercise 21.12, estimate the small-signal input and output resistances.

21.14 For the circuit of Exercise 21.12, estimate the effect on the frequency response of the circuit of using a coupling capacitor of 1 μF.

21.15 Use circuit simulation to confirm your answers to Exercises 21.12 and 21.14.

21.16 Calculate the quiescent collector current, the quiescent output voltage and the small-signal voltage gain of the following circuit.

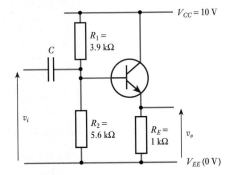

21.17 Estimate the input resistance of the circuit of Exercise 21.16 and hence determine an appropriate value for a coupling capacitor to allow satisfactory operation down to 50 Hz.

21.18 Use circuit simulation to confirm your answers to Exercises 21.16 and 21.17.

21.19 Design a simple voltage regulator to produce an output voltage of 4.0 V from a larger supply voltage. What factors determine the choice of the resistor in your circuit?

Chapter 22   Power Electronics

Objectives

When you have studied the material in this chapter you should be able to:

- describe a range of power amplifier circuits based on bipolar transistors;
- discuss methods of reducing distortion in power amplifiers and in coping with problems of temperature instability;
- explain the various classes of amplifier circuit and outline the distinction between these classes;
- explain the operation of special-purpose switching devices such as thyristors and triacs and describe their use in AC power control;
- sketch circuits of simple unregulated and regulated power supplies;
- discuss the need for voltage regulation in power supplies and describe the operation of both conventional and switching voltage regulators.

22.1   Introduction

We have seen in earlier chapters how transistors may be used to produce various types of amplifier. Such circuits usually deliver more power to their load than they absorb from their input, so they provide some degree of power amplification. However, the term **power amplifier** is normally reserved for circuits whose main function is to deliver large amounts of power to a load. Power amplifiers are used in audio systems to drive the speakers and are also used in a wide range of other applications. In some cases, as in audio amplifiers, we require a linear (or nearly linear) relationship between the input of the amplifier and the power delivered to the load. In other cases, for example when we are controlling the power dissipated in a heater, linearity is less important. In such situations, our main concern is often the efficiency of the control process. Linear amplifiers often dissipate a lot of power (in the form of heat), and this is usually undesirable. Where linearity is unimportant, we often use techniques that improve the efficiency of the control process at the expense of some distortion of the amplified signal. In some cases, we are concerned only with the amount of power that is delivered, and the nature of the waveform is unimportant. In such cases, we often use switching techniques, which can control large amounts of power and offer great efficiency.

In this chapter, we will look at power electronic circuits using both linear and switching techniques. Linear circuits can be produced using either FETs or bipolar transistors, but here we will concentrate on the latter. Although switching circuits can be constructed using transistors, special-purpose components can often perform such tasks more efficiently. Here we will look at several devices that are specifically designed for switching applications, including **thyristors** and **triacs**.

22.2   Bipolar transistor power amplifiers

When designing a power amplifier, we normally require a low output resistance so that the circuit can deliver a high output current. In Chapter 21, we looked at various transistor configurations and noted that the common-collector amplifier has a very low output resistance. You may recall that this arrangement is also known as an emitter follower, since the voltage on the emitter follows the input voltage. While the emitter follower does not produce any voltage gain, its low output resistance makes it attractive in high-current applications, and it is often used in power amplifiers.

22.2.1   Current sources and current sinks

In many cases, the load applied to a power amplifier is not simply resistive but also has an impedance that includes inductive or capacitive components. For example, a speaker connected to an audio amplifier will have inductance as well as resistance, while a long cable will add capacitance to a load. When driving reactive loads, an amplifier will need to *supply* current to the load at some times (when it is acting as a **current source**) but will need to *absorb* current from the load at other times (when it is acting as a **current sink**). These two situations are illustrated in Figure 22.1, which shows an emitter follower connected to a capacitive load.

Figure 22.1   An emitter follower with a capacitive load

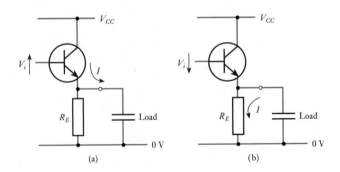

Figure 22.2 An emitter follower using *pnp* transistors

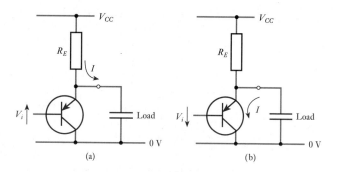

(a) (b)

Consider initially the situation shown in Figure 22.1(a), where the input is becoming more positive. The transistor drives the output positive by passing its emitter current into the load, the low output impedance of the transistor allowing the capacitor to be charged quickly. If now the input becomes more negative, as shown in Figure 22.1(b), charge must be removed from the capacitor. This cannot be done by the transistor, which can only source current in this configuration. Therefore, the charge must be removed by the emitter resistor R_E, which will be considerably greater than the output resistance of the amplifier. Therefore, this arrangement can charge the capacitor quickly but is slow to discharge it.

Replacing the transistor with a *pnp* device, as shown in Figure 22.2, produces an arrangement that can discharge the load quickly but is slow to charge it. Clearly, the rate at which the capacitor can be charged or discharged through the resistor R_E can be increased by decreasing the value of the resistor. However, decreasing R_E increases the current flowing through the transistor and therefore increases its power dissipation. In high-power applications, this can cause serious problems.

22.2.2 Push-pull amplifiers

One approach to this problem, often used in the high-power output stage of an amplifier, is to use two transistors in a **push-pull** arrangement as shown in Figure 22.3.

Here one transistor is able to source and the other to sink current, so the load can be driven from a low-resistance output in either direction. This

Figure 22.3 A simple push-pull amplifier

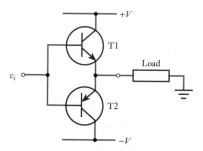

Figure 22.4 Driving a push-
pull output stage

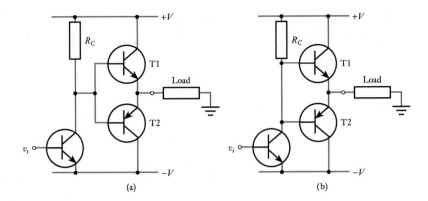

(a)   (b)

arrangement is commonly used with a **split power supply**, that is, a supply
that provides both positive and negative voltages, with the load being
connected to earth. For positive input voltages, transistor T1 will be conducting but T2 will be turned OFF, since its base junction will be reverse-
biased. Similarly, for negative input voltages T2 will be conducting and T1
will be turned OFF. Hence at any time only one of the transistors is turned
ON, reducing the overall power consumption.

A possible method of driving the push-pull stage is shown in Figure 22.4(a). Here a conventional common-emitter amplifier is used to drive
the bases of the two output transistors. This circuit is shown redrawn in
Figure 22.4(b), which is a more conventional method of drawing the
arrangement but is electrically identical.

Distortion in push-pull amplifiers

A problem with the simple push-pull arrangement described above is
that, for small values of the base voltage on either side of zero, both of the
output transistors are turned OFF. This gives rise to an effect known as
crossover distortion, as shown in Figure 22.5(a).

One solution to this problem is shown in Figure 22.5(b). Here two
diodes are used to apply different voltages to the two bases. When

Figure 22.5 Tackling crossover
distortion in push-pull amplifiers

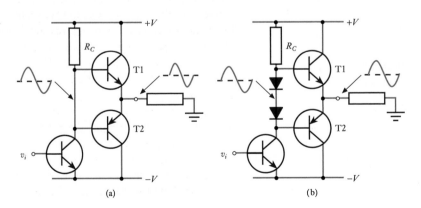

(a)   (b)

Figure 22.6 Improved push-pull output stage arrangements

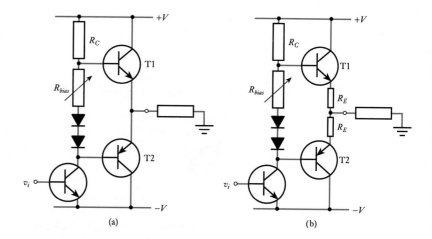

(a)   (b)

conducting, the voltage across each diode is approximately equal to the base-to-emitter voltage of each transistor. Consequently, one transistor should turn ON precisely where the other turns OFF, greatly reducing the distortion produced.

A slight problem with the arrangement of Figure 22.5(b) is that the current passing through the output transistors is considerably greater than that passing through the diodes. Consequently, the conduction voltage of the diodes is less than the V_{BE} of the transistors, and a small dead band still remains. Crossover distortion is reduced but not completely eliminated. More effective methods of biasing the output transistors are available, including that shown in Figure 22.6(a). Here a small variable resistor R_{bias} is added in series with the two diodes. The current flowing through this resistor increases the voltage between the two bases, compensating for the difference between the conduction voltages of the diodes and the turn-on voltages of the transistor. This resistance can be adjusted to minimise the crossover distortion produced.

Unfortunately, the power dissipated in the transistors causes their temperature to rise. This increases the current that they pass for a given base-to-emitter voltage, and this in turn causes them to dissipate more power. This leads to a problem of **temperature instability**, which in extreme cases can lead to **thermal runaway**. This problem can be tackled by mounting the biasing diodes close to the output transistors so that the temperature of the devices is similar. This will provide **temperature compensation**, since any thermally induced change in the turn-on voltage of the transistors should be matched by a similar change in the conduction voltage of the diodes. Problems of instability can also be tackled by the addition of small emitter resistors, as shown in Figure 22.6(b). The voltage drop across these resistors is subtracted from the base-emitter voltage of the circuit. They therefore provide negative feedback by reducing the voltage across the base-emitter junction as the current increases, thus stabilising the quiescent current. Typical values for these resistors might be a few ohms for low-power applications and much less for high-power circuits. The power dissipated in these resistors is thus fairly small, and their effect on the biasing of the circuit can be removed by adjustments to R_{bias}.

Files 22A
22B

Computer Simulation Exercise 22.1

Simulate the arrangement of Figure 22.5(a). Suitable transistors would be a 2N2222 for T1 and a 2N2907A for T2. Use +15 V and −15 V supplies, a value of 10 kΩ for R_C and a 10 Ω load. Any *npn* transistor may be used for the drive transistor, although a suitable biasing arrangement must be added to set the quiescent voltage on the bases of the output transistors to about zero. You may need to experiment with component values to achieve this. Alternatively, the drive transistor can be replaced with a sinusoidal current generator (ISIN in PSpice). If a current generator is used, this should be configured to give an offset current of 1.5 mA so that the quiescent voltage on the bases of the transistors is close to zero (this current flowing through the 10 kΩ resistor will give a voltage drop of 15 V, making the quiescent base voltage equal to about zero).

Apply a sinusoidal input to the circuit to produce a sinusoidal voltage on the bases of the transistors of 5 V peak at 1 kHz and observe the form of the output. You should observe that the output suffers from crossover distortion. Display the frequency spectrum of the output using the fast Fourier transform (FFT) feature of the simulator and observe the presence of harmonics of the input signal. Note which harmonics are present in this signal.

Modify the circuit by adding two diodes, as in Figure 22.5(b). You may use any conventional small-signal diodes – for example 1N914 devices. The drive transistor biasing arrangement, or the offset of the current generator, must now be adjusted to return the quiescent voltage on the bases of the output transistors to zero.

Again apply a sinusoidal input and observe the output voltage. Display the spectrum of this waveform and notice the effect of the diodes on the crossover distortion.

22.2.3 Amplifier efficiency

An important aspect in the choice of a technique to be used for a power output stage is its efficiency. We may define the efficiency of an amplifier as

$$\text{efficiency} = \frac{\text{power dissipated in the load}}{\text{power absorbed from the supply}} \qquad (22.1)$$

Efficiency is of importance, since it determines the power dissipated by the amplifier itself. The **power dissipation** of an amplifier is important for a number of reasons. One of the least important, except in battery-powered applications, is the actual cost of the electricity used, as this is generally negligible. Power dissipated by an amplifier takes the form of **waste heat**, and the production of excess heat requires the use of larger, and more expensive, power transistors to dissipate this heat. It might also require the use of other methods of heat dissipation, such as **heat sinks** or **cooling fans**. It may also be necessary to increase the size of the power supply to deliver the extra power required by the amplifier. All these factors increase the cost and size of the system.

Of great importance in determining the efficiency of an amplifier is its **class**, a term that describes its mode of operation. The main classes of operation are described in the following section.

22.3 Classes of amplifier

All amplifiers can be allocated to one of a number of classes, depending on the way in which the active device is operated.

22.3.1 Class A

In class A amplifiers, the active device (for example, a bipolar transistor or a FET) conducts during the complete period of any input signal. An example of such a circuit would be a conventionally biased single-transistor amplifier of the type shown in Figure 22.7.

It can be shown that, for conventional class A amplifiers, maximum efficiency is achieved for a sinusoidal input of maximum amplitude, when it reaches only 25 percent. With more representative inputs, the efficiency is very poor.

The efficiency of class A amplifiers can be improved by coupling the load using a transformer. The primary replaces the load resistor, while the load is connected to the secondary to form a **transformer-coupled amplifier**. This enables efficiencies approaching 50 percent to be achieved, but it is unattractive because of the disadvantages associated with the use of inductive components, including their cost and bulk.

Figure 22.7 A class A amplifier

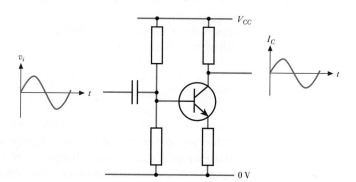

22.3.2 Class B

In a class B amplifier, the output active devices conduct for only half the period of an input signal. These are normally push-pull arrangements, in which each transistor is active for half of the input cycle. Such circuits were discussed in Section 22.2, and circuit examples were shown in Figures 22.5(b) and 22.6.

Class B operation has the advantage that no current flows through the output transistors in the quiescent state, so the overall efficiency of the system is much higher than in class A. If one assumes the use of ideal transistors, it can be shown that the maximum efficiency is about 78 percent.

22.3.3 Class AB

Class AB describes an amplifier that lies part way between classes A and B. The active device conducts for more than 50 percent of the input cycle but less than 100 percent. A class AB amplifier can be formed from a standard push-pull stage by ensuring that both devices conduct for part of the input waveform. For example, it could be achieved in the circuits of Figure 22.6 by appropriate adjustment of R_{bias}. Such an arrangement may be used to produce lower distortion than is normally associated with class B operation, without the efficiency penalty imposed by the use of class A.

The efficiency of a class AB amplifier will lie between those of class A and class B designs and will depend on the bias conditions of the circuit.

22.3.4 Class C

Following on from the definitions of classes A and B, it is perhaps not surprising that the definition of class C is that the active device conducts for *less* than half of the input cycle. Class C is used to enable the device to be operated at its peak current limit without exceeding its maximum power rating. The technique can produce efficiencies approaching 100 percent but results in gross distortion of the waveform. For these reasons, class C is used only in fairly specialised applications. One such use is in the output stage of radio transmitters, where inductive filtering is used to remove the distortion. A possible circuit for a class C amplifier is given in Figure 22.8. Often the collector resistor in this circuit is replaced by an *RC* tuned circuit.

22.3.5 Class D

In class D amplifiers, the active devices are used as switches and are either completely ON or completely OFF. A perfect switch has the characteristics of having infinite resistance when open and zero resistance when closed. If the devices used for the amplifier were perfect switches, this would result

Figure 22.8 A class C amplifier

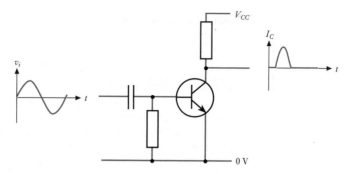

in no power being dissipated in the amplifier itself, since when a switch was ON it would have current flowing through it but no voltage across it, and when it was OFF it would have voltage across it but no current flowing through it. Since power is the product of voltage and current, the dissipation in both states would be zero. Although no real device is an ideal switch, transistors make very good switching devices, and amplifiers based on power transistors are both efficient and cost-effective. Amplifiers of this type are often called **switching amplifiers** or **switch-mode amplifiers**. Class D amplifiers may use single devices or push-pull pairs. In the latter case, only one of the two devices is ON at any time.

Switching amplifiers may be used to provide continuous control of power by switching the output voltage ON and OFF repeatedly at high speed. The power delivered to the load is controlled by varying the fraction of time for which the output is turned ON. This process is referred to as **pulse-width modulation (PWM)**, and we will look at this process in more detail in Section 22.5.3 when we look at switch-mode power supplies.

22.3.6 Amplifier classes – a summary

Class A, B, AB and C arrangements are linear amplifiers, while class D circuits are switching amplifiers. The distinction between the linear amplifiers may be seen as the differences between the quiescent currents flowing through the output devices. This in turn is determined by the biasing arrangements of the circuit.

22.4 Four-layer devices

Although transistors make excellent logical switches, they have limitations when it comes to switching high currents at high voltages. For example, to make a bipolar transistor with a high current gain requires a thin base region, which produces a low breakdown voltage. An alternative approach is to use one of a number of devices designed specifically for use in such applications. These components are not transistors, but their construction and mode of operation have a great deal in common with those of bipolar transistors.

22.4.1 The thyristor

The thyristor is a **four-layer device** consisting of a *pnpn* structure as shown in Figure 22.9(a). The two end regions have electrical contacts called the anode (*p* region) and the cathode (*n* region). The inner *p* region also has an electrical connection, called the gate. The circuit symbol for the thyristor is shown in Figure 22.9(b).

In the absence of any connection to the gate, the thyristor can be considered as three diodes in series formed by the *pn*, *np* and *pn* junctions. Since two of these diodes are in one direction and one is in the other, any applied voltage must reverse-bias at least one of the diodes and no current will flow in either direction. However, when an appropriate signal is applied to the gate, the device becomes considerably more useful.

Figure 22.9 A thyristor

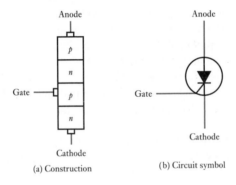

(a) Construction   (b) Circuit symbol

Thyristor operation

The operation of the thyristor is most readily understood by likening it to two interconnected transistors, as shown in Figure 22.10(a), which can be represented by the circuit of Figure 22.10(b). Let us consider initially the situation in which the anode is positive with respect to the cathode but no

Figure 22.10 Thyristor operation

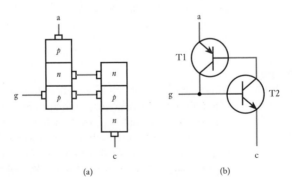

(a)   (b)

current is flowing in either device. Since T1 is turned OFF, no current flows into the base of T2, which is therefore also turned OFF. Since T2 is turned OFF, no current flows from the base of T1, so this transistor remains OFF. This situation is stable, and the circuit will stay in this state until external events change the condition of the circuit.

Consider now the effect of a positive pulse applied to the gate. When the gate goes positive, T2 will turn ON, causing current to flow from its collector to its emitter. This current will produce a base current in T1, turning it ON. This in turn will cause current to flow through T1, producing a base current in T2. This base current will tend to increase the current in T2, which in turn will increase the current in T1, and the cycle will continue until both devices are saturated. The current flowing between the anode and the cathode will increase until it is limited by external circuitry. This process is said to be *regenerative* in that the current flow is self-increasing and self-maintaining. Once the thyristor has been 'fired', the gate signal can be removed without affecting the current flow.

The thyristor will only function as a control device in one direction; if the anode is made negative with respect to the cathode, the device simply acts like a reverse-biased diode. This is why the thyristor is also called a **silicon-controlled rectifier** or **SCR**.

The thyristor may be thought of as a very efficient electrically controlled switch, with the rather unusual characteristic that when it is turned ON, by a short pulse applied to the gate, it will stay ON as long as current continues to flow through the device. If the current stops or falls below a certain **holding current**, the transistor action stops and the device automatically turns OFF. In the OFF state, only leakage currents flow and breakdown voltages of several hundreds or thousands of volts are common. In the ON state, currents of tens or hundreds of amperes can be passed with an ON voltage of only a volt or so. The current needed to turn ON the device varies from about 200 µA for a small device to about 200 mA for a device capable of passing 100 A or so. The switching times for small devices are generally of the order of 1 µs but are somewhat longer for larger devices. It should be remembered that, even though the thyristor is a very efficient switch, power is dissipated in the device as a result of the current flowing through it and the voltage across it. This power produces heat, which must be removed to prevent the device from exceeding its maximum working temperature. For this reason, all but the smallest thyristors are normally mounted on heat sinks, which are specifically designed to dispel heat.

The thyristor in AC power control

Although the thyristor can be used in DC applications, it is most often found in the control of AC systems. Consider the arrangement shown in Figure 22.11(a). Here a thyristor is connected in series with a resistive load to an AC supply. External circuitry senses the supply waveform, shown in Figure 22.11(b), and generates a series of gate trigger pulses as shown in Figure 22.11(c). Each pulse is positioned at the same point within the phase

Figure 22.11 Use of a thyristor in AC power control

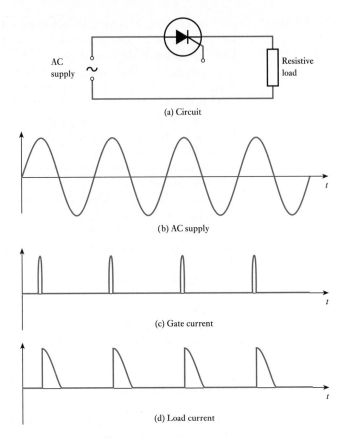

(a) Circuit

(b) AC supply

(c) Gate current

(d) Load current

of the supply, so the thyristor is turned ON at the same point in each cycle. Once turned ON, the device continues to conduct until the supply voltage and the current through the thyristor drop to zero, producing the output waveform shown in Figure 22.11(d). In the illustration, the thyristor is fired approximately halfway through the positive half-cycle of the supply, and therefore the thyristor is ON for approximately one-quarter of the cycle. Therefore, the power dissipated in the load is approximately one-quarter of what it would be if the load were connected directly to the supply. By varying the phase angle at which the thyristor is fired, the power delivered to the load can be controlled from 0 to 50 percent of full power. Such control is called half-wave control.

The gate-current pulse is generated by applying a voltage of a few volts between the gate electrode and the cathode. Since the cathode is within the supply circuit, it is common to use **opto-isolation** to insulate the electronics used to produce these pulses from the AC supply (as described in Section 4.3). To achieve full-wave control using thyristors requires the use of two devices connected in inverse parallel, as shown in Figure 22.12. This allows power to be controlled from 0 to 100 percent of full power but unfortunately requires duplication of the gate pulse-generating circuitry and the isolation network.

Figure 22.12 Full-wave power control using thyristors

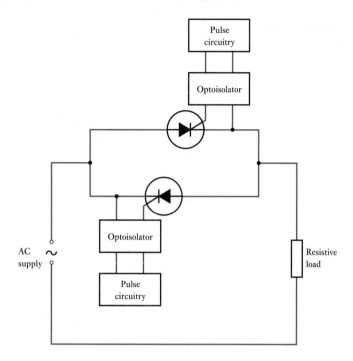

22.4.2 The triac

A more elegant solution to full-wave control of AC power is to use a **triac**. This is effectively a bidirectional thyristor that can operate during both halves of the supply cycle. It resembles two thyristors connected in inverse parallel but has the advantage that gate pulses can be supplied by a single isolated network. Gate pulses of either polarity will trigger the triac into conduction throughout the supply cycle. Since the device is effectively symmetrical, the two electrodes of the device are simply given the names MT1 and MT2, where MT simply stands for 'main terminal'. Voltages applied to the gate of the device are applied with respect to MT1. The circuit symbol for a triac is shown in Figure 22.13(a).

Gate trigger pulses in triac circuits are often generated using another four-layer device, the **bidirectional trigger diode** or **diac**. The diac

Figure 22.13 A triac and a diac

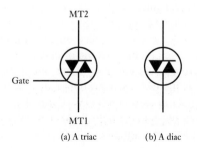

Figure 22.14 A simple lamp
dimmer using a triac

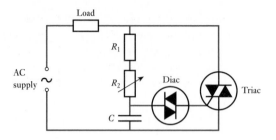

resembles a triac without any gate connection. It has the property that for
small applied voltages it passes no current, but if the applied voltage is
increased above a certain point, termed the **breakover voltage**, the device
exhibits **breakdown** and begins to conduct. Typical values for the breakover voltage for a diac are 30 to 35 V. The device operates in either direction and is used to produce a burst of current into the gate when a control
voltage, derived from the supply voltage, reaches an appropriate value. The
circuit symbol for a diac is shown in Figure 22.13(b).

Triacs are widely used in applications such as lamp dimmers and motor
speed controllers. A circuit for a simple domestic lamp dimmer is given in
Figure 22.14. Operation of the circuit is very straightforward: as the supply
voltage increases at the beginning of the cycle, the capacitor is charged
through the resistors and its voltage increases. When it reaches the breakover voltage of the diac (about 30 V), the capacitor discharges through the
diac, producing a pulse of current, which fires the triac. The phase angle at
which the triac is triggered is varied by changing the value of R_2, which
controls the charging rate of the capacitor. R_1 is present to limit the minimum resistance of the combination to prevent excessive dissipation in the
variable resistor. Once the triac has been fired it is maintained in its ON
state by the load current flowing through it, while the voltage across the
resistor–capacitor combination is limited by the ON voltage of the triac,
which is of the order of 1 V. This situation is maintained until the end of
the present half-cycle of the supply. At this point the supply voltage goes
to zero, reducing the current through the triac below its holding current and
turning it OFF. The supply voltage then enters its next half-cycle, the capacitor voltage again begins to rise (this time in the opposite sense), and the
cycle repeats. If the component values are chosen appropriately, the output
can be varied from zero to nearly full power by adjusting the setting of the
variable resistor.

The simple lamp dimmer circuit of Figure 22.14 controls the power in
the lamp by varying the phase angle of the supply at which the triac is fired.
Not surprisingly, this method of operation is called **phase control** or sometimes **duty-cycle control**. Using this technique, large transients are produced as the triac switches ON part way through the supply cycle. These
transients can cause problems of **interference**, either by propagating noise
spikes through the supply lines or by producing electromagnetic interference (EMI). An alternative method of control is to turn the triac ON for

complete half-cycles of the supply, varying the ratio of ON to OFF cycles to control the power. Switching occurs where the voltage is zero, so inter-ference problems are removed. This technique is called **burst firing** and is useful for controlling processes with a relatively slow speed of response. It is not suitable for use with lamp control, since it can give rise to flickering.

22.5 Power supplies and voltage regulators

In Section 19.8, we saw how semiconductor diodes can be used to rectify alternating voltages, and in Section 21.7 we looked at the use of bipolar transistors in voltage regulation. These techniques can be brought together, and developed, to form a range of power supplies.

22.5.1 Unregulated DC power supplies

A basic low-voltage unregulated supply takes the form of a step-down transformer, a full-wave rectifier and a **reservoir capacitor** (which is also called a **smoothing capacitor**). A typical circuit is shown in Figure 22.15.

The output voltage of an unregulated supply is determined by the input voltage and the step-down ratio of the transformer. As the load current is increased, the ripple voltage increases (as discussed in Section 19.8) and the mean output voltage falls. Unregulated supplies are used in applications where a constant output voltage is not required and where ripple is accept-able (for example when charging a battery).

Figure 22.15 A typical unregulated power supply

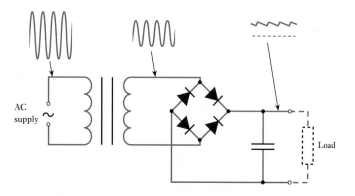

22.5.2 Regulated DC power supplies

When a more constant output voltage is required, an unregulated supply can be combined with a voltage regulator to form a regulated supply, as shown in Figure 22.16.

We considered simple voltage regulators in Section 21.7, where we looked at the circuit shown in Figure 22.17(a). You will recall that this is

Figure 22.16   A regulated
power supply

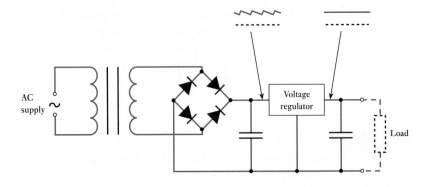

Figure 22.17   Voltage
regulators

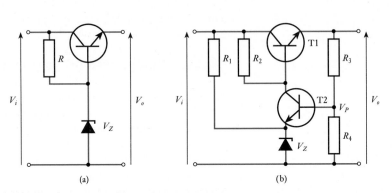

basically an emitter follower circuit where the output voltage is given by
the base voltage of the transistor (which is set by the Zener diode) minus the
fairly constant base-emitter voltage. In practice, it is common to use a slightly
more sophisticated arrangement, such as that shown in Figure 22.17(b).
This is similar to the earlier circuit, except that the output voltage is now
sampled (using the potential divider of R_3 and R_4) and a fraction of this voltage is used to provide negative feedback. The voltage on the emitter of T2
is held constant by the Zener diode. If the voltage on the base of T2 rises to
a point where the base-emitter voltage is greater than the turn-on voltage of
the transistor, this will produce a collector current that flows though R_2.
This will reduce the voltage on the base of T1 and hence reduce the output
voltage. The circuit will therefore stabilise at a point where the voltage at
the midpoint of the potential divider V_P is approximately equal to V_Z plus
the base-emitter voltage of T2 (about 0.7 V). Since V_P is determined by the
output voltage and the ratio of R_3 and R_4, it follows that

$$V_P = V_o \frac{R_4}{R_3 + R_4} = V_Z + 0.7 \text{ V}$$

and therefore

$$V_o = (V_Z + 0.7 \text{ V}) \frac{R_3 + R_4}{R_4} \tag{22.2}$$

| Example 22.1 | Determine the output voltage of the following regulator (assuming that the input voltage is sufficiently high to allow normal operation). |

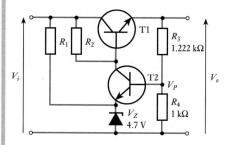

The voltage V_P on the base of T2 will be $V_Z + 0.7$. Therefore, the output voltage will be this value multiplied by $(R_3 + R_4)/R_4$. Therefore

$$V_o = (V_Z + 0.7 \text{ V})\frac{R_3 + R_4}{R_4}$$

$$= (4.7 + 0.7)\frac{1.222 \text{ k}\Omega + 1 \text{ k}\Omega}{1 \text{ k}\Omega}$$

$$= 12.0 \text{ V}$$

Voltage regulators normally take the form of a dedicated integrated circuit. These invariably use more complex circuits than that shown in Figure 22.17(b), often replacing T2 by an operational amplifier to give greater gain and better regulation. They also generally include additional circuitry to provide **current limiting** to prevent the circuit from being damaged by excessive current flow. Fixed-voltage regulator ICs are available for a wide range of standard voltages (for example +5, +15 and −15 V), while other components allow the output voltage to be set using external components (usually a couple of resistors).

Power dissipation

A disadvantage of the power supplies described so far is that they are relatively inefficient. This can be appreciated by considering the regulator of Figure 22.17(b). In order to provide a constant output voltage, the magnitude of the unregulated input voltage (V_i) must be somewhat larger than the regulated output voltage (V_o). Therefore, the voltage across the output transistor is $V_o - V_i$, and the power dissipated in the transistor is equal to this voltage times the output current. In many cases, the power dissipated in the regulator is comparable to that delivered to the load, and in high-power supplies this results in the production of large amounts of heat. A further disadvantage of these supplies is that the transformers are heavy and bulky, since the low frequency of the AC supply requires a large inductance.

Example 22.2

Compare the power dissipated in the load with that dissipated in the output transistor of the regulator when the circuit of Figure 22.17(b) is connected to a load $R_L = 5\ \Omega$, given that $V_i = 15$ V and $V_o = 10$ V.

The output voltage $V_o = 10$ V and the load resistance R_L is $5\ \Omega$, therefore the output current is

$$I_o = \frac{V_o}{R_L} = \frac{10}{5} = 2\ \text{A}$$

Therefore the power delivered to the load is

$$P_o = V_o I_o = 10 \times 2 = 20\ \text{W}$$

The current through the output transistor (T1) is equal to output current I_o, and the voltage across the transistor is given by the difference between the input voltage and the output voltage. Therefore, the power dissipated in the output transistor P_T is given by

$$P_o = (V_i - V_o)I_o = (15 - 10) \times 2 = 10\ \text{W}$$

Thus the power dissipated in the output transistor is half that of the power delivered to the load.

22.5.3   Switch-mode power supplies

One way of tackling the problems associated with high power consumption in power supplies is through the use of a **switching regulator**. The basic configuration of such a regulator is shown in Figure 22.18(a). The unregulated voltage is connected to a switch that is opened and closed at a rate of about 20 kHz (or more). While the frequency remains constant, the **duty cycle** (that is, the ratio of the ON time to the OFF time) is varied. If the switch is closed for a relatively short period during each cycle the average value of the output will be low, as shown in Figure 22.18(b). However, if the switch is closed for a larger proportion of each cycle the average value will be higher, as in Figure 22.18(c). By varying the duty cycle of the switching waveform, the average value of the output voltage can be varied from zero up to the input voltage.

A great advantage of switching regulators is that their power dissipation is very low. When an ideal switch is OFF the current through it is zero, while when it is ON the voltage across it is zero. Therefore, in either state the power dissipated in the switch is zero. Transistors are not ideal switches, but both bipolar transistors and MOSFETs have very good switching characteristics. When the transistor is turned OFF it passes negligible current, while when it is turned ON the voltage across it is small. Thus in either state the switch (and the regulator as a whole) consume very little power.

The averaging circuit normally uses an inductor–capacitor arrangement, as shown in Figure 22.19. When the switch is first closed, current starts to

Figure 22.18 A switching regulator arrangement

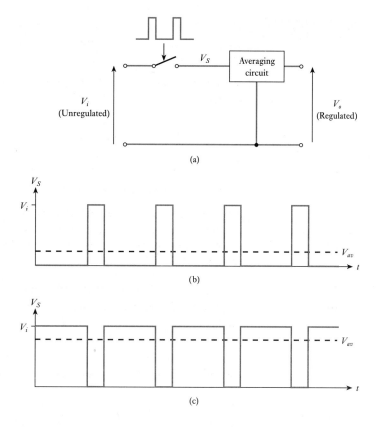

(a)

(b)

(c)

Figure 22.19 An *LC* averaging circuit

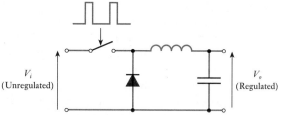

flow through the inductor and into the capacitor. The diode is reverse-biased by the applied voltage and so passes no current. Because of the nature of inductance, the current builds slowly in the circuit as energy is stored in the inductor. When the switch is now opened, the energy stored in the inductor produces an e.m.f., which acts to continue this current. This has the effect of forward-biasing the diode and further charging the capacitor as the current decays. Current is taken out of the capacitor by the load and the circuit soon reaches equilibrium, where the voltage on the capacitor is equal to the average value of the switching waveform plus a small ripple voltage.

The duty cycle of the switch in the switching regulator is controlled using feedback from the output, and Figure 22.20 shows a possible

Figure 22.20 The use of
feedback in a switching
regulator

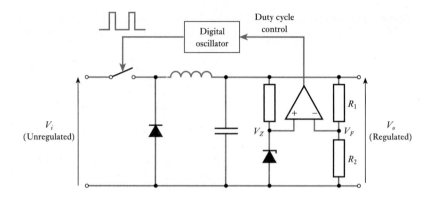

arrangement to achieve this. The potential divider formed by R_1 and R_2
produces a voltage V_F that is related to the output voltage by the expression $V_F = V_o \times R_2/(R_1 + R_2)$. This is compared with a reference voltage V_Z
produced using a Zener diode. The output from the comparator is used to
vary the duty cycle of a digital oscillator, which then controls the switch.
If V_F falls below V_Z the duty cycle of the switch will be increased to raise
the output voltage, while if V_F rises above V_Z the output will be reduced.
In this way, the feedback maintains the output such that V_F is equal to V_Z,
and thus

$$V_F = V_o \frac{R_2}{R_1 + R_2} = V_Z$$

and rearranging

$$V_o = V_Z \frac{R_1 + R_2}{R_2} \tag{22.3}$$

Switching regulators can be used to replace conventional regulators in
arrangements of the form shown in Figure 22.16. This reduces the power
consumed in the regulator and may cut its size and weight (by avoiding
the need for a large heat sink). In some cases, it may also be possible to
achieve additional weight savings by removing the transformer. Here the
AC supply is simply rectified and applied to the switching regulator. A
power supply that makes use of a switching regulator is often referred to as
a **switch-mode power supply** or as a **switch-mode power unit** (**SMPU**).
   While it is possible to construct switching regulators from discrete components, it is more common to use integrated circuit elements. These contain all of the active components within a single device. Combining such an
integrated circuit with a handful of external components allows a complete
switch-mode power supply to be constructed easily. Details of appropriate
circuits are normally given in the datasheet for the particular component.
Alternatively, complete switch-mode power supplies can be purchased as
ready-made modules.

Key points

■ Power amplifiers are designed to deliver large amounts of power to their load.

■ Power amplifiers can be constructed using either FETs or bipolar transistors. Bipolar circuits often make use of emitter follower circuits, since these have low output resistance.

■ Many power amplifiers use a push-pull arrangement using a split power supply. Where distortion is of importance, care must be taken in the design of the biasing arrangement of such circuits to reduce crossover distortion.

■ The efficiency of an amplifier is greatly affected by its class of operation:

  • In class A amplifiers, the active device conducts all the time.
  • In class B amplifiers, the active device conducts for half of the period of the input.
  • In class AB amplifiers, the active device conducts for more than half of the period of the input.
  • In class C amplifiers, the active device conducts for less than half of the period of the input.
  • Class D amplifiers are switching circuits, and the active device is always either fully ON or fully OFF.

■ While transistors make excellent logical switches, they are less good when dealing with high voltages and high currents. In such cases, we often use special-purpose devices such as thyristors or triacs.

■ A transformer, a rectifier and a capacitor can be combined to form a simple unregulated power supply. Unfortunately, such circuits suffer from variability in the output voltage and from output voltage ripple.

■ A more constant output voltage can be produced by adding a regulator. Conventional regulators are cheap and easy to use but are very inefficient. Switching regulators provide much higher efficiency but at the expense of a more complex circuit.

■ Switch-mode power supplies can provide very high efficiency combined with low volume and weight.

Exercises

22.1 What is meant by the term 'power amplifier'?

22.2 Why is efficiency of importance in power amplifiers?

22.3 Which bipolar transistor configuration is most often used in power amplifier circuits? Why is this?

22.4 Explain the operation of the simple push-pull amplifier of Figure 22.3. Why does this circuit produce crossover distortion?

22.5 How may the crossover distortion in a push-pull amplifier be reduced?

22.6 Explain what is meant by temperature instability and describe how this problem may be tackled.

22.7 Outline the distinction between the various classes of amplifier. Which forms are linear amplifiers?

22.8 Into which class does the simple push-pull amplifier of Figure 22.3 fall?

22.9 Why is a bipolar transistor not ideal for switching high currents at high voltages?

22.10 Explain briefly the operation of a thyristor.

22.11 Explain the need for opto-isolation in AC control using thyristors.

22.12 Why is a thyristor not ideal for AC power control?

22.13 How does a triac differ from a thyristor?

22.14 What is meant by 'phase control' of a triac circuit? Describe some potential problems with this approach.

22.15 What is meant by 'burst firing' of a triac circuit? Describe some potential problems with this approach.

22.16 Sketch the circuit of a simple unregulated power supply, explaining the function of each component. What limits the usefulness of such a circuit?

22.17 Sketch a simple voltage regulator that uses feedback to stabilise the output voltage. Explain the operation of your circuit.

22.18 Modify the circuit of Example 22.1 to produce an output voltage of 15 V.

22.19 A voltage regulator of the form shown in Figure 22.17 is connected to a load of 10 Ω. Calculate the power dissipated in the output transistor if the input voltage is 25 V and the output voltage is 15 V.

22.20 Explain what is meant by a switching regulator. What are the advantages of this form of regulator?

22.21 Explain the operation of the switching regulator of Figure 22.20.

Chapter 23

Electric Motors and Generators

Objectives

When you have studied the material in this chapter you should be able to:

- **discuss the various forms of electrical machine;**
- **explain how the interaction between a magnetic field and a rotating coil can be used to generate electricity;**
- **explain how the interaction between a changing magnetic field and a coil can be used to generate motion;**
- **describe the operation of various AC and DC forms of generator and motor;**
- **discuss the use of electrical machines in a variety of industrial and domestic applications.**

23.1 Introduction

An important area of electrical engineering relates to various forms of rotating **electrical machine**. These may be broadly divided into **generators**, which convert mechanical energy into electrical energy, and **motors**, which convert electrical energy into mechanical energy. In general, machines are designed to perform one of these two tasks, although in some cases a generator may also function as a motor, or vice versa, but with reduced efficiency.

Electrical machines may be divided into **DC machines** and **AC machines**, and both types operate through the interaction between a magnetic field and a set of windings. There are a great many forms of electrical machine, and this book does not aim to give a detailed treatment of each type. Rather, it sets out to describe the general principles involved, leaving readers to investigate further if they need more information on a particular type of motor or generator.

23.2   A simple AC generator

In Chapter 14, we looked at the nature of magnetic fields and at the effects of changes in the magnetic flux associated with conductors. You will recall that Faraday's law dictates that if a coil of N turns experiences a change in magnetic flux, then the induced voltage V is given by

$$V = N\frac{d\Phi}{dt} \tag{23.1}$$

where $d\Phi/dt$ is the rate of change of flux in webers/second.

In Chapter 14, we were primarily concerned with the situation where the change in magnetic flux is caused by a change in the magnetic field associated with a stationary conductor. However, a similar effect is produced when a conductor moves within a constant field. Consider for example the situation shown in Figure 23.1. Here a coil of N turns and cross-sectional area A rotates with an angular velocity ω in a magnetic field of uniform flux density B.

If at a particular time the coil is at an angle θ to the field, as shown in Figure 23.1(b), its effective area normal to the field is $A \sin \theta$. Therefore, the flux linking the coil (Φ) is given by $BA \sin \theta$, and the rate of change of this flux ($d\Phi/dt$) is given by

$$\frac{d\Phi}{dt} = BA\frac{d(\sin \theta)}{dt} \tag{23.2}$$

Now

$$\frac{d(\sin \theta)}{dt} = \frac{d\theta}{dt}\cos \theta = \omega \cos \theta \tag{23.3}$$

since $d\theta/dt = \omega$.

Therefore, combining Equations 23.1, 23.2 and 23.3, we have

$$V = N\frac{d\Phi}{dt} = NBA\frac{d(\sin \theta)}{dt} = NBA\omega \cos \theta \tag{23.4}$$

and the induced voltage varies as the cosine of the phase angle, as shown in Figure 23.2(a). Given a constant speed of rotation, the phase angle θ varies linearly with time, and hence the output voltage varies as the cosine of time, as shown in Figure 23.2(b).

Figure 23.1   The rotation of a coil in a uniform magnetic field

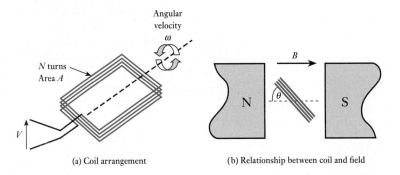

(a) Coil arrangement   (b) Relationship between coil and field

Figure 23.2   Voltage produced
by the coil of Figure 23.1

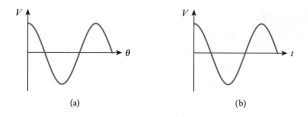

(a)   (b)

| Example 23.1 | A coil consisting of 100 turns of copper wire has an area of 20 cm². Determine the peak magnitude of the sinusoidal voltage produced across the terminals of this coil if it rotates within a magnetic field of 400 mT at a rate of 1000 rpm. |

From Equation 23.4, we know that

$$V = NBA\omega \cos\theta$$

In this expression ω is the angular frequency and in this case the cyclic frequency is 1000 rpm, which is $1000/60 = 16.7$ Hz. Since $\omega = 2\pi f$, it follows that $\omega = 2 \times \pi \times 16.7 = 105$ rad/s. Therefore, substituting into the above equation gives

$$V = NBA\omega \cos\theta$$

$$= 100 \times 400 \times 10^{-3} \times 20 \times 10^{-4} \times 105 \cos\theta$$

$$= 8.4 \cos\theta$$

Therefore, the output is a sinusoidal voltage with a peak value of 8.4 V.

23.2.1   Slip rings

One problem with the arrangement of Figure 23.1 is that any wires that are connected to the coil will become tangled as the coil rotates. A solution to this problem is to use **slip rings**, which provide a sliding contact to the coil as it rotates. This idea is illustrated in Figure 23.3. Electrical contact is made to the slip rings through **brushes**, which normally take the form of graphite blocks that are held against the rings by springs.

Figure 23.3   The use of slip
rings

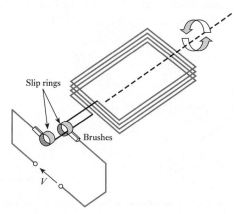

23.3   A simple DC generator

The alternating signal produced by the arrangement of Figure 23.3 could be converted to DC using a conventional rectifying arrangement (as discussed in Section 19.8). However, a more efficient (and common) approach is to rectify the output of the generator by replacing the two slip rings of Figure 23.3 with a single, split, slip ring as shown in Figure 23.4(a). A slip ring that is split in this way is referred to as a **commutator**.

The commutator is arranged so that, as the voltage produced by the coil changes polarity, the connections to the coil are reversed. Hence the voltage produced across the brushes is of a single polarity, as shown in Figure 23.4(b).

While the arrangement of Figure 23.4 produces a voltage that is unidirectional, its magnitude varies considerably as the coil rotates. This 'ripple' can be reduced by summing the output from a number of coils that are set at different angles. This is illustrated in Figure 23.5(a), which shows an arrangement with two coils set at 90° to each other. The coils are connected in series, and the commutator now has four segments to allow the connections to the coils to be changed as the coils rotate. The resultant output voltage is shown in Figure 23.5(b). This process can be extended by using additional coils to further reduce the variation in the output voltage.

Figure 23.4   Use of a commutator

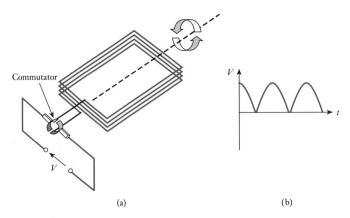

(a)   (b)

Figure 23.5   A simple generator with two coils

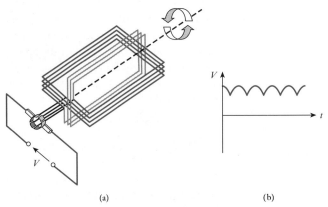

(a)   (b)

Figure 23.6 The use of an iron core and shaped pole pieces

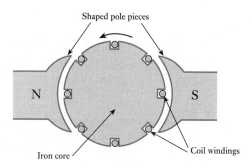

Shaped pole pieces

N   S

Iron core   Coil windings

The ripple voltage produced by the generator can be further reduced by winding the coils on a cylindrical iron core and by shaping the pole pieces of the magnets as shown in Figure 23.6. This has the effect of producing a high and approximately uniform magnetic field in the small air gap. The coils are wound within slots in the core, and the arrangement of coils and core is known as the **armature**. The armature of Figure 23.6 has four coils.

23.4 DC generators or dynamos

Having looked at the basic principles of a simple DC generator, we are now in a position to consider a more practical arrangement. DC generators, or **dynamos** as they are sometimes called, can take a number of forms depending on the method used to produce the magnetic field. **Permanent-magnet generators** are available, but it is more common to generate the magnetic field electrically using **field coils**. The current that flows in these coils can be supplied by an external energy source (in the case of a **separately excited generator**), but it is more usual to supply this from the current produced by the generator itself (as in a **self-excited generator**). Since the field coils consume only 1 or 2 percent of the rated output current, this loss of power is generally quite acceptable.

In addition to using multiple armature windings, it is common to use multiple poles, and Figure 23.7 shows a typical four-pole arrangement. The poles are arranged as alternating north and south poles and are held in place by a steel tube called the **stator**, which also forms the outer casing of the unit. Field coils are wound around each pole piece, these being connected in series and wired to produce the appropriate magnetic polarity. The generator in Figure 23.7 has eight slots in the armature, although many devices will have twelve or more. A typical device would produce an output ripple of about 1 or 2 percent.

23.4.1 Field coil excitation

In some generators the field coils are connected in series with the armature coils (a **series-wound DC generator**), while in others the field coils are

Figure 23.7 A DC generator or dynamo

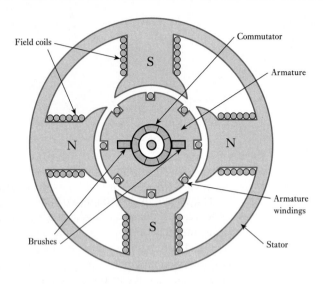

Figure 23.8 Connections for a shunt-wound generator

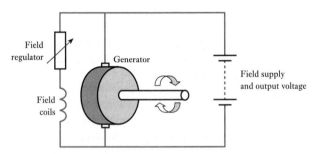

connected in parallel with the armature coils (a **shunt-wound DC generator**). A third variant has two sets of field windings, one connected in series and one in parallel, and such a machine is called a **compound DC generator**.

One of the most common forms of DC generator is the shunt-wound generator. This is an example of a self-excited generator, and in this case the field coils are connected in parallel with the armature. Such an arrangement is shown in Figure 23.8. In a typical configuration, the generator might be used to charge a battery. The voltage across the armature forms the output voltage of the generator, and this is connected across the battery. This voltage is also used to drive the field coils, the current in the coils being controlled by a field regulator, which might be a simple variable resistance.

23.4.2 DC generator characteristics

The various forms of DC generator have slightly different electrical characteristics, and these differences will often determine the device chosen for

Figure 23.9 Shunt-wound
generator characteristics

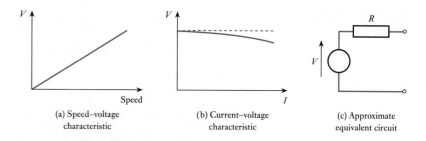

(a) Speed–voltage
characteristic

(b) Current–voltage
characteristic

(c) Approximate
equivalent circuit

a particular situation. We will not look at generator characteristics in detail
but simply note that the output voltage increases with the speed of rotation
of the armature and that in many DC generators this is a nearly linear relationship. Generators are often run at a constant speed (although this is *not*
the case in all situations), with the characteristics of the generator chosen to
give the required output voltage.

The voltage produced by a generator is also affected by the current taken
from the device. As the current increases the output voltage falls, as we
would expect for a voltage source that has output resistance. This fall is
caused partly by the resistance in the armature and partly by an effect
known as **armature reaction**, where the armature current produces a magnetic flux that opposes that produced by the field coils.

Figure 23.9 shows typical characteristics for a shunt-wound DC generator. Figure 23.9(a) shows the relationship between the output voltage
and the speed of rotation for zero output current; Figure 23.9(b) shows the
effect of current on the output voltage; and Figure 23.9(c) shows a simple
equivalent circuit of the generator.

23.5 AC generators or alternators

AC generators, or **alternators**, are in many ways similar to DC generators
in that power is produced by the effect of a changing magnetic field on a
set of coils. However, as we noted in Section 23.2, alternators do not
require commutation, and this allows some simplification of their construction. Since the power required to produce the electric field is much less than
that delivered by the generator, it makes sense to reverse the construction
and to rotate the field coils while keeping the armature windings stationary.
Note that the *armature windings* are the coils that produce the output e.m.f.
of the generator. In the case of a DC generator, the armature windings are
mounted on the rotating part of the machine (the **rotor**), so the rotor is also
known as the armature. In an AC synchronous generator, the large and
heavy armature coils are mounted in the stationary part of the machine (the
stator). In this case, the field coils are mounted on the rotor and direct current is fed to these coils by a set of slip rings.

As with DC motors, multiple poles and sets of windings are used to
improve efficiency. In some cases, three sets of armature windings are

Figure 23.10   A four-pole alternator

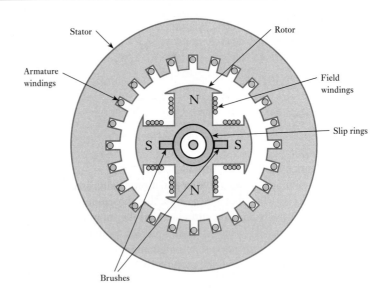

spaced 120° apart around the stator to form a **three-phase generator**. Figure 23.10 shows a section through a simple four-pole alternator.

In the alternator, the e.m.f. generated is in sync with the rotation of the rotor, and for this reason it is referred to as a **synchronous generator**. If the generator has a single pair of poles, then the frequency of the output will be the same as the rotation frequency. Therefore, to produce an output at 50 Hz using a two-pole generator, the speed of rotation would need to be $50 \times 60 = 3000$ rpm. Generators with additional pole pairs produce an output at a correspondingly higher frequency, since each turn of the rotor represents several cycles of the output. In general, a machine with N pole pairs produces an output at a frequency of N times the rotational frequency.

Example 23.2	A four-pole alternator is required to operate at 60 Hz. What is the required rotation speed?

A four-pole alternator has two pole pairs. Therefore the output frequency is twice the rotation speed. Therefore, to operate at 60 Hz, the required rotation speed must be $60/2 = 30$ Hz. This is equivalent to $30 \times 60 = 1800$ rpm.

23.6   DC motors

We noted in Chapter 14 that current flowing in a conductor produces a magnetic field about it. This process is illustrated in Figure 23.11(a). When the current-carrying conductor is within an externally generated magnetic field, the induced field interacts with the external field and a force is exerted on the conductor. This is shown in Figure 23.11(b). If the conductor is able to move, this force will cause motion in the direction indicated in

Figure 23.11 Fields and forces associated with a current-carrying conductor

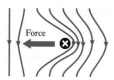

(a) The magnetic field about a current flowing into the page

(b) The effects of an external magnetic field

Figure 23.11(b). Therefore, if a conductor lies in a magnetic field, motion of the conductor in the field will generate an electric current, while current in the conductor will generate motion.

The reciprocal nature of the properties of conductors in magnetic fields means that many electrical machines will operate as either generators or motors. For example, the DC generators described in Section 23.4 will also function as DC motors, although machines designed as motors will generally be more efficient in this role than machines designed as generators. However, the diagram of Figure 23.7 could equally well represent a four-pole DC motor.

As in the case of generators, motors vary in the number of magnetic poles they possess and in the way that their windings are configured. Shunt-wound, series-wound and compound motors are available, and each has slightly different characteristics.

23.6.1 Shunt-wound DC motor

The shunt-wound DC motor is widely used since it has several attractive features. Its speed of rotation is largely dependent on the applied voltage, while the torque it applies is related to the current. Thus if a constant voltage is applied to such a motor the speed of rotation will tend to remain fairly constant, while the current taken by the motor will vary depending on the load applied. In practice, the speed is affected to some extent by the load applied to the motor, and therefore the speed drops slightly with increasing torque. Figure 23.12(b) shows typical characteristics for a shunt-wound DC motor.

Figure 23.12 Characteristics of a shunt-wound DC motor

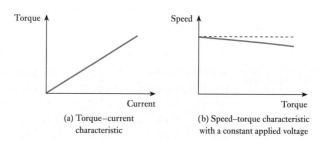

(a) Torque–current characteristic

(b) Speed–torque characteristic with a constant applied voltage

23.7   AC motors

AC motors can be divided into two main forms: synchronous motors and induction motors. High-power versions of either type invariably operate from a three-phase supply, but single-phase versions are also widely used, particularly in a domestic setting. Here we will look at examples of both synchronous and induction motors, and at both three-phase and single-phase versions.

23.7.1   Synchronous motor

Just as DC generators can also be used as DC motors, the synchronous AC generator (or alternator) of Section 23.5 can also be used as a **synchronous motor**. As the name implies, such a device operates at a speed determined by the frequency of the AC input. When used with a conventional AC supply, they are therefore **constant-speed motors**.

In a three-phase synchronous motor, the three sets of stator coils produce a magnetic field that rotates around the rotor at a speed determined by the frequency of the AC supply and the number of poles in the stator. The DC field current, which is fed to the rotor through the slip rings, turns the rotor into an electromagnet (possibly with several poles), which is dragged around by the rotating magnetic field. Thus the rotor speed is determined by the speed of rotation of the magnetic field.

Single-phase motors do not have the benefit of multiple input phases to generate the rotating magnetic field. Various techniques can be used to overcome this problem, and this gives rise to several variants on the basic design. These techniques will not be discussed here.

Unfortunately, because the rotor in a synchronous motor is dragged around by the rotating magnetic field, torque is only produced when the rotor is in sync with this field. When the motor is energised from rest the rotating magnetic field is established almost immediately, but the rotor is initially stationary. Consequently, the magnetic field races past the rotor rather than causing it to turn. Basic synchronous motors therefore produce no starting torque and are not self-starting. To overcome this problem, motors incorporate some form of starting mechanism to get them moving from rest. In some cases, this involves configuring the motor to operate as an induction motor (as discussed below) until it gets up to speed. Once the motor has reached synchronism it switches to operate as a synchronous motor, which in some situations gives greater efficiency than can be gained from an induction motor.

Example 23.3

What is the speed of rotation of an eight-pole synchronous motor when used with a single-phase 50 Hz supply?

An eight-pole motor has four pole pairs, so the magnetic field rotates at four times the supply frequency, which in this case is $4 \times 50 = 200$ Hz. Since the rotor is dragged around by this rotating magnetic field, this is also the speed of rotation of the motor. Therefore, the motor turns at 200 Hz = 200 × 60 = 12,000 rpm.

23.7.2   Induction motors

Perhaps the most important forms of AC motor are the various types of **induction motor**. These differ from synchronous motors in that field current is not fed to the rotor through slip rings but is *induced* in the rotor by transformer action. The most common form of induction motor is the **cage rotor induction motor**, which is also known as the **squirrel-cage induction motor**. This uses a stator of a similar form to that in the synchronous motor (or the synchronous generator of Figure 23.10) but replaces the rotor and slip rings with an arrangement of the form shown in Figure 23.13. This can be seen as a series of parallel conductors that are shorted together at each end by two conducting rings.

As in the three-phase synchronous motor, in a three-phase induction motor the stator coils produce a rotating magnetic field that cycles around the rotor at a frequency determined by the supply frequency and the number of poles in the stator. A stationary conductor in the stator will see a varying magnetic field, and this changing flux will induce an e.m.f. in much the same way that an e.m.f. is induced in the secondary of a transformer. This induced e.m.f. results in a current flowing in the rotor, and this in turn produces a magnetic field. This *rotor* field interacts with the *stator* field in the same way as in a synchronous motor, and again the rotor is dragged around by the rotating field in the stator. However, in the case of an induction motor, the stator will always turn slightly *more slowly* than the stator field. This is because, if the rotor turned in sync with the stator field, there would be no change in the flux associated with the rotor, and hence no induced current. The slight speed difference is called the **slip** of the motor. This increases with the applied load and might be a few percent at full load.

Figure 23.13   A squirrel-cage induction motor

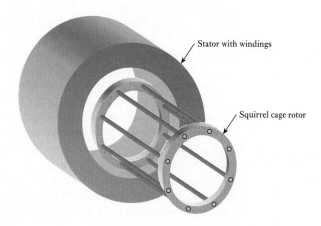

Three-phase induction motors have an advantage over synchronous motors in that they are self-starting.

In domestic situations three-phase supplies are usually not available, and several techniques can be used to produce the required rotating magnetic field from a single-phase supply. This gives rise to several forms of induction motor, such as the **capacitor motor** and the **shaded-pole motor**. Such motors are inexpensive and are widely used in domestic appliances. However, further discussion of the operation of these motors is beyond the scope of this book.

23.8 Universal motors

While most motors are designed to operate on either AC or DC, some motors can operate on either. These **universal motors** resemble series-wound DC motors but are designed for both AC and DC operation. Typically operating at high speeds (usually greater than 10,000 rpm), these motors offer a high power-to-weight ratio, making them ideal for portable appliances such as hand drills and vacuum cleaners.

23.9 Electrical machines – a summary

It can be seen that there are a great many forms of electrical machine, and this chapter has given only a brief overview of their characteristics and variety.

While both DC and AC generators have been described, power generation is dominated by synchronous AC machines. Such devices range from small alternators used in automotive applications to the large generators used in power stations. While all synchronous generators tend to be relatively efficient, this efficiency increases with size, with large generators converting more than 98 percent of their mechanical input power into electricity.

Both DC and AC motors are widely used, although often in different situations. Where moderate or large amounts of mechanical power are required AC motors are more common, particularly where variable-speed operation is not required. In industrial applications, three-phase induction motors are the dominant devices, while domestic appliances such as washing machines and dishwashers will normally use single-phase induction motors. DC motors are widely used in low-power applications, particularly in situations where variable-speed control is required. The simple relationship between speed and voltage in many DC motors makes them very easy to control. DC motors are also used in high-power applications that require variable speed, such as traction applications. However, the development of high-powered electronic speed controllers for AC motors has reduced their use in such applications.

Having discussed both DC and AC motors in this chapter, we must remember the existence of the stepper motor (as discussed in Chapter 4), which offers alternative characteristics for use in appropriate applications.

Key points

- Electrical machines can be broadly divided into generators, which convert mechanical energy into electrical energy, and motors, which convert electrical energy into mechanical energy.

- In most cases, generators can also function as motors, and vice versa.

- Electrical machines can be divided into DC machines and AC machines.

- All electrical machines operate through the interaction between a magnetic field and a set of windings.

- The rotation of a coil in a uniform magnetic field produces a sinusoidal e.m.f. This principle is at the heart of an AC generator or alternator.

- A commutator can be used to convert the above sinusoidal e.m.f. into a unipolar form. This is the basis of a DC generator or dynamo.

- While the magnetic field in an electrical machine can be produced by a permanent magnet, it is more common to produce this electrically using field coils.

- In DC generators, the armature coils normally rotate within stationary field coils. In AC generators, the field coils normally rotate within stationary armature coils.

- DC motors are often similar in form to DC generators.

- Some forms of AC generator can also be used as motors.

- There are many forms of AC motor, the most widely used being the various types of induction motor.

- Many types of motor are not inherently self-starting, and some form of starting mechanism must be incorporated.

Exercises

23.1   What is meant by the term 'electrical machine'?

23.2   A coil of 50 turns with an area of 15 cm² rotates in a magnetic field of 250 mT at 1500 rpm. What is the peak magnitude of the sinusoidal voltage produced across its terminals?

23.3   Explain the function of slip rings.

23.4   What is the function of a commutator?

23.5   How may the ripple voltage produced by a DC generator be reduced?

23.6   How are the field coils in a DC generator normally excited?

23.7   Describe the characteristics of a shunt-wound dynamo.

23.8   What is meant by 'armature reaction'?

Exercises continued

23.9 How does the construction of a typical alternator differ from that of a dynamo?

23.10 What is meant by the term 'armature', and what form does this take in a dynamo and in an alternator?

23.11 How does a *synchronous* generator get its name?

23.12 A six-pole alternator is required to operate at 50 Hz. What is the required rotation speed?

23.13 How does a DC motor differ from a DC generator?

23.14 Briefly describe the characteristics of a shunt-wound DC motor.

23.15 Why is a synchronous motor so called?

23.16 What is the speed of rotation of a twelve-pole synchronous motor when used with a single-phase 50 Hz supply?

23.17 How does the construction of an induction motor differ from that of a synchronous motor?

23.18 What is meant by the 'slip' of an induction motor?

23.19 Explain what is meant by a universal motor?

23.20 What sort of generator would normally be used in a power station?

23.21 What sort of motor would typically be used in a domestic washing machine?

Chapter 24

Positive Feedback, Oscillators and Stability

Objectives

When you have studied the material in this chapter you should be able to:

- **describe the use of positive feedback in the production of both sine-wave and digital oscillators;**
- **explain the conditions required for a circuit to oscillate;**
- **sketch simple circuits of sine-wave and digital oscillators;**
- **discuss the problems of amplitude stability in such circuits;**
- **describe the use of crystals in producing highly stable oscillators;**
- **explain the effects of positive feedback on the stability of a circuit.**

24.1 Introduction

In Chapter 7, we looked at feedback in general terms and spent some time looking at the use and characteristics of negative feedback. In this chapter, we will continue our study in this area by looking at some of the features of positive feedback.

Positive feedback is used in a range of both analogue and digital circuits to produce a variety of effects. In this chapter, we will concentrate on the most common use of positive feedback, which is in the production of **oscillators**.

While positive feedback is often used intentionally to achieve particular circuit characteristics, it can also occur *unintentionally* in circuits. This is particularly common in circuits that make use of negative feedback. In this situation, the presence of feedback can adversely affect the operation of the circuit and can have dramatic effects on its stability.

24.2   Oscillators

When considering generalised feedback arrangements in Chapter 7, we derived an expression for the gain of a feedback system of the form shown in Figure 24.1. You will recall that this is referred to as the **closed-loop** gain G of the system and is given by the expression

$$G = \frac{A}{(1 + AB)}$$

where A represents the **forward gain** and B the **feedback gain** of the arrangement. If the **loop gain** AB is negative and its magnitude is less than or equal to 1, then the overall gain is greater than the forward gain and we have positive feedback. You will note that if $AB = -1$ the closed-loop gain is theoretically infinite. Under these circumstances, the system will generally produce an output even in the absence of any input. This situation is used in the production of oscillators, and $AB = -1$ represents the condition for oscillation to occur.

Figure 24.1   A generalised feedback

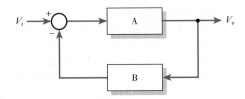

When considering feedback in Chapter 7, we considered A and B to represent simple (real) voltage ratios. Using this approach, a non-inverting amplifier has a positive gain and an inverting amplifier has a negative gain. Thus the condition $AB = -1$ can be satisfied if the magnitude of B is equal to $1/A$ and if either A or B (but not both) is 'inverting'. The inversion of a sine wave represents a phase shift of 180°, and an alternative way of describing the condition for oscillation is thus that the product AB must have a magnitude of 1 and a phase angle of 180° (or π radians).

These requirements are expressed by the **Barkhausen criterion**, which, using our notation, may be represented by the condition that for oscillation to occur:

1. The magnitude of the loop gain AB must be equal to 1.
2. The phase shift of the loop gain AB must be 180°, or 180° plus an integer multiple of 360°.

The second condition is slightly more complex than our original requirement, since it acknowledges that shifting a sine wave by a complete cycle leaves it unchanged. Thus, if a phase shift of 180° will cause oscillation, then a phase shift of 180° plus any multiple of 360° will have the same effect.

In order to make a useful oscillator, a frequency-selective element is added to ensure that the condition for oscillation is met at only a single frequency. The circuit then oscillates continuously at that frequency.

Figure 24.2 An *RC* ladder network

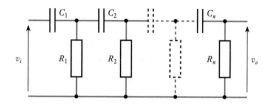

24.2.1 The *RC* or phase-shift oscillator

A simple way of producing a phase shift of 180° at a single frequency is to use an *RC* ladder network, as shown in Figure 24.2. Here a number of *RC* stages are cascaded, each producing an additional high-frequency cut-off. From the discussion in Chapter 17, we know that a single *RC* stage of this type produces a maximum phase shift of 90°, but this maximum value is achieved only at infinite frequency. We therefore require at least three stages to produce a phase shift of 180° at any non-infinite frequency.

If we adopt a ladder with three identical *RC* stages, then standard circuit analysis reveals that the ratio of the output voltage to the input voltage is given by the expression

$$\frac{v_o}{v_i} = \frac{1}{1 - \dfrac{5}{(\omega CR)^2} - j\left(\dfrac{6}{\omega CR} - \dfrac{1}{(\omega CR)^3}\right)} \tag{24.1}$$

The magnitude and phase angle of this ratio is clearly dependent on the angular frequency ω. We are interested in the condition where the phase shift is 180°. This implies that the gain is negative and real, and that the imaginary part of the ratio is zero. This condition is met when

$$\frac{6}{\omega CR} = \frac{1}{(\omega CR)^3}$$

or

$$6 = \frac{1}{(\omega CR)^2}$$

This can be rearranged to give

$$\omega = \frac{1}{CR\sqrt{6}}$$

and therefore

$$f = \frac{1}{2\pi CR\sqrt{6}}$$

Substituting for $(\omega CR)^2$ in Equation 24.1 gives

$$\frac{v_o}{v_i} = \frac{1}{1 - 5 \times 6} = -\frac{1}{29}$$

Figure 24.3 An *RC* or phase-shift oscillator

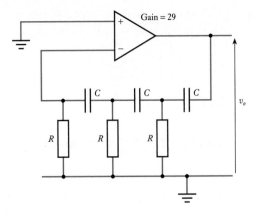

If we use the *RC* ladder network as our feedback path, it is clear that $B = -1/29$. In order for the loop gain AB to be equal to -1, we therefore require the forward gain of the arrangement A to be $+29$. Oscillators based on this principle are called **RC oscillators** or sometimes **phase-shift oscillators**. Figure 24.3 shows their basic form.

It can be seen from Figure 24.3 that the phase-shift oscillator consists of an inverting amplifier (the input is applied to the inverting input) and a feedback network with a phase shift of $180°$. The same result may be achieved using a non-inverting amplifier and a feedback network with a phase shift of $0°$. This approach is used in the Wien-bridge oscillator.

24.2.2 Wien-bridge oscillator

The Wien-bridge oscillator uses a series/parallel combination of resistors and capacitors for the feedback network, as shown in Figure 24.4. If we consider that R_1 and C_1 together constitute an impedance $\mathbf{Z}_1$, and that R_2 and

Figure 24.4 The Wien-bridge network

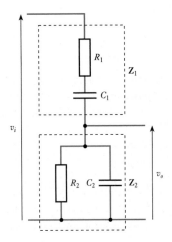

C_2 are represented by an impedance $\mathbf{Z}_2$, it is clear that the output of the network is related to the input by the expression

$$\frac{v_o}{v_i} = \frac{\mathbf{Z}_2}{\mathbf{Z}_1 + \mathbf{Z}_2}$$

Since

$$\mathbf{Z}_1 = R_1 + \frac{1}{j\omega C_1}$$

and

$$\mathbf{Z}_2 = \frac{1}{\dfrac{1}{R_2} + j\omega C_2}$$

if we make $R_1 = R_2$ and $C_1 = C_2$, it is relatively straightforward to show that

$$\frac{v_o}{v_i} = \frac{1}{3 - j\left(\dfrac{1 - \omega^2 R^2 C^2}{\omega C R}\right)} \tag{24.2}$$

In order for the phase shift of this network to be zero, the imaginary part must also be zero. This is true when

$$\omega^2 R^2 C^2 = 1$$

that is, when

$$\omega = \frac{1}{RC}$$

Substituting for ω in Equation 24.2 gives

$$\frac{v_o}{v_i} = \frac{1}{3}$$

Thus at the selected frequency the network has a phase shift of zero and a gain of 1/3. Further investigation of Equation 24.2 will show that the gain is a maximum at this point and that this is therefore the resonant frequency of the circuit. To form an oscillator, this network must be combined with a non-inverting amplifier with a gain of 3, making the magnitude of the loop gain unity. Figure 24.5 shows a possible arrangement using the non-inverting amplifier circuit discussed in Chapter 8.

Figure 24.5(a) shows the basic non-inverting amplifier circuit with the resistors chosen to give a gain of 3. Figure 24.5(b) shows the same circuit redrawn in a more convenient form. Figure 24.5(c) shows the oscillator

Figure 24.5   The Wien-bridge oscillator

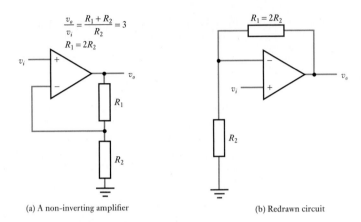

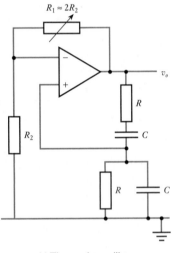

(a) A non-inverting amplifier   (b) Redrawn circuit

(c) The complete oscillator

formed by adding the feedback network. From the above, it is clear that the frequency of oscillation of the circuit is given by

$$f = \frac{1}{2\pi CR} \tag{24.3}$$

File 24A

Computer Simulation Exercise 24.1

Simulate the Wien-bridge network of Figure 24.4 and measure its gain and phase response using component values of $R = 1\,\text{k}\Omega$ and $C = 1\,\mu\text{F}$. Measure the response over a frequency range from 10 Hz to 10 kHz and determine the frequency at which the output reaches its maximum amplitude. Measure the voltage gain and the phase angle at this frequency and compare these with the values given above.

24.2.3 Amplitude stabilisation

In the phase-shift and Wien-bridge circuits discussed above, the loop gain of the circuit is determined by component values in the oscillator. If the gain thus set is too low, the oscillations will die; if it is too high, the oscillations will grow until limited by circuit constraints.

In Figure 24.5(c), R_2 has been shown as a variable resistor to allow it to be adjusted to the correct value. In practice, the gain must be set such that the magnitude of the loop gain is slightly greater than unity to ensure that any oscillation grows rather than decays and to allow for any downward fluctuation in the gain of the amplifier.

Several methods exist for limiting the magnitude of the oscillation. In the circuit shown in Figure 24.5(c), the amplitude is restricted simply by the limitations on the output swing of the amplifier. Fortunately for this application, operational amplifiers have non-linear gain characteristics, and the gain tends to drop as the amplitude approaches the supply rails. Thus, if the gain is set to slightly greater than that required to maintain the oscillation for small signals, as the signal amplitude increases it will enter a region where the gain falls and the magnitude will stabilise at that value. While this is a simple method, it does produce some distortion (resembling the clipping shown in Figure 5.17) because the amplifier is being used in its non-linear region.

A possible solution is to replace the variable resistor R_1 in the circuit of Figure 24.5(c) with a suitable thermistor (as discussed in Section 3.3). The resistor values are chosen such that, when the thermistor is at normal room temperature, the gain is slightly greater than that required for oscillation, and thus the amplitude of the output increases. This increases the power dissipated in the thermistor, causing it to heat up. The increase in temperature causes the resistance of the thermistor to fall, reducing the gain of the circuit. The amplitude of the oscillation therefore stabilises at a point where the magnitude of the loop gain is exactly unity. This limits the amplitude of the output signal without causing distortion.

Although the use of a thermistor is a possible solution to this problem, there are several more elegant solutions. However, the detailed design of oscillators is beyond the scope of this book and we will not discuss this further.

24.2.4 Digital oscillators

The oscillators considered so far are intended to produce a sinusoidal output (although as we have seen this is often slightly distorted due to amplitude stabilisation problems). Positive feedback is also widely used in a range of digital applications. These include a range of digital oscillator circuits.

A simple digital oscillator is illustrated in Figure 24.6, which shows a **relaxation oscillator**. To understand the operation of this circuit, imagine

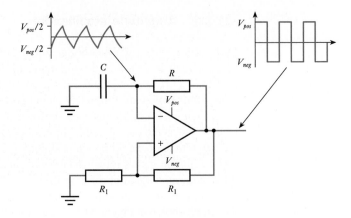

that initially (when power is applied to the circuit) the inputs have a slight
bias such that the non-inverting input is more positive than the inverting
input. This offset will be amplified by the op-amp, and its output will
become large and positive. The actual voltage produced will depend on the
nature of the op-amp (as discussed in Chapter 8), but for our current purposes it is sufficient to assume that the output will be close to the positive
supply voltage V_{pos}. The potential divider formed by the two resistors
sets the voltage on the non-inverting input, and since these are of equal
value (R_1) the voltage on the non-inverting input will be about $V_{pos}/2$. If we
assume that initially the capacitor is uncharged, the voltage on the inverting input will be zero, and thus the voltage difference between the two
op-amp inputs will maintain the output at its maximum positive value.

If the output is positive, a positive voltage is applied across the RC combination, which will charge the capacitor and cause the voltage across it
to increase exponentially towards V_{pos}. The voltage across the capacitor sets
the voltage on the inverting input, so this also increases. However, when
this voltage becomes greater than the voltage on the non-inverting input
(which is $V_{pos}/2$), the polarity of the input voltage to the op-amp will be
reversed, and its output will go close to the negative supply voltage V_{neg}.
This in turn will change the voltage on the non-inverting input to $V_{neg}/2$,
which will tend to force the output to be even more negative. The voltage
across the RC network is now reversed, and the capacitor will start to
charge exponentially towards V_{neg}. This continues until the voltage on the
inverting input reaches $V_{neg}/2$, when the output will reverse once more and
the cycle will start again. This produces a continuous oscillation of the
output between the two supply voltages. The frequency of oscillation is
determined by the rate at which the capacitor charges, and this is set by the
time constant of the arrangement, which is equal to CR.

In practice, the output produced by the relaxation oscillator of Figure 24.6 is not a perfect square wave, since the slew rate of the operational amplifier limits the speed at which the output can change. The slew
rate of an operational amplifier was discussed in Section 8.5.

File 24B

24.2.5 Crystal oscillators

The **frequency stability** of an oscillator is largely determined by the ability of the feedback network to select a particular operating frequency. In a **resonant circuit**, this ability is described by its quality factor or Q, which determines the ratio of its resonant frequency to its bandwidth (this topic was discussed in Section 17.9 when we looked at filters). A circuit with a very high Q will be very frequency-selective and will therefore tend to have a stable frequency. Networks based on resistors and capacitors have relatively low values of Q. Those based on combinations of inductors and capacitors are better in this respect, with Q values of up to several hundred. These are suitable for most purposes but are not adequate for some demanding applications, such as the measurement of time. In such cases, it is normal to use a frequency-selective network based on a crystal.

Some materials have a **piezoelectric** property in that deformation of the substance causes them to produce an electrical signal. The converse is also true: an applied electric field will cause the material to deform. A result of these properties is that, if an alternating voltage is applied to a crystal of one of these materials, it will vibrate. The mechanical resonance of the crystal, caused by its size and shape, produces an electrical resonance with a very high Q. Resonant frequencies from a few kilohertz to many megahertz are possible with a Q as high as 100,000.

These piezoelectric resonators are commonly referred to simply as **crystals** and are most commonly made from **quartz** or some form of **ceramic** material. The circuit symbol for a crystal is shown in Figure 24.7(a). Functionally, the device resembles a series *RLC* resonant circuit (with a

Figure 24.7 A crystal

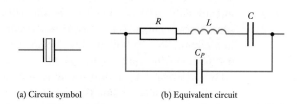

(a) Circuit symbol   (b) Equivalent circuit

Figure 24.8   A crystal oscillator

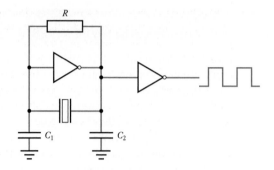

small amount of parallel capacitance C_p), and Figure 24.7(b) shows a simple equivalent circuit. The devices have a pair of resonant frequencies: at one (the parallel resonant frequency) the impedance approaches infinity, while at the other (the series resonant frequency) it drops almost to zero. Over the remainder of the frequency range the device looks like a capacitor. The parallel resonance occurs at a slightly higher frequency than the series resonance, but the frequency difference is normally so small that it may be ignored. The presence of these two forms of resonance allows the device to be used in a number of different circuit configurations.

Crystal oscillators are widely used in a range of analogue and digital applications. They form the basis of the time measurement in digital watches and clocks and are used to generate the timing reference (clock) in most computers. Figure 24.8 shows the circuit of a simple **digital oscillator** based on a crystal. This is a form of **Pierce oscillator**, where a logical inverter is used to provide a high-gain inverting amplifier and the crystal provides positive feedback at its resonant frequency. The second inverter in this circuit 'squares up' the output from the oscillator and also acts as a buffer, increasing the circuit's ability to drive a load.

24.3   Stability

In the previous section, we used a general expression for the gain of a feedback network, namely

$$G = \frac{A}{(1 + AB)}$$

and considered the conditions necessary to produce oscillation. However, so far we have only considered the situation where a circuit designer intentionally sets out to produce this effect. Unfortunately, oscillation sometimes occurs unintentionally as a result of unwanted positive feedback in a circuit. To see how this can occur, we need to look in more detail at the nature of the *gains* in the circuit.

In our earlier discussions, we have assumed that the forward and feedback gains of our circuit (A and B) can be represented by simple, real numbers. When we design a circuit to make use of negative feedback (as in most amplifiers), the selected component values will set A and B such that $| 1 + AB |$ is positive and greater than unity. This gives the various advantages associated with the use of negative feedback, as discussed in Section 7.6. However, from our study of frequency response in Chapter 17, we know that the gain of an amplifier has not only a *magnitude* but also a *phase angle*. We also know that the gain of all amplifiers falls at high frequencies, and that associated with this fall in gain is an increasing phase shift. In almost all cases, the phase shift will become greater than 180° as the frequency is increased. Since a phase shift of 180° corresponds to an inversion of a sine wave, the effective gain of the circuit changes polarity. Thus a circuit designed to take advantage of negative feedback may, at high frequencies, see the effects of positive feedback instead. Not only does this remove the beneficial effects of negative feedback (such as its effects on gain, frequency response and input/output impedance) but it may also result in the circuit becoming unstable and starting to oscillate. Thus when designing a feedback circuit one must consider not only its performance within its operational frequency range but also its stability.

The stability of an amplifier is determined by the term $(1 + AB)$. If this term is positive, we have negative feedback and the stability of the circuit is assured. If, as a result of a phase shift, the term $(1 + AB)$ becomes less than 1 (because AB becomes negative), the feedback becomes positive and all the advantages of negative feedback are lost. In the extreme case, if $(1 + AB)$ becomes equal to 0 the closed-loop gain of the arrangement is infinite and the system becomes unstable and will oscillate. This corresponds to the circuit satisfying the conditions for oscillation set out in the Barkhausen criterion discussed earlier.

The condition that $(1 + AB) = 0$ represents the case where $AB = -1$, or, in other words, where the loop gain has a magnitude of 1 and a phase of 180°. Indeed, the amplifier will remain stable even if the phase shift is equal to 180°, provided that the magnitude of the loop gain is less than unity. The task of the designer is thus to ensure that the loop gain of the amplifier falls below unity *before* the phase shift reaches 180°.

24.3.1 Gain and phase margins

In practice, it is advisable to allow some margin for variability in the phase and gain values. This leads to the concept of the **phase margin**, which is the angle by which the phase is less than 180° when the loop gain falls to unity, and the **gain margin**, which is the amount (in dB) by which the loop gain is less than 0 dB (that is unity gain) when the phase reaches 180°. These quantities may be illustrated using a Bode diagram, as shown in Figure 24.9.

The above discussion makes clear why designers of operational amplifiers such as the 741, described in Chapter 8, choose to add a single dominant time constant to the amplifier to roll off the gain as shown in

Figure 24.9 Gain and phase margins

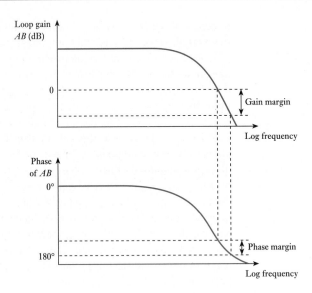

Figure 8.11. This ensures that the gain falls to less that 0 dB well before the phase shift reaches 180°. This produces large gain and phase margins and ensures good stability.

24.3.2 Unintentional feedback

Stability can also be affected by unintended feedback in a circuit. For example, the presence of stray capacitance or stray inductance in a circuit may introduce additional feedback paths that do not form part of the original design. If these represent positive feedback then they can cause instability in a similar manner to that described above. These problems are more severe in high-frequency applications, where small amounts of capacitance can have a dramatic effect. Such problems must be tackled by careful design to minimise these spurious effects.

Key points

- Positive feedback is used in a range of both analogue and digital circuits.

- One of the primary uses of positive feedback is in the production of oscillators.

- The requirements for oscillation are that the loop gain AB must have a magnitude of 1 and a phase of 180° (or 180° plus some integer multiple of 360°).

- This condition can be satisfied by using an arrangement that produces a phase shift of 180° (at a particular frequency) in association with a non-inverting amplifier – as in a phase-shift oscillator.

- Alternatively, it can be satisfied using an arrangement that produces a phase shift of 0° in association with an inverting amplifier – as in the Wien-bridge oscillator.

- Sine-wave oscillators present the problem of maintaining the gain of the circuit at 1 without causing distortion of the output.

- Positive feedback is also used in digital applications, such as the production of digital oscillators.

- Where good frequency stability is required, circuits normally make use of crystals.

- While positive feedback is a useful tool in circuit design, it can also pose problems. At high frequencies, negative feedback arrangements can exhibit positive feedback, which may lead to instability. Unwanted feedback can also cause problems.

Exercises

24.1 State the Barkhausen criterion for oscillation.

24.2 Since an RC network can produce up to 90° of phase shift, why can a phase-shift oscillator not use just two stages in its ladder network?

24.3 Calculate the frequency of oscillation of a phase-shift oscillator that uses a three-stage ladder network, each with $R = 1\text{k}\Omega$ and $C = 1\ \mu\text{F}$.

24.4 Simulate the RC ladder network of Figure 24.2 using three stages, each with $R = 1\ \text{k}\Omega$ and $C = 65\ \text{nF}$, and measure the gain and frequency response of this arrangement. Measure the frequency at which the phase shift is equal to 180° and the gain at this frequency, and confirm that these are as expected.

24.5 A Wien-bridge oscillator of the form shown in Figure 24.5 is constructed using $R = 100\ \text{k}\Omega$ and $C = 10\ \text{nF}$. Calculate the frequency of oscillation.

24.6 Why does amplitude stabilisation present a problem in simple sine-wave oscillators?

24.7 Explain how a thermistor might be used to stabilise the output amplitude of an oscillator.

24.8 A relaxation oscillator of the form shown in Figure 24.6 is constructed with $R = 1\ \text{k}\Omega$, $C = 10\ \mu\text{F}$ and $R_1 = 10\ \text{k}\Omega$. Estimate the frequency of oscillation of this circuit by considering the charging rate of the capacitor.

24.9 Use simulation to confirm your answer to the previous exercise. You may find it useful to start with the circuit of Computer Simulation Exercise 24.2.

24.10 Why are crystal oscillators used in digital watches rather than circuits based on RC or RL techniques?

24.11 Calculate the percentage accuracy required in the frequency of oscillation of a clock that must keep time to within 1 second per month.

24.12 Why does the phase response of an amplifier have implications for its stability at high frequencies?

24.13 Explain what is meant by the gain margin and the phase margin of a circuit.

24.14 How can stray capacitance affect the stability of a circuit?

Chapter 25

Digital Components

Objectives

When you have studied the material in this chapter you should be able to:

- discuss the use of both field-effect and bipolar transistors in the construction of logic circuits;
- discuss the characteristics that describe the performance of a logic gate and determine its compatibility with other gates;
- explain how the noise performance of a logic gate can be quantified;
- outline the characteristics of a range of popular logic families;
- describe the operation and characteristics of basic CMOS and TTL logic gates.

25.1 Introduction

In Chapters 9 and 10, we looked at a range of digital applications based on the use of standard logic gates. In these chapters, we considered these gates as 'black boxes' and did not concern ourselves with how these functions are physically implemented. There are sound reasons for considering gates in this way, since in most cases we buy and use gates as standard components, or they are integrated into more complex arrangements. For this reason, we are rarely required to design logic gates and are not normally concerned with their internal circuitry. However, logic gates are *not* ideal components, and their external characteristics are dictated by their internal construction. For this reason, it is sometimes useful to have a broad idea of how the various types of gate operate so that we can understand their external behaviour better.

In this chapter, we will concentrate on the construction of small- and medium-scale integration devices (these terms will be discussed in Chapter 27). This means that we will be looking at the techniques used to construct devices containing a handful of individual gates, or simple circuits such as flip-flops, counters or registers. Figure 25.1 illustrates some typical logic

Figure 25.1 Typical logic device pin-outs

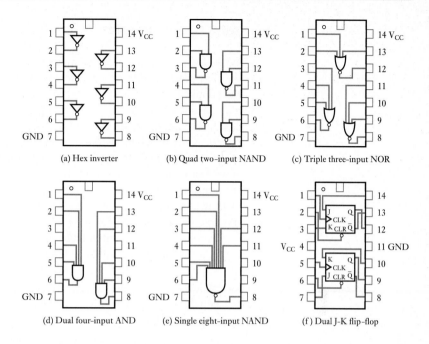

(a) Hex inverter   (b) Quad two–input NAND   (c) Triple three–input NOR

(d) Dual four–input AND   (e) Single eight–input NAND   (f) Dual J-K flip-flop

devices and their pin connections (pin-outs). In Chapter 27, we will look at the construction of more sophisticated components (such as micro-computers, memories and programmable logic devices) and will see that these use many of the same techniques.

The relevance of looking in detail at gate operation will vary from reader to reader. Those who are likely to be concerned with detailed logic design will probably be more interested in the internal operation of the gates than those who envisage themselves as system integrators. For this reason, this chapter is written to allow readers to select the material that they need while ignoring that which is not relevant. Information on the internal operation of the most common forms of logic gate is given in Sections 25.6 and 25.7, and readers may skip these sections if they wish without affecting their understanding of the remainder of the chapter.

25.2 Gate characteristics

In Chapter 9, we noted that there are many different forms of logic gate (for example, AND, OR and inverter (NOT) gates). Gates also differ in the circuitry used to implement them, and these implementation differences give the gates very different characteristics. In order to understand these characteristics, we will start by looking at the simplest form of logic gate – the inverter or NOT gate.

25.2.1 The inverter

When looking at the characteristics of linear amplifiers in earlier chapters, we have seen that all real amplifiers have an output swing that is limited by the supply voltages used. A typical inverting linear amplifier might have a characteristic as shown in Figure 25.2. If we wish to use such a device as a linear amplifier, we must ensure that the input is restricted so that the operation is maintained within the linear range of the device.

It is also possible to use the amplifier of Figure 25.2 as a logical device. If we restrict the input signal so that it is always *outside* the linear region of the amplifier, we are left with two allowable ranges for the input voltage, as shown in Figure 25.3. We may consider these ranges as representing two possible input states, '0' and '1'.

Clearly, when the input voltage corresponds to state '0' the output voltage is at its maximum value, and when the input state is '1' the output is at its minimum value. If we choose component values appropriately, we can arrange that the maximum and minimum output voltages lie within the voltage ranges defined at the input to represent '0' and '1', as shown in Figure 25.4(a).

Figure 25.2 An inverting linear amplifier

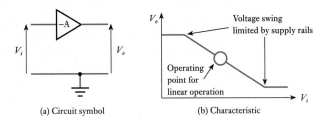

(a) Circuit symbol   (b) Characteristic

Figure 25.3 Use of an inverting amplifier as a logical device

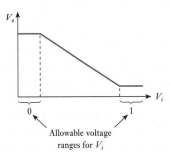

Figure 25.4 Transfer characteristics for logical inverters

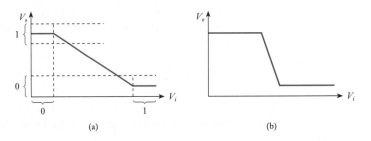

(a)   (b)

From Figure 25.4(a), it is clear that when the input is '0' the output is '1', and vice versa. The circuit therefore has the characteristics of a logical inverter. Since the input and output voltages are compatible, the output of this arrangement could be fed to the input of a similar gate.

When designing linear amplifiers, we aim to produce an extended linear region to permit a large output swing. When producing a logical inverter, we wish the linear portion of the characteristic to be as small as possible to reduce the region of uncertainty. Such circuits therefore have a very high gain and a rapid transition from one state to the other, as shown in Figure 25.4(b).

Logical inverters can be produced using a range of circuit techniques, and slight modifications to these basic circuits will produce the functions of the other basic gates. These in turn can be combined to form more complex digital functions. We will look at examples of logic circuits later in this chapter.

25.2.2 Logic levels

The voltage ranges representing '0' and '1' in Figure 25.4(a) represent the **logic levels** of the circuit. In many gates, logical 0 is represented by a voltage close to zero, but the range of allowable voltages varies considerably. There is also great variation in the voltage used to represent logical 1: this might be 2–4 V in some components, but 12–15 V in others. In order for one logic gate to work with another, the logic levels used must be compatible.

25.2.3 Noise immunity

Noise is present in all real systems. This has the effect of adding random fluctuations to the voltages representing the logic levels. To enable the system to tolerate a certain amount of noise, the voltage ranges defining the '0' and '1' states at the output of a gate are more tightly constrained than those at the input. This ensures that small perturbations of an output signal caused by noise will not take the signal outside the defined ranges of the input of another gate. Thus the circuit is effectively immune to small amounts of noise, but it may be affected if the magnitude of the noise is large enough to take the logic signal, in either logic state, outside the allowable logic bands. The maximum noise voltage that can be tolerated by a circuit is termed, logically enough, the **noise immunity** V_{NI} of the circuit.

25.2.4 Transistors as switches

The functions of the various logic gates are invariably implemented using some form of transistor. In Chapters 20 and 21, we looked at the characteristics of FETs and bipolar transistors, and in each case we considered their use as logical switches. Both forms of transistor make good switches, but neither is ideal, and their characteristics are somewhat different.

The FET as a logical switch

MOSFETs are the dominant form of field-effect transistor for digital applications. While in analogue applications it is common to describe such devices as FETs, in digital systems it is more common to talk of **MOS devices**, describing their method of construction rather than their principle of operation.

MOSFETs make excellent switches, and the vast majority of modern digital circuitry is based on these devices. The major advantages of MOS technology over circuits based on bipolar transistors are that MOS devices are simpler and less expensive to fabricate. Each MOS gate requires a much smaller area of silicon, allowing a greater number of devices to be produced on a given chip. When used in CMOS gates, MOSFETs can also be used to produce logic circuits with extremely low power consumption. This reduces the amount of waste heat that must be dissipated and allows greater packing densities.

In the early days of digital technology, MOS circuits tended to be slower than those based on bipolar components. However, modern devices can be extremely fast, and MOSFETs have largely replaced bipolar transistors in digital applications.

When we use a MOSFET as a logical switch, we ensure that it is driven into one of two states. In the first, the device is effectively 'turned ON' and the channel from the drain to the source has a relatively low resistance and resembles a closed switch. In the other, the device is 'turned OFF' and the effective resistance of the device is so high that it resembles an open switch.

Figure 25.5(a) illustrates the use of a MOSFET as a logical switch. The input to this arrangement is restricted to being either close to zero or close to the supply voltage V_{DD}. When the input voltage is close to zero, the MOSFET is turned OFF and negligible drain current flows. The output voltage is therefore close to the supply voltage V_{DD}. When the input voltage is close to V_{DD}, the MOSFET is turned ON and the output voltage is pulled

Figure 25.5 A logical inverter based on a MOSFET

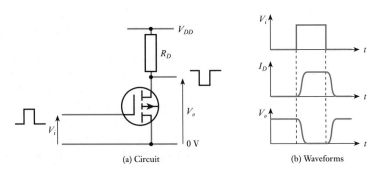

(a) Circuit (b) Waveforms

Figure 25.6 Rise and fall times

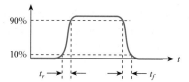

down close to zero. Therefore, the circuit acts as a simple logic inverter, with voltages close to V_{DD} representing logical 1, and voltages close to 0 V representing logical 0.

Figure 25.5(b) shows the relationship between a pulse applied to the input of the inverter and the corresponding drain current and output voltage. The figure shows that there is a delay between a change in the input voltage and the response of the output. Because the waveforms produced are not 'square', we generally quantify the time taken for them to change by defining the **rise time** t_r as the time it takes for the waveform to increase from 10 percent to 90 percent of the height of the step and the **fall time** t_f as the time taken for the waveform to fall from 90 percent to 10 percent of the height of the step. These two measures are shown in Figure 25.6.

We noted in Section 20.8 that the circuit of Figure 25.5(a) is not normally used in integrated circuits because the resistor is 'expensive' in terms of circuit area. It also produces a circuit that is slow to respond, since when the MOSFET turns OFF the drain resistor produces a relatively high output resistance, making it slow to charge circuit capacitances. This problem is overcome in CMOS gates by using a 'push-pull' arrangement, which provides a low resistance output in either state. We will look at CMOS gates in Section 25.6.

File 25A

Computer Simulation Exercise 25.1

Use simulation to investigate the characteristics of the circuit of Figure 25.5(a). A suitable arrangement would use an IRF150 MOSFET with $R_D = 10\ \Omega$ and $V_{CC} = 5$ V. Apply a suitable input waveform and plot V_i, I_D and V_o against time. Repeat this procedure using a 20 Ω drain resistor and compare your results. What is the effect of further increasing the drain resistance?

The bipolar transistor as a logical switch

Bipolar transistors also make good logical switches, and a simple switching arrangement is illustrated in Figure 25.7(a). When the input voltage is close to zero, the transistor is turned OFF and negligible collector current flows. The output voltage is therefore close to the supply voltage V_{CC}. R_B is chosen so that when the input voltage is high the transistor is turned ON and the output voltage is equal to the **saturation voltage** of the device, which is generally about 0.1 V. Therefore, the circuit acts as a simple logic

Figure 25.7 A logical inverter
using a bipolar transistor

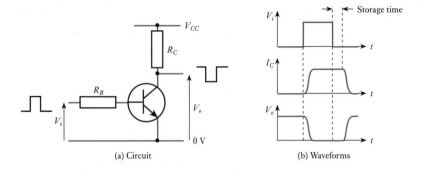

(a) Circuit   (b) Waveforms

inverter, with voltages close to V_{CC} representing logical 1 and voltages close to 0 V representing logical 0.

Figure 25.7(b) shows the relationship between a pulse applied to the input of the inverter and the corresponding collector current and output voltage. As in the MOSFET, there is a delay between a change in the input voltage and the response of the transistor. However, in this case the time taken for the device to turn OFF is much greater than the time taken for it to turn ON. This increase in switching time results from the **saturation** of the transistor and represents the time taken to remove excess charge stored in the base region (the so-called **storage time**). The presence of storage time imposes major restrictions on the speed of operation of circuits in which bipolar transistors are allowed to saturate. Some switching circuits increase speed by preventing their transistors from entering their saturation region, and in such arrangements the time taken to turn the transistor OFF is comparable to that required to turn it ON.

File 25B

Computer Simulation Exercise 25.2

Use simulation to investigate the characteristics of the circuit of Figure 25.7(a). A suitable arrangement would use a 2N2222 transistor with $R_B = 10\ \text{k}\Omega$, $R_C = 1\ \text{k}\Omega$ and $V_{CC} = 5$ V. Apply a suitable input waveform and plot V_i, I_C and V_o against time. Repeat this procedure using a base resistor R_B of 1 kΩ (to increase the base current) and compare your results.

25.2.5   Timing considerations

Propagation delay time

Logic gates invariably consist of a number of transistors (either MOSFET or bipolar), each producing a slight delay as signals pass through them. Inevitably, the resultant delay is different for changes in each direction, and two **propagation delay times** are used to describe the speed of response of

Figure 25.8 Propagation delay times

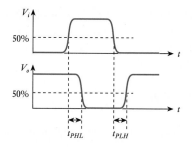

the circuit. These are t_{PHL}, the time taken for the output to change from high to low, and t_{PLH}, the time taken for the output to change from low to high. In some cases, a single value is used corresponding to the average time for the two transitions. This average propagation delay time t_{PD} is given by

$$t_{PD} = \frac{1}{2}(t_{PHL} + t_{PLH})$$

Since in general the input waveform will not be a perfect square pulse, t_{PHL} and t_{PLH} are measured between the points at which the input and output signals cross a reference voltage corresponding to 50 percent of the voltage difference between the logic levels. This is illustrated in Figure 25.8.

The speed of operation of a logic circuit is greatly affected by the load that it must drive. When an integrated circuit is connected to a printed circuit board (PCB), the conductive track joined to each output of the device represents a capacitive load (in addition to the load represented by the components it is linked to). In order to change from one output state to another the circuit must charge or discharge this stray capacitance, and the output stage of the device must be designed to supply sufficient current to do this. Circuits that are only connected to other circuitry *within* the integrated circuit will have a much smaller load capacitance, which might be only 1 percent of that of an external connection. Consequently, gates and other logic circuitry can operate much faster when they are an integral part of a larger system (such as a microprocessor or memory circuit) than when they must drive external connections.

Set-up time

In logic circuits that have clock input signals (such as flip-flops), it is often necessary for control inputs to be applied a short while *before* an active transition of the clock to ensure correct operation. The time for which the control input is stable before the clock trigger occurs is termed the **set-up time** t_S. Device manufacturers normally specify a minimum value for this quantity, which ranges from less than 1 ns to more than 50 ns.

Hold time

It is also often required that a control signal (in circuits such as flip-flops) should not change for a short interval *after* the active transition of a clock input. The time for which a control input is stable after a clock trigger occurs is termed the **hold time** t_H. Typical minimum values might range from 0 to 10 ns.

25.2.6 Fan-out

In many cases, it is necessary to connect the output of one logic gate to the input of a number of gates. Since each input draws current from this output, there will be a limit to the number of gates that a single output can supply. This is termed the **fan-out** of the circuit. The fan-out is clearly determined by the output resistance of the gate and by the input resistance of the gates it is driving. Because of the very high input resistance of MOS devices, it is the input capacitance that is of importance. Circuits based on this technology generally have a greater fan-out than those based on bipolar transistors, but in some cases (such as CMOS) the speed of the gate is reduced as the output is more heavily loaded.

25.3   Logic families

We noted earlier that logic devices differ in the voltage ranges that they use to represent logical 0 and logical 1. They also differ in other characteristics, such as their switching speed, their current-driving capabilities and their noise immunity. In order to assure correct operation when gates are inter-connected, they are normally produced as 'families' of components with similar electrical characteristics. There are several **logic families** in general use, the most significant of which are:

- complementary metal oxide semiconductor (CMOS);
- transistor–transistor logic (TTL); and
- emitter-coupled logic (ECL).

By far the most widely used logic families are **complementary metal oxide semiconductor logic** and **transistor–transistor logic**. For this reason, towards the end of this chapter we will look briefly at the operation of these two forms of logic circuit. However, before becoming involved in the detailed operation of logic gates we will consider their general characteristics.

25.4   Logic family characteristics

In the last section we identified three families of logic circuit, and we will now look briefly at each of these. In doing so, we will concentrate on their key characteristics and consider features that might determine their suitability for a given application.

25.4.1 Complementary metal oxide semiconductor

Early versions of logic based on MOSFETs used entirely *p*-channel devices (to produce what were called PMOS circuits) or entirely *n*-channel devices (to produce NMOS). Both of these technologies suffered from high output resistance (as discussed in Section 20.8), making them slow in operation. Both PMOS and NMOS are now obsolete, having been replaced by CMOS circuits.

Complementary MOS or CMOS is so called since it combines both *n*-channel and *p*-channel devices in each gate. Its logic levels are close to the supply rail voltages, and CMOS can be used with a wide range of power supply voltages. There are many different forms of CMOS, but most CMOS circuits operate using a single supply voltage of from 5 to 15 V, although there is a move towards lower supply voltages of perhaps 2.5 or even 1.2 V. CMOS gates have very good noise immunity, typically 30 percent (or more) of the supply voltage. The speed of operation and the noise immunity increase with the supply voltage, as does the power consumption.

CMOS gates have a very high input resistance, the input impedance resembling a small capacitance to ground. They also have a relatively low output resistance (for reasons that are explained in Section 25.6). This allows outputs to charge the input capacitance of other gates rapidly, producing a relatively fast response (with t_{PD} down to about 1 ns in some cases) and a large fan-out (of perhaps 50). However, the speed of operation is affected by the supply voltage used and the number of gates connected to the output.

One of the dominant characteristics of CMOS is its low power consumption. Indeed, when a CMOS gate is static (that is, it remains in a particular state) its power consumption is negligible (perhaps a few nW per gate). However, a small amount of power is dissipated each time the gate changes state, so the average power consumption is proportional to the rate at which it is 'clocked' – the clock speed. When clocked at 1 MHz, a typical gate might consume a modest 1 mW, making it ideal for battery-powered applications.

The high speed of CMOS circuitry combined with its low power dissipation and excellent noise immunity make it ideal for the production of small- and medium-scale ICs. It is also the dominant technology for new, highly integrated circuits.

25.4.2 Transistor–transistor logic

Unlike CMOS, TTL is based on the use of bipolar transistors. It has relatively high power consumption and is rarely used for the construction of highly integrated components such as memories or microcomputers. However, TTL is one of the most widely used logic families for the production of small- and medium-scale integrated circuits.

As with CMOS, TTL is not one family but a group of families sharing some characteristics but varying quite considerably in terms of speed and power consumption (see Section 25.7). So-called 'standard' TTL is largely obsolete now, having been replaced by much more advanced variants.

TTL gates operate from a +5 V supply and have typical voltage levels of 3.4 V (for logical 1) and 0.2 V (for logical 0). The *minimum* noise immunity of a gate is 0.4 V, although the *typical* noise immunity is considerably greater at 1–1.6 V. Power consumption in logic devices is normally given in terms of the 'power per gate'. This gives an indication of the power consumed by an individual gate (such as a single NAND gate) and is proportionately greater for more complex circuits. For 'standard' TTL, power consumption is about 10 mW per gate, although low-power versions may dissipate as little as 1 mW per gate. The power consumption tends to increase with the speed of the device, with high-speed versions consuming up to 22 mW per gate.

Standard TTL has a propagation delay time t_{PD} of about 9 ns, this being dominated by the **storage time** of the saturated transistors (as discussed in Section 25.2.4). High-speed versions use various techniques to remove this saturation effect and produce delays of as little as 1.5 ns. Low-power devices might have a delay of about 33 ns.

A standard TTL gate can drive up to ten other gates (that is, it has a fan-out of 10).

25.4.3 Emitter-coupled logic

ECL removes the problem of storage time by keeping its bipolar transistors within their active region. This produces very fast gates with propagation delay times down to 1 ns or less. Unfortunately, this is achieved at the expense of power consumption, which is high at about 60 mW per gate. It also suffers from a low noise immunity, typically about 0.2–0.25 V.

ECL was once the preferred option when building very high-speed systems, but it has now been largely superseded by advanced CMOS devices. These approach the speed of ECL while offering improved power consumption and better noise performance. However, ECL is commonly used in certain specialised applications such as RF frequency synthesisers.

25.5 A comparison of logic families

The previous section has outlined some of the characteristics of the technologies that are most widely used in the production of logic gates. It is not within the scope of this text to give detailed descriptions of the operation and characteristics of these technologies, but it is perhaps useful to summarise some of the results in tabular form. Table 25.1 gives a comparison of the three logic families discussed above in respect of five parameters. In each case, a number of device families are available, and the figures attempt

Table 25.1 A comparison of logic families

Parameter	CMOS	TTL	ECL
Basic gate	NAND/NOR	NAND	OR/NOR
Fan-out	>50	10	25
Power per gate (mW)	1 @ 1 MHz	1–22	4–55
Noise immunity	Excellent	Very good	Good
t_{PD} (ns)	1–200	1.5–33	1–4

to represent the range of the parameters across these series. Consequently, the data should be used with care. Often the values obtained will depend on other factors, and the numbers given should be taken simply as a guide for comparison and not as detailed data on a particular device family. Note that the values given relate to the use of the technologies for the production of simple, small-scale or medium-scale integrated circuits. We will leave discussion of the use of such technologies in very-large scale integrated circuits (such as microprocessors or gate arrays) until Chapter 27.

25.6 Complementary metal oxide semiconductor

CMOS logic is based on the use of MOS transistors (or if you prefer FETs). In Chapter 20, we considered the use of MOSFETs as logical switches and in Figure 20.20 we looked at two circuits that use these devices in inverters. The circuits shown use *n*-channel MOSFETs and are therefore examples of simple **NMOS logic gates**. Unfortunately, these gates suffer from a high output resistance in one of their output states (when the MOS-FET is turned OFF), which limits their speed of operation. This problem is common to all amplifiers that use a single output transistor (remember that the load MOSFET in Figure 20.20(b) is acting as a resistor). In Section 22.2, we discussed a similar problem in the design of power amplifiers and looked at a method of tackling this using a push-pull arrangement. This technique can also be applied to the design of logic gates, and such an arrangement is shown in Figure 25.9.

The circuit uses both an *n*-channel and a *p*-channel device and is therefore described as complementary MOS logic, or simply CMOS. Here V_{DD}

Figure 25.9 A CMOS inverter

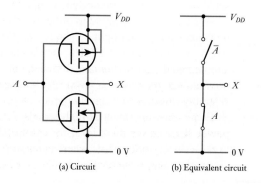

(a) Circuit   (b) Equivalent circuit

represents logical 1 and 0 V represents logical 0. Being of different polarities, the two transistors respond in the opposite sense to voltages applied to their gates. While a gate voltage of V_{DD} will turn ON the n-channel transistor, it will turn OFF the p-channel device. Similarly, a voltage of 0 V will turn OFF the n-channel transistor and turn ON the p-channel device. Since the gates of the two MOSFETs are joined, input voltages of either logic level will turn one device ON and the other OFF. This arrangement produces a low output resistance in either state, which can charge load capacitances more quickly, producing a faster switching time. The low output resistance also gives a high fan-out of up to about fifty gates (although switching times are increased as more gates are connected to the output).

25.6.1   Propagation delay

Early CMOS gates were relatively slow, with propagation delays considerably greater than the TTL gates of the day. In recent years, the speed of operation of CMOS has increased considerably, with modern, advanced CMOS gates having delay times of a few nanoseconds. This is comparable with the fastest of the newer forms of TTL. CMOS gates that are not required to drive external loads can be physically smaller and have greatly reduced delay times. This permits the construction of CMOS microprocessors and other logic circuits that can operate with clock speeds of several gigahertz.

25.6.2   Power dissipation

Since one of the two transistors is always turned OFF, there is no DC path between the supply rails, and the only current drawn from the supply is that which is fed to the output. The high input resistance of the gates makes this output current negligibly small, except when the input capacitance of a gate is being charged or discharged after an output has changed. Power is also consumed when the circuit switches from one state to another, as, for a short period, both transistors are conducting at the same time. The resultant **power consumption** is therefore generally negligible when the circuit is static but increases with the switching rate.

When gates are required to drive external circuits, typical values for the power consumption might be about 10 nW per gate when static and about 1 mW when clocked at 1 MHz. Gates that are connected to other gates in an integrated circuit with no external connections require less drive capability and can therefore be physically smaller. Such devices will have considerably lower power consumption.

Even when operating at high speeds, CMOS gates consume very little power. This makes them ideal for applications in which power consumption is critical, for example where battery operation is required. Low power dissipation also reduces the amount of waste heat that must be removed, allowing more circuitry to be integrated into a single circuit. This permits the production of integrated circuits of great complexity, which would not be possible using other circuit techniques.

25.6.3   CMOS gates

The simple inverter of Figure 25.9 can be modified to provide additional logic functions. Examples of two-input NAND and NOR gates are shown in Figure 25.10. Gates of this form are invariably formed in integrated circuits, so the circles around the individual transistors have been removed. Like the inverter, these circuits provide both an active pull-up and an active pull-down of the output, giving a low output resistance, and they provide no DC path between the supply rails when in either output state.

Figure 25.10   Two-input CMOS gates

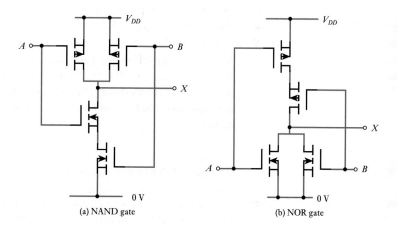

(a) NAND gate   (b) NOR gate

25.6.4   CMOS inputs

CMOS inputs resemble a small capacitor of the order of 1 pF. Because of their high input resistance, CMOS gates are very sensitive to **static electricity**, which can easily destroy them (particularly before being installed in circuits). Normal precautions include storing devices in conductive enclosures and minimising handling.

In certain instances, a circuit may not need to use all the inputs of a particular gate. For example, it may be convenient to OR together three signals using a four-input OR gate because such a device is already available in the circuit. Unused CMOS inputs should *not* be left unconnected, first because such inputs are prone to damage from static electricity, and second because unconnected inputs tend to float midway between the two supply rails, making them susceptible to noise. All unused inputs should be tied high or low, or joined to other inputs.

25.6.5 CMOS outputs

CMOS outputs have a typical output resistance of about 250 Ω (for a supply voltage of 5 V). Since the input resistance of gates is so high, a large number of devices can be connected to a single output. The main restriction on the fan-out is that the propagation delay increases with the number of gates driven.

25.6.6 Power-supply voltages

Most CMOS gates operate using a single supply voltage of from 5 to 15 V, although most gates are usable over a range of from 3 to 18 V. The speed of operation increases with the supply voltage, as does the power dissipation. The trend is towards a reduction in the supply voltage, and newer families are optimised for use with a 3.3 V or 2.5 V supply, while some can be used down to 1.2 V.

25.6.7 Logic levels and noise immunity

The output logic levels of CMOS gates are very close to the supply rails and can normally be assumed to be equal to 0 V (V_{SS}) and the positive supply voltage (V_{DD}).

The allowable input ranges for the two logic levels differ slightly depending on the nature of the input circuit, but typical ranges are shown in Figure 25.11. Here the output of one gate is applied to the input of another. While the output logic levels of the first are likely to be very close to V_{DD} and 0 V, the input of the second will accept any voltage in the range $0 - V_{DD} \times 0.3$ as a logical 0 and any voltage in the range $V_{DD} \times 0.7 - V_{DD}$ as a logical 1. This means that small amounts of noise added to the output voltage will not result in the input misinterpreting the signal. It can be seen that the noise immunity of each logic level is equal to 30 percent of the supply voltage. Some input configurations give an improved noise immunity of about 50 percent of the supply voltage.

Figure 25.11 CMOS input and output logic levels

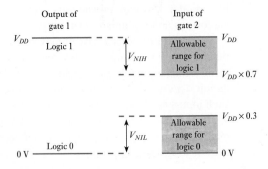

25.6.8   CMOS families

The first manufacturer to produce CMOS logic was RCA, who described them as the **4000 series**, having numbers 4000, 4001, etc. Some manufacturers have adopted the same numbering system, while others have devised their own related numbering schemes. Motorola, for example, produces components in the MC14000 and MC14500 series. Many years ago, the original 4000 series devices were replaced by an improved 4000B series.

Most manufacturers have now moved away from the original 4000 series parts and have produced a range of circuits that follow the circuit functions and pin assignments of the **74XX TTL family** of devices (see Section 25.7). These are given part numbers such as 74CXX, 74HCTXX, 74ACXX, 74ACTXX, 74AHCXX, 74AHCTXX, 74LVCXX, 74ALVCXX and 74AL74AVC, where in each case the 'C' stands for CMOS. The 'XX' signifies a two- or three-digit code that represents the function of the device. For example, a 74AC00 contains four two-input NAND gates, while a 74AC163 is a 4-bit synchronous binary counter. Parts with 'A' in their names are 'advanced' devices with improved speed and lower power consumption. 'V' and 'LV' parts are optimised for low-voltage operation. 'T' indicates that the devices are unlike those of the other CMOS families in that they are designed to operate with the supply voltages and logic levels of TTL gates. This enables them to be used easily with TTL components, allowing them to act as direct, low-power replacements for the corresponding TTL parts. Conventional CMOS circuits of the 4000 series or the 74CXX types cannot normally be used directly with TTL parts, since their logic levels are different.

25.7   Transistor–transistor logic

The essential form of TTL gates is illustrated in Figure 25.12, which shows two basic circuits. Figure 25.12(a) shows a simple TTL inverter. When the input A is taken *low* the resistor R pulls the base of the input transistor

Figure 25.12   TTL inverter and NAND gates

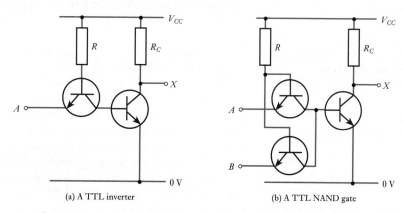

(a) A TTL inverter   (b) A TTL NAND gate

Figure 25.13 An integrated circuit TTL NAND gate

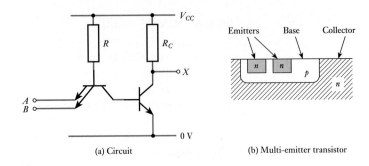

(a) Circuit   (b) Multi-emitter transistor

above its emitter voltage and turns it ON, driving its collector down towards ground. This takes the base of the output transistor low and thus turns it OFF, causing its collector to rise close to V_{CC}. When the input A is taken *high* the input transistor pulls the base of the output transistor high, turning it ON. This causes its output voltage to drop to close to ground. Thus when the input is low the output is high, and when the input is high the output is low, and we have the characteristics of an inverter.

The addition of a second input transistor, as in Figure 25.12(b), converts the inverter into a NAND gate. If either input A or input B is *low*, the associated input transistor is turned ON and the output is driven high. However, if both inputs are *high*, the base of the output transistor is pulled high and the output goes low.

It can be seen that the input transistors of the circuit of Figure 25.12(b) have common connections to their bases and their collectors. When using integrated circuit techniques, the circuit can be improved by combining the functions of the input transistors into a single device. This is shown in Figure 25.13(a). Since the components in this figure are formed in a single integrated circuit, the circles around the transistors have been omitted. The multi-emitter transistor is produced by forming a number of emitter regions in a single base region, as illustrated in Figure 25.13(b), and it can be extended to produce almost any number of inputs.

A range of logical functions can be produced using circuits similar to those shown above. Integrated circuit TTL gates improve these basic circuits by using additional components to increase speed and drive capabilities (as discussed below).

25.7.1   Standard TTL

As discussed earlier, the term 'TTL' describes not one but a range of logic families of very different characteristics. While often replaced by CMOS in modern designs, the more advanced TTL families are still widely used, particularly in the construction of relatively simple, low-volume applications. A wide range of manufacturers produce circuits of this form, and standardisation has been very successful in providing a common specification for

Figure 25.14 A 7400 TTL two-input NAND gate

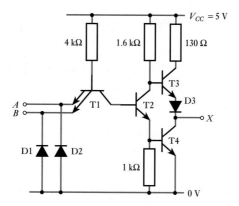

such devices. The 'standard' commercial TTL family of components contains a broad spectrum of circuits, each of which is specified by a generic serial number starting with the digits '74'. The two-digit suffix is followed by a two- or three-digit code that represents the function of the device (as described above for CMOS gates). The family is often called the **74XX family**, or simply the **74 family**. Devices in the 74 family are specified for operation over a temperature range from 0 to 70 °C. A companion family has corresponding numbers starting with 54, and these may be used over a range from −55 to 125 °C.

Real TTL components use slightly more sophisticated circuits than those shown earlier. Figure 25.14 shows one of the four two-input NAND gates found in a 7400. It can be seen that this is similar to the circuit of Figure 25.13(a), but it has an additional push-pull output stage (as discussed in Chapter 22). This arrangement is referred to as a **totem-pole** output, and it provides increased current drive capability and a reduced output resistance. The diodes D1 and D2 are **input clamp diodes**, which prevent negative-going noise spikes from damaging the input transistor.

In fact, the 74XX family of components are effectively obsolete, having been replaced with more advanced circuits that give faster operation with lower power consumption. A major disadvantage of the 'standard' components is that their transistors saturate, leading to the problems of storage time discussed in Section 25.2.4. More modern variants adopt techniques to prevent saturation, thus dramatically reducing their switching time.

25.7.2 Propagation delay

Standard TTL gates of the form shown in Figure 25.14 are relatively slow, with propagation delay times of about 9 ns. However, more modern TTL families can be much faster, with delays down to about 1 ns, making them comparable with the fastest CMOS parts.

25.7.3 Power-supply voltage

Unlike CMOS, which can operate with a wide range of supply voltages, TTL is restricted to working with a nominal supply voltage of 5 V, which for ordinary commercial parts should be within the range 4.75 to 5.25 V.

25.7.4 Logic levels and noise immunity

The standard totem-pole output produces a logical 1 voltage of about 3.4 V and a logical 0 voltage of about 0.2 V. Therefore, the logic levels are *not* equal to the supply-line voltages (0 and V_{CC}).

The operation of the circuit dictates that the noise immunity of TTL has a minimum value of about 0.4 V for each logic level, which is considerably less than for a CMOS gate operating with a similar supply voltage. However, the average value for the noise immunity is between 1 and 1.6 V.

25.7.5 Other output configurations

While the totem-pole arrangement is the most common output configuration for TTL gates, some devices use alternative output circuits, such as the **open collector output**. An example of such a device is the 7401, which contains four two-input NAND gates with open collector outputs. The circuit of one of the gates in this component is shown in Figure 25.15. It is clear that the circuit is identical to that of the 7400 (shown in Figure 25.14), except that the output circuit has been simplified. In order for this circuit to function, an external **pull-up resistor** must be connected from the output to the positive supply. One advantage of this arrangement is that the pull-up resistor can be connected to an external supply voltage, allowing the device to drive higher-power devices directly. Some open collector gates can switch up to 30 V at currents of up to 40 mA. Unfortunately,

Figure 25.15 The 7401 two-input NAND gate with open collector output

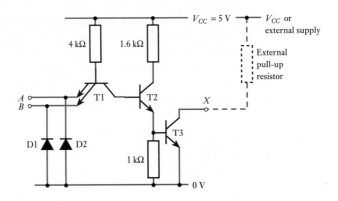

open collector gates are slightly slower in operation than conventional totem-pole devices.

While conventional gates have two output states, namely 0 and 1, some devices have a third output state corresponding to the output being allowed to 'float'. In this third state, the output goes into a high-impedance condition where its output level will be determined by whatever is connected to the gate externally. For obvious reasons, such devices are called **three-state logic gates**. The output of the gate is 'enabled' or 'disabled' by a **control input**, which is usually given the symbol C on simple gates. Three-state gates can be used in the creation of bus systems in which the outputs of several devices are connected together on to a common line. Each device can then place data on the line, provided that the output of only one device is enabled at any time.

25.7.6 TTL inputs

If unused inputs of TTL devices are left disconnected, they will act as if they were connected to a logical 1. Such inputs are said to be **floating**. Although unused inputs will 'float' to a logical 1, it is inadvisable to leave such inputs disconnected, even if the application requires the input to be at logical 1. Unconnected inputs represent a high impedance to ground, making them very sensitive to electrical noise, which could cause them to switch between states. It is much wiser to 'tie' such inputs high or low, as required. Inputs that are required to be at logical 1 should not be tied directly to the positive supply rail but should be connected through a resistor. A typical value for such a resistor might be $1 \text{ k}\Omega$. If appropriate, several inputs can be connected together to the same resistor. Inputs that are required to be low may be tied directly to ground (0 V).

25.7.7 Other TTL families

In addition to 'standard' 54XX and 74XX devices (which are now largely obsolete), there are related families with modified characteristics. These are defined by adding alphabetic characters after the '54' or '74' suffix to specify the family. For example, a 74LS00 is a low-power version of the 7400, and the 74AS00 is a high-speed version.

As noted in the last section, parts with a 'C' in the names (for example the 74ACXX, 74HCTXX and 74ACTXX families) are *not* TTL logic families, the 'C' in each name standing for CMOS.

Key points

- Physical logic gates are not ideal components, and their characteristics (such as speed and power consumption) depend on the technology used to implement them.

- Logic gates are manufactured in a range of logic families. Using gates in a single family will normally guarantee compatibility, but members of different families will not normally interface directly.

- The ability of a gate to ignore noise on its inputs is termed its 'noise immunity'.

- Both MOSFETs and bipolar transistors may be used as logical switches, and both are used in logic circuits.

- All logic gates exhibit a propagation delay when responding to changes in their inputs.

- The most widely used logic families are complementary metal oxide semiconductor (CMOS) logic and transistor–transistor logic (TTL).

- CMOS is available in a range of forms offering high speed or very low power consumption.

- TTL logic is also produced in many versions, each optimised for a particular characteristic.

Exercises

25.1 Sketch the transfer function of a logical inverter.

25.2 What is meant by the noise immunity of a logic gate?

25.3 What are the logic levels of the simple inverter shown in Figure 25.5?

25.4 What are the major advantages of MOS technology over circuits based on bipolar transistors?

25.5 Why is the turn-on time of a bipolar transistor less than the turn-off time?

25.6 Explain the terms 'PMOS', 'NMOS' and 'CMOS'.

25.7 What points in the input and output waveforms are used to measure the propagation delay times?

25.8 Define the terms 'set-up time' and 'hold time'.

25.9 What is meant by the fan-out of a logic gate?

25.10 Which two logic families are most widely used for the production of integrated logic circuits?

25.11 What is the normal supply voltage for a CMOS gate, and what are its normal logic levels?

25.12 Why is the power consumption of CMOS gates greatly affected by its clock frequency?

25.13 What is the normal supply voltage for TTL gates? What are their normal logic levels?

25.14 How many TTL gates can be safely connected to the output of a single gate?

25.15 How do ECL gates overcome the problem of saturation time in bipolar transistors?

25.16 How do CMOS gates achieve a lower output resistance than NMOS gates?

25.17 Why are special precautions needed when handling CMOS components?

25.18 A four-input CMOS NAND gate is to be used in an application that requires a three-input gate. What should be done to the unused input in this circuit?

25.19 A four-input CMOS NOR gate is to be used in an application that requires a three-input gate. What should be done to the unused input in this circuit?

25.20 What factor differentiates 74HCTXX and 74ACTXX devices from other CMOS gates?

25.21 What is the difference between a 7400 and a 5400 device?

25.22 Explain the function of the input clamp diodes in the circuit of Figure 25.14.

25.23 Why might an open-collector device be used in preference to a totem-pole output device?

25.24 What is meant by a three-state output gate?

25.25 A four-input TTL NAND gate is to be used in an application that requires a three-input gate. What should be done to the unused input in this circuit?

25.26 A four-input TTL NOR gate is to be used in an application that requires a three-input gate. What should be done to the unused input in this circuit?

25.27 How does a 74C00 differ from a 7400 gate?

Chapter 26

Data Acquisition and Conversion

Objectives

When you have studied the material in this chapter you should be able to:

- explain the need for techniques to convert analogue signals into a digital form, and vice versa;
- identify the major components of a typical data-acquisition system;
- discuss the characteristics and limitations of sampling as a method of obtaining a picture of a time-varying signal;
- describe the general characteristics of analogue-to-digital and digital-to-analogue converters;
- explain techniques that allow a number of analogue inputs or outputs to be used with a single data converter.

26.1 Introduction

We have seen in earlier chapters that the effects of noise are often less of a problem in digital systems than in those using analogue techniques. Digital data can also be easily processed, transmitted and stored. For these reasons, we often choose to represent analogue quantities in a digital form, which raises the question of how we translate from one form to the other.

The process of taking analogue information, often from a number of sources, and converting it into a digital form is often termed **data acquisition**. It consists of several stages. This chapter begins by looking at the process of sampling a changing analogue quantity to determine its time-varying nature. It then discusses the hardware required to convert these samples into a digital form and the reconstruction of this digital information into an analogue signal. Finally, it considers the process of combining information from a number of sources into a single system input, and the converse problem of generating a number of analogue output signals from a single information source.

Sampling

In order to obtain a picture of the changes in a varying quantity, it is necessary to take regular measurements. This process is referred to as **sampling**. Clearly, if a quantity is changing rapidly we will need to take samples more frequently than if it changes slowly, but how can we determine the sampling rate required to give a 'good' representation of a signal? It would seem obvious that the required sampling rate would be determined by the most rapidly changing, in other words the highest-frequency, components in a signal, but how do we decide how fast we need to sample to get a 'good picture'?

Fortunately, an answer to this question is available in the form of **Nyquist's sampling theorem**. This states that the sampling rate must be greater than twice the highest frequency present in the signal being sampled. It also states that under these circumstances none of the information in the signal is lost by sampling. In other words, it is possible to reconstruct completely the original signal from the samples.

In general, the waveform to be represented will contain components of many frequencies. In order to sample it reliably, we need to know the highest frequency present. Let us assume that we know that a certain signal contains no components above a frequency of F Hz. According to Nyquist's theorem, provided that we sample this waveform at a rate greater than $2F$, we will obtain sufficient information to reconstruct the original signal completely. This minimum sampling rate is often called the **Nyquist rate**. This process is illustrated in Figure 26.1.

While in practice the waveform we will wish to sample will consist of many frequency components, for simplicity Figure 26.1(a) shows a sine wave of frequency F. Figure 26.1(b) shows the results of sampling this signal at a rate greater than the Nyquist rate. Given these samples, it is possible to reconstruct the original waveform, since any other line drawn through the sample points would have frequency components above F. Since we know that in this case the signal has no components above this frequency, the original waveform is the only possible reconstruction. Since this sampling rate allows the reconstruction of a signal of frequency F, it will also allow reconstruction of any signal that contains no components above this frequency.

Figure 26.1(c) shows the results of sampling the waveform at a frequency below the Nyquist rate. Here the samples can be reconstructed in a number of ways, including that shown in the figure. This is clearly not the original waveform. Thus, if a signal is sampled below the Nyquist rate it will not, in general, be possible to reconstruct the original signal. The waveform generated appears to have been produced by a signal of a lower frequency than the original. This effect is known as **aliasing** and resembles a *beating* between the signal and the sampling waveform.

It should be pointed out that the Nyquist rate is determined by the highest frequencies present in a signal, *not* by the highest frequencies of interest. If a signal contains unwanted high-frequency components, they must be removed before sampling or they will result in spurious signals in the

Figure 26.1 The effects of sampling at different rates

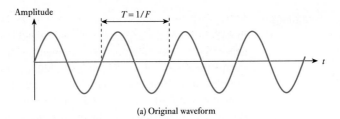

(a) Original waveform

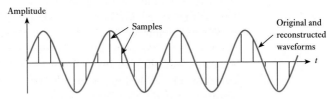

(b) Waveform sampled above the Nyquist rate

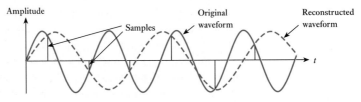

(c) Waveform sampled below the Nyquist rate

frequency band of interest. It is normal to use filters to remove signals that are above the range of interest to prevent this effect. These are referred to as **anti-aliasing filters**. For example, although human speech contains frequencies up to above 10 kHz, it has been found that good intelligibility can be obtained using only those components up to about 3.4 kHz. Therefore, to sample such a signal for transmission over a channel of limited bandwidth, it would be normal to filter the speech signal to remove frequencies above 3.4 kHz and then to sample the waveform at about 8 kHz. This is somewhat above the Nyquist rate (which would be 6.8 kHz) to allow for the fact that the filters are not perfect and some frequency components will be present a little above 3.4 kHz. It is common to sample at about 20 percent above the Nyquist rate. A typical anti-aliasing arrangement might use a sixth-order **Butterworth filter** (see Chapter 17).

26.3   **Signal reconstruction**

In many cases, it is necessary to reconstruct an analogue signal from a series of samples that have been transmitted, processed or stored. Reconstruction requires the removal of the step transitions in the sampled signal and can be seen as removing the high-frequency signal components that these represent. This can be achieved using a low-pass filter to remove these unwanted frequencies. Such a filter is called a **reconstruction filter** and would normally have a similar characteristic to the anti-aliasing filter used before sampling. Therefore, a typical reconstruction arrangement would use a sixth-order Butterworth filter.

26.4 Data converters

The process of sampling an analogue signal involves taking an instantaneous reading of its magnitude and converting this into a digital form. Similarly, the process of reconstruction requires us to take digital values and convert these back into their analogue equivalents. These two operations are performed by **data converters**, which can be divided into **analogue-to-digital converters (ADCs)** and **digital-to-analogue converters (DACs)**.

A range of converters are available, each providing conversion to a particular **resolution**. This determines the number of steps or **quantisation levels** that are used. An n-bit converter produces or accepts an n-bit parallel word and uses 2^n discrete steps. Thus an 8-bit converter uses 256 levels, and a 10-bit converter uses 1024 levels. It should be noted that the resolution of a converter may be considerably greater than its accuracy. The latter is a measure of the error associated with a particular level rather than simply the number of levels used. In many applications a simple 8-bit conversion is sufficient, this giving a resolution of about 0.25 percent. However, in situations where greater accuracy is required, converters of up to 20-bit resolution or more are available; 20-bit conversion gives a resolution of about one part in one million and is sufficient for almost all applications.

Conversions of either form take a finite time, which is referred to as the **conversion time** or **settling time** of the converter. The times taken for conversion differ greatly depending on the form of the converter used, although, in general, digital-to-analogue conversion is faster than the inverse operation.

Although both ADCs and DACs can be constructed fairly simply from basic components, in practice integrated circuit converters are almost always used. These are generally inexpensive, although the cost increases with resolution and speed.

26.4.1 Digital-to-analogue converters

DACs are available with a range of resolutions, and in general conversion time increases with resolution. A typical general-purpose 8-bit DAC would have a settling time of between 100 ns and 1 µs, while a 16-bit device might have a settling time of a few microseconds. However, for specialist applications high-speed converters have settling times down to a few nanoseconds. It is sometimes more convenient to specify the number of samples that can be converted in a second rather than the settling time. Converters used for generating the video signals used in graphics display systems might have a resolution of 8 bits and a maximum **sampling rate** of above 100 MHz, corresponding to a settling time of less than 10 ns.

26.4.2 Analogue-to-digital converters

While we are generally not concerned with the internal operation of a data converter, several different forms of ADC are available, and these tend to have different characteristics. Most general-purpose devices use a **successive approximation** approach, which produces fast conversions in a relatively simple circuit. Typical successive approximation converters might have settling times of from 1 to 10 μs for 8-bit conversion, increasing to perhaps 10 to 100 μs for a 12-bit device. High-speed variants are available with considerably improved conversion times.

In applications requiring very high-speed operation, it is common to use what are termed **flash converters**. These can produce sample rates in excess of 150 million conversions per second with conversion times of only a few nanoseconds. However, such devices are significantly more complex, and hence expensive, than other techniques.

ADCs generally have one or more lines to control the operation of the device. For example, a 'start convert' input might instruct the component to start the conversion process, while an 'end of conversion' signal might be used by the converter to tell an external system that the conversion is complete.

26.5 Sample and hold gates

With rapidly changing quantities it is often useful to be able to *sample* a signal and then *hold* its value constant. This may be required when performing analogue-to-digital conversion so that the input signal does not change during the conversion process, upsetting the operation of the converter. It may also be necessary when performing digital-to-analogue conversion to maintain a constant output voltage during the conversion period of the DAC.

The basic principle of a sample and hold gate is illustrated in Figure 26.2(a). A capacitor is connected to an analogue input signal by an electrically operated switch. When the switch is closed, the capacitor quickly charges or discharges so that its voltage, and hence the output voltage, equals the input voltage. If the switch is now opened, the capacitor simply holds its current charge and its voltage remains constant. The circuit is used to take a *sample* of a varying voltage by closing the switch and then to *hold* that value by opening the switch.

In practice, the simple circuit above has a couple of weaknesses. Firstly, when the switch is closed the capacitor represents a very low impedance to

Figure 26.2 Sample and hold
gates

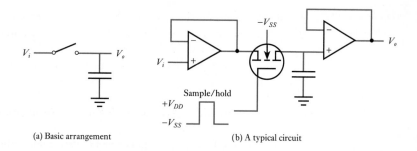

(a) Basic arrangement   (b) A typical circuit

the source and thus loads it heavily, possibly distorting the input value. If
the source has a fairly high output resistance, it may take some time for the
capacitor to charge up, reducing the speed at which samples can be taken.
Secondly, in practice the capacitor will be connected to a load that will tend
to discharge the capacitor because of its finite input resistance. This will
cause the output voltage to fall as the capacitor is slowly discharged. To
overcome these problems, buffer amplifiers are normally used at the input
and the output. Figure 26.2(b) shows a simple circuit using operational
amplifiers as buffers and a FET as the electrically operated switch.

While sample and hold gates can be constructed using discrete components, they are usually produced in integrated circuit form. Typical integrated components require a few microseconds to sample the incoming
waveform, which then decays (or **droops**) at a rate of a few millivolts per
millisecond. Higher-speed devices, such as those used for video applications, can sample an input signal in a few nanoseconds but are designed to
hold the signal for a shorter time. Such high-speed devices may experience
a droop rate of a few millivolts per microsecond.

26.6   Multiplexing

Although it is quite possible to have a system with a single analogue input
or a single analogue output, it is more common to have multiple inputs and
outputs. Clearly, one solution to this problem is to use a separate converter
for each input and output signal, but often a more economical solution is to
use some form of **multiplexing**. The principle of signal multiplexing is
illustrated in Figure 26.3.

A number of analogue input signals can be connected to a single ADC
using an **analogue multiplexer**. This is a form of electrically controlled
switch based on the use of **analogue switches**, as discussed in Section 20.8.
Each analogue signal is connected in turn to the ADC for conversion, the
sequence and timing being determined by control signals from the system.
This is illustrated in Figure 26.3(a).

For certain applications the arrangement of Figure 26.3(a) is unsuitable,
since each analogue input signal is sampled at a different time. This may
make it impossible to obtain detailed information as to the relationship
between the signals, such as their phase difference. The problem can be

Figure 26.3 Input and output multiplexing

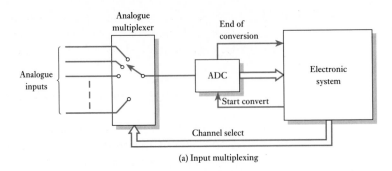

(a) Input multiplexing

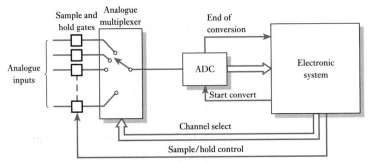

(b) Input multiplexing with sample and hold gates

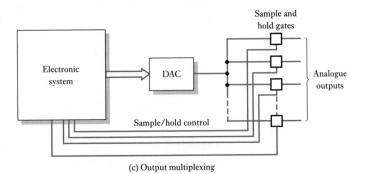

(c) Output multiplexing

overcome by sampling all the inputs simultaneously using a number of sample and hold gates, as shown in Figure 26.3(b). Once the input signals have been sampled, they can be read sequentially without losing the time relationship between the channels.

Figure 26.3(c) shows an arrangement whereby a number of output channels are produced from a single DAC. Here the converter is sent data relating to each channel in turn. When the conversion is complete, a control signal is used to activate the appropriate sample and hold gate. The gate samples the output from the DAC and reproduces this value at its output. The system sets the values of each output channel in turn, updating the values as necessary.

In practice, anti-aliasing and reconstruction filters would be used on the inputs and output of the arrangements shown in Figure 26.3. These have been omitted from the diagram to aid clarity.

26.6.1 Single-chip data-acquisition systems

The combination of an ADC and a multiplexer in a single integrated circuit is often termed a **single-chip data-acquisition system**. Although a slight exaggeration, the combination of the two functions is often convenient. These components, which are often specifically designed for use with microprocessor-based systems, provide all the control lines necessary for easy interfacing.

Key points

- Converting an analogue signal to a digital form is achieved by sampling the waveform and then performing analogue-to-digital conversion.

- As long as the signal is sampled at a rate above the Nyquist rate, no information is lost as a result of the sampling operation.

- When sampling signals that have a broad frequency spectrum, it is necessary to use anti-aliasing filters to remove components that are at frequencies above half the sampling rate.

- When reconstructing analogue signals from a series of samples, filters are used to remove the high-frequency components associated with the sampling process.

- A wide range of DACs and ADCs are available. These differ in their resolution, accuracy, speed and cost.

- Sample and hold gates may be required to hold an input signal constant while it is sampled or to hold an output signal constant between the times at which it is updated.

- In systems with a number of analogue inputs or outputs, multiplexing can be used to reduce the number of data converters required.

Exercises

26.1 Explain the function of sampling in data acquisition.

26.2 What is meant by the term 'Nyquist rate'? A signal has a frequency spectrum that extends as high as 4 kHz. What is the minimum rate at which the signal may be sampled to obtain a good representation of its form? What would be the effect of sampling below this rate?

26.3 Describe the use of anti-aliasing and reconstruction filters.

26.4 A signal has a frequency range from 20 Hz to 20 kHz. However, for a particular application only those frequency components up to 10 kHz are of importance. Explain how this signal may be sampled to minimise the amount of data produced while maintaining the amount of useful information. What would be an appropriate sampling rate in your arrangement?

26.5 Explain the terms 'resolution' and 'accuracy' as they apply to data converters.

26.6 How many quantisation levels are used by a 12-bit data converter?

26.7 Give examples of control signals that might be used with a typical analogue-to-digital converter.

26.8 Explain the use of sample and hold gates in analogue input/output systems. What is meant by the term 'droop' when applied to such gates?

26.9 Explain the function of multiplexing in a data-acquisition system.

26.10 What is meant by a 'single-chip data-acquisition system'?

Chapter 27

Implementing Digital Systems

Objectives

When you have studied the material in this chapter you should be able to:

- explain the classification of integrated circuits according to their level of integration;
- describe the nature and uses of the various forms of semiconductor memory;
- discuss the use of array logic in allowing widespread use of VLSI technology;
- outline the implementation of combinational logic functions using simple programmable logic devices (PLDs);
- list the various elements in a simple microcomputer system and explain how such a system might produce the functions of combination or sequential logic circuits;
- describe the general form of a programmable logic controller (PLC);
- suggest a range of implementation strategies that might be appropriate for the production of a digital system.

27.1 Introduction

In Chapter 25, we saw how a number of basic gates may be implemented in a single integrated circuit (IC). In the early days of digital electronics, the number of components that could be combined in a single 'chip' was limited. However, over the years the densities of ICs have increased considerably, and today it is possible to place millions of both active and passive components on a single piece of silicon only a few millimetres square. Electronic devices may be classified by their **integration level**, and Table 27.1 shows a common way of defining the various levels of integration.

In Chapter 25, we looked at technologies for implementing SSI and MSI devices and looked at the basic forms of these types of integrated circuit. More highly integrated devices make use of the same underlying technologies

Table 27.1 Integration levels for electronic devices

Integration level	Number of transistors
Zero-scale integration (ZSI)	1
Small-scale integration (SSI)	2–30
Medium-scale integration (MSI)	$30–10^3$
Large-scale integration (LSI)	$10^3–10^5$
Very large-scale integration (VLSI)	$10^5–10^7$
Ultra large-scale integration (ULSI)	$10^7–10^9$
Giga-scale integration (GSI)	$10^9–10^{11}$
Tera-scale integration (TSI)	$10^{11}–10^{13}$

but are far more complex in their structure and operation. The production of these circuits is totally dominated by CMOS technology, which is evolving and growing ever more powerful year by year.

27.1.1   The evolution of integrated circuit complexity

Back in 1965, Gordon Moore, one of the founders of Fairchild's Semiconductor Division, noted an exponential growth in the number of transistors that could be placed in a single integrated circuit and suggested that this trend would continue. **Moore's law**, as it is generally known, predicts that the number of transistors that can be integrated in a single device will double every couple of years. This prediction has proved remarkably accurate, as can be seen from Figure 27.1, which shows the number of transistors in a series of processors produced by a single manufacturer, Intel.

Using modern device production techniques, it is possible to combine millions of gates in a single integrated circuit. However, there are practical limits to the number of separate gates that may usefully be put in a single package. One of the major constraints is simply the number of pins that are required to connect to the inputs and outputs of the gates. A circuit with 1000 separate gates would require several thousand pins and would inevitably occupy a large amount of space on a circuit board. The external interconnections between these pins would also require a large amount of board area.

Figure 27.1   Integration densities of Intel processors

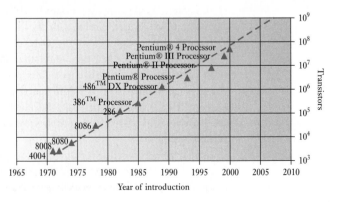

To take full advantage of large-scale integration, it is necessary to implement not only the gates required by a circuit but also their interconnections. If this is done, then only the circuit's inputs and outputs need to be brought to the outside world, rather than connections to each node of the circuit. This produces an arrangement in which most of the interconnections are internal, greatly reducing stray capacitance and increasing the speed of operation of the circuit (as discussed in Chapter 25). Internally connecting the gates in a package permits complex circuits to be implemented in a single device but results in a device that is dedicated to a particular function.

Complex integrated circuits are relatively inexpensive to mass produce but are very expensive to design. Consequently, it is quite feasible to design a special-purpose device for an application requiring a million components (where the design costs can be distributed between the million units), but it is unattractive to do so for a project requiring only a handful of systems. One way of tackling this problem is to produce a standardised component that can be produced in high volume and then customised (or programmed) for a particular application.

In this chapter, we will look at two forms of highly integrated, general-purpose device, namely **array logic** and **microprocessors**. Having looked at each of these in turn, we will then consider a packaged form of the microprocessor, the **programmable logic controller** or **PLC**, before looking at the process of choosing a technology for a particular application. However, since all highly integrated digital systems make use of some form of memory, we will start by looking at the nature and forms of **semiconductor memory**.

27.2 Semiconductor memory

Memory is used for a variety of purposes in electronic systems. We are perhaps most familiar with its use in microcomputers – to store computer programs and associated data. However, memory is also used in other components such as array logic. Memory devices may also be used alone to implement simple combinational logic functions directly.

In some cases, memory elements are used to store information that will never be changed, while in others they are used for data that is constantly changing. For example, the program to control an engine management system in a car will be fixed, while the data it uses to describe the speed and temperature of the engine will change regularly. These different requirements have led to various forms of memory, which may be divided into two broad categories, namely RAM and ROM.

27.2.1 RAM

Data that must be changed frequently is normally stored in **random-access memory** or **RAM**, this being the name given to memory that can be both *written* and *read* quickly. Here we use the word **write** to describe the process of placing information into the memory and the word **read** to describe the process of accessing this information. Note that in most cases the action of reading data does not change the contents of the memory elements. Thus the information can be read again and again if necessary.

RAM is implemented using one of two circuit techniques. **Static RAM** uses a bistable arrangement similar to that described in Section 10.3. Information written into such a device is retained indefinitely provided that power is maintained. **Dynamic RAM** stores information by charging or discharging an array of capacitors. Dynamic RAM requires far fewer components for each bit of information stored, permitting more storage elements to be integrated within a single chip. However, it suffers from the disadvantage that the charges on the capacitors tend to decay over time, making it necessary to **refresh** the devices periodically by applying an appropriate sequence of control signals.

One characteristic of RAM is that it is **volatile**. That is, it loses its contents when power is removed. In many applications this is unimportant, since when power is lost the electronic system itself stops functioning. However, in some applications the contents of memory must not be lost when power is removed. For example, storage of the program for a control system must be non-volatile, since it must be present when the system is first turned on. For this reason, programs are normally stored in ROM, as described below, rather than RAM.

There are many situations where a non-volatile read/write memory is required. In such cases it is normal to use some form of **battery backup** to protect the contents of the RAM from power failure. Fortunately, when CMOS memory is not being clocked its power consumption is extremely low (as discussed in Chapter 25), allowing even a small battery to maintain its contents for extended periods.

27.2.2   ROM

ROM is **read-only memory**, that is, it can be read from but cannot be written to. Such devices are **non-volatile** and are thus suitable for storing programs or any non-changing data. There are many forms of ROM. Some must be programmed by the device manufacturer, while others can be programmed by the user, often using special equipment. Table 27.2 lists the acronyms of several forms of ROM and gives their meanings.

The general term 'ROM' is applied to all forms of read-only device. Some of these are **mask programmed**, which means that the device is

Table 27.2 Acronyms for various types of ROM

ROM	Read-only memory
PROM	Programmable read-only memory
EPROM	Erasable and programmable read-only memory
EEPROM	Electrically erasable and programmable read-only memory

programmed photolithographically by the chip manufacturer as the last stage of production. A designer using this approach must supply the manufacturer with the required memory contents and pay a large fee for the production of the mask. However, when the mask has been made the unit cost of the device is low. This is the most attractive option for high-volume production, but it is unsuitable for use during development and for low-volume applications.

An alternative for small-scale projects is to use one of the range of **programmable read-only memories** or **PROM**s. These are available in a number of variants, but all allow the user to program the device, saving the high cost of mask production. In fact, the term 'PROM' is normally reserved for small, fusible link devices that once programmed cannot be modified.

For flexible system development, it is advantageous to have a memory device that can be programmed, and then reprogrammed if necessary. These features are provided by **erasable and programmable read-only memories** or **EPROM**s. Although the term can be applied to a number of components, this description is usually applied to memory that is erased by exposure to ultraviolet (UV) light. The chips are fitted with a transparent window, which allows UV light to reach the silicon surface. Programming is normally performed using an **EPROM programmer**, which provides the appropriate voltages and control signals. After use, EPROMs can be erased using a UV source in 20 to 30 minutes.

EPROMs have the disadvantage that they must normally be removed from their circuit and placed in a special eraser and a programmer to allow them to be modified. For this reason, they have been largely replaced by the EEPROM, which can be modified electrically without the need for a UV source. This allows the contents to be modified while the chip is in place. It might seem that the EEPROM should be classified as a RAM, since it can be written to (programmed) as well as read. However, it should be noted that RAM can normally be written and read in a fraction of a microsecond. An EEPROM can be read at this speed, but it may require several milliseconds to write a single byte. The EEPROM is thus a read quickly but write slowly device, and it is better described as a ROM than a RAM. The relatively slow programming speed of EEPROMs can be rather inconvenient for large memories. These problems are overcome by **flash** memory, which provides the ability to electrically program and reprogram a device at very high speeds. This permits even the largest devices to be programmed in a few seconds.

27.2.3 Memory organisation

Memory elements are often incorporated into sophisticated electronic devices such as microprocessors and array logic devices. Here they may be arranged as an array of words or as a linear array of individual bits. In either case, the array will have associated address logic that allows a given

Figure 27.2 A typical memory device

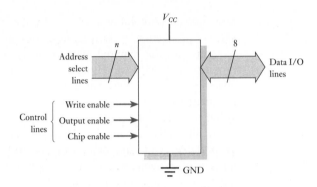

combination of input signals (the memory **address**) to select a particular memory location. This, combined with appropriate read/write logic, allows information to be written into, or read from, the selected location.

Memory is also available in dedicated integrated circuits that combine the memory elements with address decoding and control circuitry. Figure 27.2 shows the connections to a typical memory device. The number of input/output lines (I/O lines) will depend on the organisation of the memory. In many devices words are stored in bytes, and such devices would have eight I/O lines for reading and writing data (as in Figure 27.2). Other devices store data in longer or shorter words, while others store data in an array of individual bits of information.

The number of address select lines will depend on the size of the memory. The number of memory locations in a memory device is invariably a power of 2, since n address lines can specify 2^n memory addresses. Because the numbers concerned are large, it is normal to express memory sizes in **kilobytes** (kbytes), where 1 kbyte is equal to 1024 bytes (which is 2^{10}). A similar notation is used for larger memories, where a block of 2^{20} (1,048,576) bytes of memory is referred to as a **megabyte** (Mbyte). Thus a device with 65,536 8-bit memory registers would be called a 64-kbyte memory and would have sixteen address lines, while a device with 1,048,576 memory locations would be called a 1-Mbyte device and would have twenty address lines.

Where several memory devices must be used together to provide sufficient memory, external address decoding logic is used to partition the memory. For example, 4 Mbytes of memory might be constructed using four 1-Mbyte memory devices. In this case, a 22-bit address would be needed to uniquely specify each of the 4,194,394 locations. Here the twenty least significant bits of the address would be used for the address lines of each of the four memory devices, and the two most significant bits of the address would be decoded to enable one of the devices (using its **chip enable** line). The timing of read and write operations is controlled by the **write enable** and **output enable** control lines.

27.3   Array logic

The term array logic is applied to a range of technologies, but here we will concentrate on two major forms: programmable logic devices (PLDs) and field programmable gate arrays (FPGAs).

27.3.1   Programmable logic devices

PLDs contain a large number of logic gates in a single package, but they allow a user to determine how they are interconnected. This technology is also known as **uncommitted logic**, since the gates are not committed to any specific function at the time of manufacture. There are many forms of PLD, and, unfortunately, these are known by a plethora of names, for example:

- PLA – programmable logic array;
- PAL – programmable array logic;
- GAL – generic array logic;
- EPLD – erasable programmable logic device;
- PROM – programmable read-only memory; and
- CPLD – complex programmable logic device.

In this text, it would not be appropriate to look in depth at the operation of each of these forms of PLD, but we will consider the basic principles involved. We will do this by considering a simple form of PLD – the programmable logic array.

Programmable logic array (PLA)

In Chapter 9, we saw that any combinational logic function can be represented by a truth table, and that the outputs can be expressed as a number of terms that are ORed together. For example, a system with four inputs, A, B, C and D, might have outputs X, Y and Z, where

$$X = \overline{A}\,\overline{B}\,\overline{C}D + \overline{A}\,\overline{B}CD$$

$$Y = \overline{A}\,\overline{B}CD + ABC\overline{D}$$

$$Z = \overline{A}\,\overline{B}\,\overline{C}D + \overline{A}\,\overline{B}CD + ABC\overline{D}$$

One way of implementing such a system is to use a number of inverters to produce the inverted input signals ($\overline{A}$, $\overline{B}$, $\overline{C}$ and $\overline{D}$) and then to use a series of AND gates and OR gates to generate and combine the various terms. A PLA has a structure that allows such functions to be produced easily.

    The structure of a simple PLA is shown in Figure 27.3. This shows an arrangement with four inputs (A, B, C and D), which are inverted to produce four pairs of complementary inputs. Each of these eight signals is then connected to the inputs of a number of AND gates through an array of **programmable links**. These links determine the pattern of connections between the input signals and the AND gates. In this way, each AND gate is used to detect the input pattern corresponding to an individual term. A

Figure 27.3   The structure of a
simple PLA

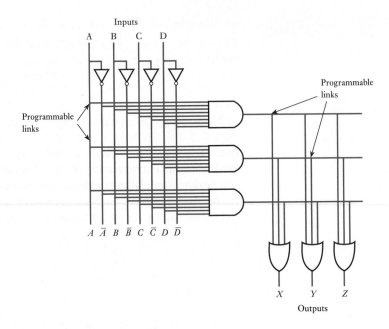

Figure 27.4   A configured PLA

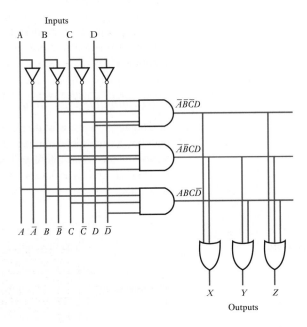

second array of links is used to connect the outputs of the AND gates to a
collection of OR gates. These OR gates combine the relevant terms to produce the various outputs. This process is illustrated in Figure 27.4, which
shows the earlier simplified PLA configured to implement the system given
in the above example. Here most of the programmable links connecting the
inputs to the AND gates have been removed, leaving only those connecting

the required signals to each gate. Similarly, the links connected to the inputs to the OR gates have been programmed to produce the required three output signals.

A PLA would normally have more inputs and outputs than the simplified example shown above, and it would also have a greater number of AND gates, allowing more complex functions to be implemented.

The architecture of other PLDs is somewhat different from that of the PLA shown above, but the user is generally not concerned with the internal construction of the device. A wide range of standard components are available with a considerable spread of capabilities and features. A small, simple device might have sixteen input/output lines and come in a twenty-pin package. Larger devices might provide forty or more input/output lines and would clearly use a larger package. More sophisticated devices include bistables, registers and multiplexers within their structure, while complex parts might include arithmetic logic units and blocks of memory. Not surprisingly, sophisticated devices of this type are referred to as **complex programmable logic devices** or **CPLDs**.

27.3.2 Field programmable gate arrays

The term **gate array** describes a component in which an array of identical cells is laid out in a rectangular matrix, together with a network of programmable interconnections. These interconnections are programmed by the chip manufacturer, producing a device tailored to a particular application. A **field programmable gate array** (**FPGA**) is a user-programmable version of the gate array.

The cells in an FPGA vary considerably from one device to another and may contain a range of elements such as registers, look-up tables and various amounts of memory. The cells are arranged in a grid together with a series of routing pathways for interconnecting the cells. This structure is illustrated in Figure 27.5. Since the cells have much greater functionality

Figure 27.5 A simplified FPGA structure

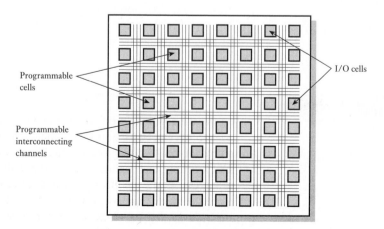

than simple logic gates, the capabilities of these devices are much greater than those of simple PLDs.

27.3.3 PLD and FPGA programming

Before use, PLDs and FPGAs must be **programmed** (or **configured**) to establish the appropriate pattern of interconnections between the various circuit elements. In the case of PLDs, this task is performed by a **PLD programmer**, which reads and interprets a **link map** supplied by the user. The method used to configure FPGAs depends on the form of the device.

The programmable links in PLDs and FPGAs may take a number of forms. Simple PLDs often use **fuses** to define the pattern of interconnections. These are initially all intact but may be blown selectively (by passing a high current through them) to leave only those connections that are required. An alternative approach uses **anti-fuses**, which are initially open-circuit but can be joined electrically. More modern devices use links based on the same technology as EEPROMs, allowing them to be programmed and then erased and reprogrammed if necessary. A further technique uses an electrically operated switch at each interconnection, with the state of the switch being controlled by a static random-access memory (SRAM) cell. Since SRAM is volatile, the configuration pattern must be rewritten each time power is applied to the device. This is normally achieved by placing the pattern in an associated non-volatile memory device. In a variation on this arrangement, the switches are controlled by a non-volatile memory such as flash so that the pattern of interconnections remains when the power is removed.

Configuration data for both PLDs and FPGAs are produced by a dedicated computer-aided design (CAD) package that interprets a description of the desired functionality. This description may be written in a manufacturer-specific **hardware description language** (**HDL**) or in a generic system description language such as **VHDL**. Alternatively, the functionality may be described in a schematic form. Once the required functionality has been defined, the CAD software will attempt to fit this into a particular device. A range of components of varying complexity are available, so if a particular function cannot be accommodated in one device, another may be used.

27.4 Microprocessors

Earlier we noted the problem of creating a single, high-volume VLSI component that can be used in a wide range of applications. In array logic, this problem is tackled by producing a device with a large number of undedicated functions that can be configured for a given application. An alternative strategy is adopted in the case of the microprocessor, which uses a unit that is capable of executing a range of instructions and a program that is used to adapt this to a given situation.

Microcomputers, under the control of a program, can sense input signals, use this information in calculations and then produce relevant output signals. Therefore, given an appropriate program, they can be made to perform the functions of any combinational or sequential logic circuit. This potentially allows large amounts of logic circuitry to be replaced by a single microcomputer chip, with a considerable cost saving. Moreover, since the operation of the microcomputer is controlled by a program, its operation can be modified without changing the physical structure of the system. This flexibility is invaluable in allowing products to be updated easily and cost-effectively. These advantages have led to the widespread use of microcomputers in a range of applications from washing machines to aircraft autopilots, and from coffee makers to the controllers of atomic power stations.

It should be noted that both array logic and microprocessors are *programmable*, but that the meaning of this term is slightly different in these two cases. A PLD or FPGA is programmed by altering the configuration of the device to suit a particular application. A microcomputer is programmed by giving it a list of instructions that enable it to perform a given task.

The physical components of a computer are collectively known as the **hardware** of the system, while the programs that control it are referred to as the **software**.

27.4.1 Microcomputer systems

While we often talk of functions being performed by a *microprocessor*, it is more accurate to say that a *microcomputer* is being used. A microprocessor is simply one element in a microcomputer, and a range of other components are needed to make a functioning unit. All computers, from mainframes to microcomputers, consist of a number of primary elements, and these are shown in Figure 27.6. The **central processing unit** or **CPU** is the heart of the computer. It is responsible for executing the various instructions in a program and for performing the operations that this involves. The CPU is often referred to as the **processor**, and in the case of a microcomputer this is a **microprocessor**. A major element in the CPU is

Figure 27.6   The essential elements of a computer

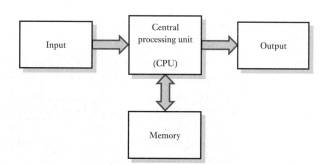

the **arithmetic logic unit** or **ALU**, which is responsible for performing the arithmetic and logical operations required by the program.

Memory is used to store both **programs** (a sequence of instructions) and **data** (the information used or produced by a program). A typical small computer might have several thousand bytes of memory, while large computers might have millions, or sometimes thousands of millions, of bytes.

The most powerful computer would be of little practical use without some method of communicating with it. This is achieved by the **input/ output section (I/O section)** of the computer, the form of which varies considerably depending on the nature of the information concerned. In a desktop computer, the input/output section would consist largely of circuitry designed to communicate with the keyboard, monitor and printer and links to other machines. On a small microcomputer in a washing machine, the input and output information would relate to such items as the water temperature and the motor speed, and the input/output sections would be quite different.

In some cases, the processor, memory and input/output sections of a computer are combined in a single integrated circuit. Such a device is not a microprocessor but a **single-chip microcomputer**. Such components are particularly useful in control and instrumentation applications, where the requirements for memory are normally modest.

Communication in the microcomputer

Communication between the various sections of the computer takes place over a number of **buses**. These are parallel data highways that permit information to flow in one, or both, directions. Figure 27.7 shows the bus structure of a typical microcomputer.

The buses may be considered to be a collection of parallel conductors (wires). Three buses are used to carry data, address information and control signals. For example, if the processor wished to write a data word into a particular memory location, it would place the data on the **data bus**, the address where the information was to be stored on the **address bus**,

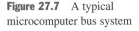

Figure 27.7 A typical microcomputer bus system

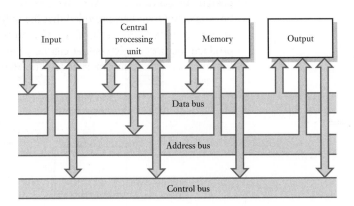

and various control signals to synchronise the storage operation on the **control bus**.

The number of lines in the data bus is equal to the **word length** of the device and thus determines the number of bits of data that can be moved about the machine at any one time. Thus in an 8-bit microprocessor the data bus would be eight bits wide, whereas in a 16-bit computer it would be sixteen bits wide.

The number of lines in the address bus limits the number of memory locations that can be specified by the processor. This is called the **addressing range** of the device. An 8-bit address bus would be able to specify only 2^8 (256) addresses, which is rather limiting for the majority of applications. Most 8-bit computers use a 16-bit address bus, giving an addressing range of 2^{16} or 65,536. Thus, in 8-bit microprocessors, addresses are usually represented by two bytes of information. Most 16-bit machines are designed for more demanding applications, which often require a greater addressing range than is possible with a 16-bit address bus. Many machines use a 20-bit bus, giving an addressing range of over one million, while others use a 24- or even a 32-bit address bus. A 32-bit bus gives an addressing range of over 4,000,000,000 locations (4 Gbytes), which is sufficient for the majority of applications. Many 32-bit processors use a 64-bit address bus and have an addressing range of 64 Gbytes.

The lines of the control bus are used by the processor to produce actions in external components and to synchronise these operations. The exact nature of these lines varies between machines.

Registers

The memory section of the computer consists of a large number of memory registers, which can be used to store both data and programs. The processor and I/O sections also contain registers for a range of purposes. The register is thus a fundamental building block in a computer system.

In Section 10.3, we looked at the use of D flip-flops in the construction of memory registers. In a microprocessor, communication between registers takes place over a bus system, which imposes some restrictions on their design. In Section 25.7, we saw how gates with three-state outputs can be connected together, provided that the output of only one gate is enabled at any time. By using D flip-flops with three-state outputs, it is possible to produce memory registers that can be connected directly to a bus system. Figure 27.8 shows such an arrangement. Here an 8-bit register is connected

Figure 27.8 A simple 8-bit register

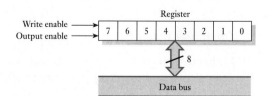

to a data bus. The register has two control inputs: one to write data from the bus into the register (this corresponds to the clock input to the flip-flops) and the other to enable the outputs of the register to drive the bus (this is connected to the three-state control of the gates). The inputs and outputs of each flip-flop are connected together to the corresponding bits of the data bus.

Communication between a number of registers is achieved simply by enabling both the output of one register and the input of another. Since all the registers in a system are connected by the same data bus, only one piece of information can be communicated at any one time. If many pieces of data are to be transferred, this will require several operations. This arrangement is thus a **sequential** communication system.

Input and output registers

Input and output registers differ from other computer registers in that they have connections to the 'outside world'. For example, an input register will have an input line connected to each bit of the register, and when the processor *reads* from this register it will 'see' the current state of these inputs. Similarly, output registers have output lines connected to each bit such that, when the processor *writes* a value into the register, this sets the state of each of these outputs.

Program storage

A computer program is a list of instructions to the processor. All microprocessors have a set of instructions that they can execute, and these make up what is called the **instruction set** of the machine. Each type of processor has its own instruction set, and generally programs written for one machine will not operate on another.

A typical microprocessor will have in its repertoire instructions for transferring data between registers, transferring data between registers and memory, performing various arithmetic and logical operations, performing comparisons and tests on register contents, and controlling the sequence of program execution. Each instruction has within it an indication of the operation to be performed (the **operation code** or **opcode**) plus additional information that is needed to carry out this operation. This additional information might be the address to be used for an operation (for example, the address of a register to be accessed) or an absolute value (for example a number to be added or subtracted). This additional information is termed the **operand** of the instruction.

Computer memory

Computers invariably use both random-access and read-only memory. In computers that are dedicated to control applications, the control program will normally be stored in ROM (since this is non-volatile) and will be configured so that the program is executed automatically when the system is switched on. In executing the program, the computer will make use of RAM to store the results of calculations and other variable data. In general-purpose computers (such as desktop PCs), a range of different programs are used, so these are normally stored on some secondary storage device (such as a hard disk) and are loaded into RAM when needed. Therefore, general-purpose computers have a relatively small amount of ROM (to get the machine up and running and to load further software) and large amounts of RAM (which is used to store programs and data).

Using microcomputers

Through the use of complex software, microcomputers can perform very sophisticated operations, including applications such as word processing and video games. However, in addition to their uses in general-purpose computers (such as desktop computers and laptops), they are also widely used in control and instrumentation applications. Modern cars, for example, often make use of several microprocessors in their engine management, anti-lock braking and navigation systems. In such applications, the input/output signals are not associated with keyboards and displays but with various sensors and actuators. It is also important to remember that micro-processors can be used in relatively simple applications to replace conventional logic circuits.

Combination logic circuits produce one or more binary outputs that are determined by the state of one or more binary inputs. Since a microcomputer can read the values from a series of inputs (using an input register) and can set the values on a series of outputs (using an output register), it can be programmed to produce any required relationship between the input and the output. It can thus replace any combination of conventional logic gates. With slightly more complex programs, it can also replace sequential logic arrangements.

27.5 Programmable logic controllers

Programmable logic controllers (PLCs) are self-contained microcomputers that are optimised for industrial control. They consist of one or more processors together with power supply and interface circuitry in a suitable housing. A range of input and output modules are normally available to allow such units to be used in a wide range of situations without the need for any electronic design or construction. Facilities are also provided for programming and for general system development.

PLCs were introduced in the 1970s as a way of producing and market-ing computers in large quantities in an industrial area that was character-ised by the diversity of its applications. At that time, many simple control systems were based on the use of electromagnetic relays, and PLCs were seen initially as a replacement for this form of circuitry. Designers working with relays were used to producing their designs using a graphic notation based on **ladder diagrams** (or **ladder logic**). To simplify the adoption of a new technology, manufacturers of PLCs added a user interface to their products, which enabled them to be programmed in a manner that was intuitive to engineers familiar with ladder diagrams. Initially, the controllers had a functionality that was limited to the simple logic functions that could be produced using relays. As time passed, more elaborate features were added to meet the diverse needs of the control engineer. These included sophisticated displays, data logging and communications facilities, and the ability to program the devices using a range of graphical or text-based languages.

PLCs are extensively used in machine and process control, and they have been designed specifically for high reliability and dependability. The availability of off-the-shelf hardware and interfacing software reduces development time and costs and makes PLCs particularly suited to low-volume applications.

27.6 Selecting an implementation method

In most cases, the method chosen to implement a digital system is likely to be determined by the complexity of the functions to be produced. Where very limited logical operations are required, it may be possible to produce these using simple circuits based on the use of conventional logic gates. Applications that require only a handful of gates will normally use standard CMOS logic or TTL devices. Unfortunately, even relatively simple logic arrangements can produce circuits requiring many devices, and it soon becomes economical, in terms of both cost and space, to use some form of array logic. The choice between the different forms of array logic is likely to be determined by the number and range of functions required and the programming support available.

For complex digital applications simple PLDs will not suffice, and the designer must choose between using a more complex programmable device (such as a CPLD or an FPGA) or a microcomputer.

27.6.1 Hardware versus software

In simple applications, hardware-based solutions are preferable since they require no software development and therefore have a lower development cost. However, as the complexity of the system increases, the potential advantages of a software-based (that is, computer-based) approach become considerable. One of the greatest advantages of computer-based systems is their flexibility. This allows a single standard computer board to be used for a range of applications, reducing the range of subsystems that must be produced. This saves on both design time and inventory costs (the cost of holding stocks of components). It also allows the operation of a system to be updated simply by changing its operating program without having to redesign the hardware. Set against these advantages is the high cost of software development.

Despite the obvious advantages of computer-based systems, it should be noted that modern PLDs and FPGAs offer much of the flexibility of computer-based systems within a hardware-based implementation. In such systems, much of the design complexity is implemented in the logic device. This enables the functionality of the system to be modified simply by changing the device configuration. This allows such systems to be upgraded in much the same way as a computer-based system. When complex PLDs and FPGAs are used in this way, the task of configuring the device is actually very similar to that of the production of software. One can view microprocessors and the various forms of array logic as devices that each implement a potentially complex set of instructions defined by the programmer. The primary difference between the two approaches is that in a computer they are executed in a serial manner, while in array logic they are executed in parallel.

The parallel operation of array logic means that they are able to perform many processing tasks much faster than a computer-based system. However, beyond a certain level of complexity the cost of implementing systems using array logic becomes prohibitively expensive. In such cases, a computer-based system is the only practical solution, allowing very complex control algorithms to be constructed in software rather than by adding more complicated hardware.

When the characteristics of an application suggest a computer-based solution, the designer is then left with the task of choosing an appropriate method of implementing this system. Complex, low-volume applications might suggest the use of a conventional desktop computer, provided that such a computer can satisfy the space and environmental constraints. When systems must be produced in slightly greater numbers, it would be normal to purchase a ready-made computer, perhaps in the form of a single-board computer or a programmable logic controller. PLCs can be purchased as complete off-the-shelf systems, greatly reducing the design effort. However, the use of a single-board computer may give more flexibility to configure the system to meet a particular requirement.

Buying a ready-made system greatly reduces the development costs, but it often does this at the expense of a higher unit cost. When producing systems in high volumes, it will be cost-effective to design the system at the chip level rather than to buy a ready-made system. This allows the designer to tailor the unit to the specific application, but requires a greater level of technical effort and experience.

Key points

- Device technologies may be categorised into various levels of integration, ranging from 'zero-scale integration' through to 'giga-scale' and 'tera-scale' techniques.

- The available circuit complexity doubles every couple of years.

- Semiconductor memory can be divided into random-access memory, which is volatile, and read-only memory, which is non-volatile.

- In situations where large numbers of gates are required, the use of standard logic gates is impracticable. In such cases, it is common to use a more highly integrated approach.

- Array logic integrates large numbers of gates within a single package. They can be mass produced, since a single component can be programmed for a wide range of applications.

- Many forms of programmable logic device (PLD) are available. These differ in their complexity and in the range of functions that they offer.

- Field programmable gate arrays offer enhanced functionality through the use of an array of logic cells.

- PLDs and FPGAs are normally programmed using a sophisticated software package that defines the internal configuration needed to produce a certain functionality. For this reason, the designer does not require a detailed knowledge of the internal structure of the device.

- An alternative approach to the implementation of complex digital systems is through the use of microprocessors.

- A microprocessor represents one of the main components of a computer – the processor or CPU.

- A chip that contains the processor, memory and input/output sections of a computer is termed a 'single-chip microcomputer'.

- Computers consist largely of a set of registers, which communicate via a series of data buses.

- The functionality of a computer is defined by a sequence of instructions (software). This can enable a computer to perform highly complex functions – or to replace relatively simple logic circuitry.

- A programmable logic controller (PLC) is a self-contained microcomputer that is optimised for industrial control.

- The method selected to implement a digital system is likely to depend on its complexity. Simple systems might be produced using a handful of conventional gates, while more complicated arrangements might suggest the use of a gate array. Where considerable complexity is involved, it will often be necessary to use some form of VLSI device in the form of a logic array or a microprocessor.

Exercises

27.1 Explain what is meant by 'integration level' in connection with digital integrated circuits and define the terms 'SSI', 'MSI', 'VLSI' and 'GSI'.

27.2 Why is it unattractive to produce a VLSI device containing 1000 separate gates?

27.3 Explain what is meant by the terms 'uncommitted logic' and 'array logic'.

27.4 Explain the basic form of a programmable logic array (PLA).

27.5 Describe the process used to program a PLD for a given application.

27.6 How does an FPGA differ from a PLA?

27.7 What is a microprocessor?

27.8 Sketch a simple block diagram indicating the main elements of a computer and showing the information flow between them.

27.9 Explain what is meant by a single-chip microcomputer.

27.10 What is the addressing range of a computer that uses 24-bit addresses?

27.11 Explain why registers require three-state operation to be used on a computer bus.

27.12 Explain the terms 'instruction set', 'opcode' and 'operand'.

27.13 Describe the basic characteristics of a programmable logic controller (PLC).

27.14 What would be a suitable implementation method for a system requiring about six basic gates?

27.15 What would be a suitable implementation method for a system requiring about twenty basic gates?

27.16 Discuss the relative merits of hardware and software approaches to implementing a complex digital system.

Appendix A

Symbols

Below are the principal symbols used in the text and their meanings.

Symbol	Meaning	Symbol	Meaning
α	temperature coefficient of resistance	G	overall gain
β	bipolar transistor DC current gain (equivalent to h_{FE})	g_m	transconductance
		H	magnetic field strength
ε	permittivity	h_{fe}	bipolar transistor small-signal current gain (common-emitter)
ε_0	absolute permittivity, permittivity of free space	h_{FE}	bipolar transistor DC current gain (common-emitter)
ε_r	relative permittivity		
ζ	damping factor	h_{ie}	bipolar transistor small-signal input resistance (common-emitter)
ϕ	phase difference		
Φ	magnetic flux	i_b, i_c, i_e	small-signal base, collector and emitter currents
μ	permeability		
μ_0	permeability of free space	I_B, I_C, I_E	DC base, collector and emitter currents
μ_r	relative permeability	I_{BB}, I_{CC}, I_{EE}	base, collector and emitter supply currents
ρ	resistivity of a material	I_{CEO}	leakage current, collector to emitter with base open circuit
σ	conductivity of a material		
T	time constant	i_d, i_g, i_s	small-signal drain, gate and source currents
ω	angular frequency of a sine wave	I_D, I_G, I_S	DC drain, gate and source currents
ω_0	angular corner frequency	I_{DD}, I_{GG}, I_{SS}	drain, gate and source supply currents
ω_0	angular resonant frequency	I_{DSS}	drain-to-source saturation current
ω_c	angular cut-off frequency	I_p	peak current of a sine wave
ω_n	undamped natural frequency	I_{pk-pk}	peak-to-peak current of a sine wave
A_i, A_p, A_v	current, power and voltage gains	I_{rms}	root-mean-square current
B	bandwidth	I_s	reverse saturation current
B	magnetic flux density	I_{SC}	short-circuit current
C	capacitance	k	Boltzmann's constant
D	electric flux density	L	inductance
e	electronic charge	M	mutual inductance
E	electric field strength	P_{av}	average power
E_m	dielectric strength	P_i, P_o	input and output power
F	magnetomotive force	Q	reactive power
f_0	corner frequency	Q	quality factor
f_0	resonant frequency	R	resistance
f_c	cut-off frequency	r_d	drain resistance
f_T	transition frequency	R_i, R_o	input and output resistance

Symbol	Meaning	Symbol	Meaning
R_L	load resistance	V_{br}	breakdown voltage
R_M	meter resistance	V_{DD}, V_{GG}, V_{SS}	drain, gate and source supply voltages
R_s	source resistance	v_{ds}, v_{gs}	small-signal drain-to-source and gate-to-source voltages
R_{SE}	meter series resistance		
R_{SH}	meter shunt resistance	V_{DS}, V_{GS}	DC drain-to-source and gate-to-source voltages
S	apparent power		
S	reluctance	v_i, v_o	small-signal input and output voltages
T	absolute temperature	V_i, V_o	input and output voltages
T	period of a repetitive waveform	V_{NI}	noise immunity
t_f	fall time	V_{OC}	open-circuit voltage
t_H	hold time	V_p	peak voltage of a sine wave
t_{PD}	propagation delay time	V_P	pinch-off voltage
t_{PHL}	propagation delay for transitions from high to low	V_{pk-pk}	peak-to-peak voltage of a sine wave
t_{PLH}	propagation delay for transitions from low to high	V_{pos}, V_{neg}	positive and negative supply voltages for an op-amp
t_r	rise time	V_{rms}	root-mean-square voltage
t_S	set-up time	V_S	source voltage
V_+, V_-	non-inverting and inverting op-amp input voltages	V_T	threshold voltage
		V_Z	Zener breakdown voltage
V_{BB}, V_{CC}, V_{EE}	base, collector and emitter supply voltages	X_C	reactance of a capacitor
		X_L	reactance of an inductor
		$\mathbf{Z}$	impedance

Appendix B

SI Units and Prefixes

Below are a series of physical quantities and their associated SI units.

Quantity	Quantity symbol	Unit	Unit symbol
Capacitance	C	farad	F
Charge	Q	coulomb	C
Conductance	G	siemens	S
Current	I	ampere	A
Electric field strength	E	volts per metre	V/m
Electric flux	Ψ	coulomb	C
Electric flux density	D	coulombs per square metre	C/m^2
Electromotive force	E	volt	V
Energy	W	joule	J
Force	F	newton	N
Frequency	f	hertz	Hz
Frequency (angular)	ω	radians per second	rad/s
Impedance	Z	ohm	Ω
Inductance (self)	L	henry	H
Inductance (mutual)	H	henry	H
Magnetic field strength	H	amperes per metre	A/m
Magnetic flux	Φ	weber	Wb
Magnetic flux density	B	tesla	T
Period	T	second	s
Permeability	μ	henries per metre	H/m
Permittivity	ε	farads per metre	F/m
Potential difference	V	volt	V
Power (active)	P	watt	W
Power (apparent)	S	volt ampere	VA
Power (reactive)	Q	volt ampere (reactive)	var
Reactance	X	ohm	Ω
Resistance	R	ohm	Ω
Resistivity	ρ	ohm metre	Ωm
Temperature	T	kelvin	K
Time	t	second	s
Torque	T	newton metre	Nm
Velocity	V	metres per second	m/s

iator

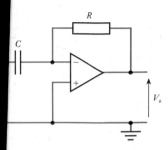

$$V_o = -RC\frac{\mathrm{d}V_i}{\mathrm{d}t}$$

Input impedance determined by C.
Low output resistance.
Virtual earth circuit.
Sensitive to noise – resistor in parallel with C
reduces noise.
See Section 18.4

Integrator

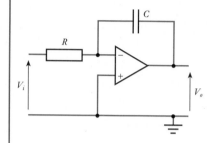

Notes

$$V_o = -\frac{1}{RC}\int_0^t V_i\,\mathrm{d}t$$

Input impedance determined by R.
Low output resistance.
Virtual earth circuit.
DC input will produce a ramp output. Offset
voltages can be a problem.
See Section 18.4

ator with reset

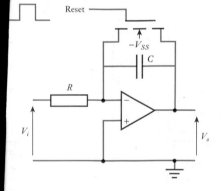

$$V_o = -\frac{1}{RC}\int_0^t V_i\,\mathrm{d}t$$

Circuit behaviour largely as an integrator.
FET acts as a switch, discharging C when closed.
Constant V_i and regular pulses on reset will
produce sawtooth waveform.
See Section 20.8

Sample and hold gate

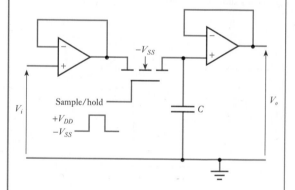

Notes

$$V_o = V_i \text{ (at time of sample)}$$

Very high input resistance.
Very low output resistance.
For best performance, use FET input op-amp for
second amplifier to minimise discharge of C and
maximise hold time.
See Section 26.5

Below is a list of the most commonly used unit prefixes.

Prefix	Name	Meaning (multiply by)
E	eta	10^{18}
P	peta	10^{15}
T	tera	10^{12}
G	giga	10^{9}
M	mega	10^{6}
k	kilo	10^{3}
h	hecto	10^{2}
da	deca	10^{1}
d	deci	10^{-1}
c	centi	10^{-2}
m	milli	10^{-3}
μ	micro	10^{-6}
n	nano	10^{-9}
p	pico	10^{-12}
f	femto	10^{-15}
a	atto	10^{-18}

Appendix C — Op-amp Circuits

The following are examples of basic operational amplifier circuits. These are included for illustrative purposes rather than as a source of definitive circuits. In each case, component values must be chosen with care, taking into account the guidance given in Section 8.6.

Non-inverting amplifier

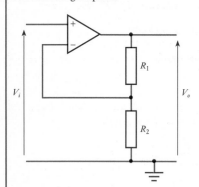

Notes

$$\frac{V_o}{V_i} = \frac{R_1 + R_2}{R_2}$$

High input resistance.
Low output resistance.
Good voltage amplifier.
See Section 8.3.1

Inverting amplifier

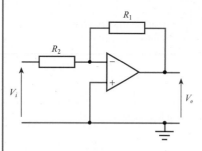

Notes

$$\frac{V_o}{V_i} = -\frac{R_1}{R_2}$$

Input resistance set by R_2.
Low output resistance.
Virtual earth amplifier.
See Section 8.3.2

Unity-gain buffer amplifier

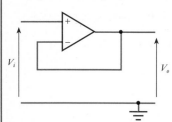

Notes

$$\frac{V_o}{V_i} = 1; \quad V_o = V_i$$

Very high input resistance.
Very low output resistance.
Excellent buffer amplifier.
See Section 8.4.1

Current-to-voltage converter

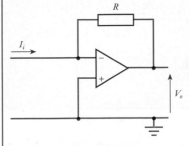

Notes

$$V_o = -I_i R$$

Very low input resistance.
Low output resistance.
Virtual earth circuit.
Also called a trans-resistive or trans-impedance amplifier.
See Section 8.4.2

Differential amplifier (subtr

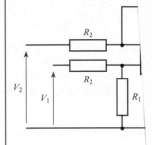

Notes

$$V_o = (V_1 - V_2)\frac{R_1}{R_2}$$

Input resistance gen
input.
Low output resistanc
If $R_1 = R_2$ then $V_o = ($
See Section 8.4.3

Inverting summing amplifier (adder)

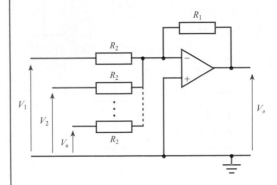

Notes

$$V_o = -(V_1 + V_2 + \ldots + V_n)\frac{R_1}{R_2}$$

Input resistance set by R_2.
Low output resistance.
Virtual earth amplifier.
Any number of inputs.
See Section 8.4.4

Non-inverting summing amplifie

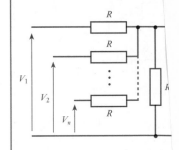

Notes

$$V_o = (V_1 + V_2 + \ldots + V_n)$$

Input resistance determin
Low output resistance.
Any number of inputs.

Differen

V_i

Notes

Integr

$+V_{DD}$
$-V_{SS}$

Note

A low-pass filter

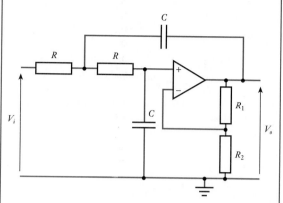

Notes

    Two-pole filter.
Values shown give a Butterworth response.

$$f_o = \frac{1}{2\pi CR}$$

Filter characteristics (and cut-off frequency) affected by gain set by R_1 and R_2.
See Section 17.10

A high-pass filter

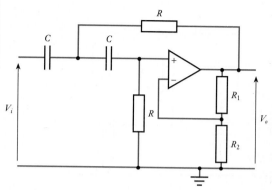

Notes

    Two-pole filter.
Values shown give a Butterworth response.

$$f_o = \frac{1}{2\pi CR}$$

Filter characteristics (and cut-off frequency) affected by gain set by R_1 and R_2.
See Section 17.10

A band-pass filter

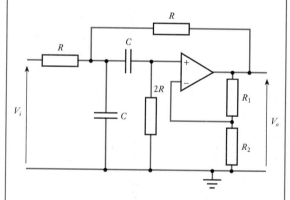

Notes

    Two-pole filter.
Values shown give a Butterworth response.

$$f_o = \frac{1}{2\pi CR}$$

Filter characteristics (and centre frequency) affected by gain set by R_1 and R_2.
See Section 17.10

A band-stop filter

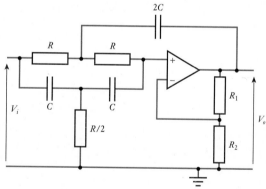

Notes

    Two-pole filter.
Values shown give a Butterworth response.

$$f_o = \frac{1}{2\pi CR}$$

Filter characteristics (and centre frequency) affected by gain set by R_1 and R_2.
See Section 17.10

A Wien-bridge oscillator

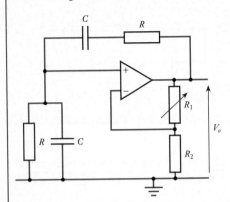

Notes

$$f = \frac{1}{2\pi CR}$$

Normally $R_1 \approx 2R_2$.
If gain is too low, oscillation will stop.
If gain is too high, output will saturate and distort.
More sophisticated circuits use automatic gain control.
See Section 24.2.2

Appendix D

Complex Numbers

Real, imaginary and complex numbers

Readers will be familiar with the problem of solving quadratic equations of the form

$$ax^2 + bx + c = 0$$

For example, the equation

$$x^2 + x - 6 = 0$$

can be rewritten as

$$(x - 2)(x + 3) = 0$$

which yields the solution that $x = 2$ or $x = -3$.

Unfortunately, some equations, for example

$$x^2 + 1 = 0$$

cannot be solved using *real* numbers. To overcome this problem, mathematicians define an *imaginary* number i, which has the property that

$$i^2 = -1$$

$$i = \sqrt{-1}$$

and this allows all forms of quadratic equation to be solved. While the symbol 'i' is widely used in mathematics, in engineering we generally use the symbol 'j' for this quantity, since 'i' is widely used to represent current.

The existence of imaginary numbers permits us to use several different forms of number, namely **real numbers** (1, 2, 3, etc.); **imaginary numbers**, which are a product of j and a real number (j1, j2, j3, etc.); and **complex numbers**, which are formed by adding a real to an imaginary number (for example, $3 + j4$).

A complex number x would therefore be of the form

$$x = a + jb$$

where a and b are real numbers. Here a represents the *real* part of x, and b (not jb) represents the *imaginary* part of x. This is written as

$$\text{Re}(x) = a$$

$$\text{Im}(x) = b$$

Graphical representation of complex numbers

Complex numbers are two-dimensional quantities that can be represented as a point on a rectangular co-ordinate plane called a **complex plane**. This has a real horizontal axis and an imaginary vertical axis, and a complex number can be represented by a line as shown in Figure D.1. This form of representation is known as an **Argand diagram**, and this figure shows the **rectangular form** of $x = a + jb$.

An alternative method of representing x is shown in Figure D.2, where the number is defined by the length r and the angle of rotation θ of a line. This is termed the **polar form** of the complex number. Comparing Figures D.1 and D.2, it is clear that the conversion from the rectangular form to the polar form is straightforward, since by Pythagoras

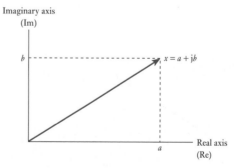

Figure D.1 Rectangular representation of a complex number

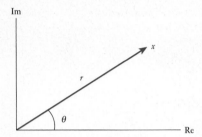

Figure D.2 Polar representation of a complex number

$$r = \sqrt{a^2 + b^2}$$

and

$$\theta = \tan^{-1}\frac{b}{a}$$

r is called the *magnitude* of the complex number x and may be written as $|x|$. θ is the *angle* of the complex number and may be written as $\angle\theta$. Therefore, the polar form of a complex number can be expressed as

$$x = r\angle\theta \quad \text{or} \quad x = |x|\angle\theta$$

Conversion from the polar form to the rectangular form is also straightforward, as illustrated in Figure D.3. Clearly, the real part of a complex number x with magnitude r and phase angle θ is equal to $r\cos\theta$, and the imaginary part is equal to $r\sin\theta$. Consequently, x may be written as

$$x = r\cos\theta + jr\sin\theta$$

A further form can be obtained by using Euler's formula, which says that

$$e^{j\theta} = \cos\theta + j\sin\theta$$

Therefore, an alternative form of x is given by

$$x = r\cos\theta + jr\sin\theta$$
$$= re^{j\theta}$$

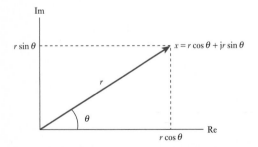

Figure D.3 Conversion from polar to rectangular form

This is called the **exponential form** of the complex number and is shown in Figure D.4.

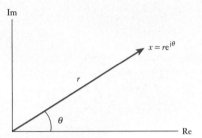

Figure D.4 Exponential representation of a complex number

The complex conjugate

The **conjugate** of a complex number x is formed by negating the imaginary part of the number and is given the symbol x^*. Therefore, if $x = a + jb$, then

$$x^* = a - jb$$

The relationship between x and x^* is shown in Figure D.5. From the figure it is clear that when using polar notation the magnitude of x^* is equal to the magnitude of x, but that the angle is reversed. Therefore, if $x = r\angle\theta$, then $x^* = r\angle-\theta$. Similarly, when using the exponential form, if $x = re^{j\theta}$, then $x^* = re^{-j\theta}$.

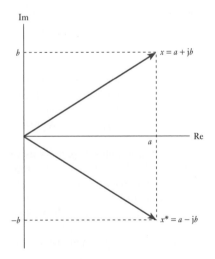

Figure D.5 The relationship between a complex number x and its conjugate x^*

Complex arithmetic

To add (or subtract) complex numbers, we simply add (or subtract) their real parts and their imaginary parts. For example, if $x = a + jb$ and $y = c + jd$, then

$$x + y = (a + jb) + (c + jd)$$

$$= (a + c) + j(b + d)$$

The multiplication of complex numbers is also straightforward, provided that we remember that $j^2 = -1$. If x and y are as before, then

$$xy = (a + jb)(c + jd)$$

$$= ac + jad + jbc + j^2bd$$

$$= ac + jad + jbc - bd$$

$$= (ac - bd) + j(ad + bc)$$

It is interesting to note that the multiplication of a complex number with its conjugate produces a real number. For example, if $x = a + jb$, then

$$xx^* = (a + jb)(a - jb)$$

$$= a^2 - jab + jab - j^2b^2$$

$$= a^2 + b^2$$

The division of complex numbers is simplified by the use of the conjugate. If, as before, $x = a + jb$ and $y = c + jd$, then

$$\frac{x}{y} = \frac{(a + jb)}{(c + jd)}$$

The presence of an imaginary element in the denominator is inconvenient, but it can be removed by multiplying top and bottom by y^*.

$$\frac{x}{y} = \frac{(a + jb)}{(c + jd)}$$

$$= \frac{(a + jb)}{(c + jd)} \frac{(c - jd)}{(c - jd)}$$

$$= \frac{(ac + bd) + j(bc - ad)}{c^2 + d^2}$$

$$= \frac{ac + bd}{c^2 + d^2} + j\frac{bc - ad}{c^2 + d^2}$$

While multiplication and division of complex numbers are straightforward (as described above), it is often simpler to perform these tasks using the polar form of the number, since

$$A\angle\alpha \times B\angle\beta = AB\angle(\alpha + \beta)$$

$$\frac{A\angle\alpha}{B\angle\beta} = \frac{A}{B}\angle(\alpha - \beta)$$

Multiplication and division are also easy using the exponential form, since

$$Ae^{j\alpha} \times Be^{j\beta} = ABe^{j(\alpha+\beta)}$$

$$\frac{Ae^{j\alpha}}{Be^{j\beta}} = \frac{A}{B}e^{j(\alpha-\beta)}$$

For this reason, it is common to perform addition or subtraction of complex numbers using the rectangular form but to use the polar or exponential form when performing multiplication or division. Fortunately, converting between these different forms is straightforward (as described above).

Answers to Selected Exercises

Chapter 2

2.4	5 mA
2.5	6 kΩ
2.6	25 W
2.7	10 nW
2.8	50 Ω
2.9	12 Ω
2.10	7.9 kΩ
2.11	600 Ω
2.12	20 Ω, 50 Ω
2.13	1.51 kΩ, 208 Ω
2.14	6 V, 10 V, 8 V
2.15	16 V, 4 V, −10 V
2.16	1 ms
2.17	50 kHz

Chapter 3

3.7	138.5 Ω, 1.385 V
3.18	1 V, 3.85 mV/ °C

Chapter 5

5.7	2
5.8	4096
5.9	0.024 percent
5.10	7 bits
5.16	9 kHz

Chapter 6

6.6	25 V
6.7	0.1
6.8	9.12 V
6.9	18.6
6.10	24 μW, 83 mW, 3.5×10^3
6.12	10.8 V
6.13	439
6.14	2.42 nW, 667 mW, 2.8×10^8
6.16	13.2 V
6.20	10 dB, 0 dB, 3 dB, 120 dB
6.21	100, 10, 0.032, 0.178
6.22	30 dB
6.23	22 dB
6.24	7.07
6.25	24 kHz
6.26	5 MHz
6.27	10 V

Chapter 7

7.14	0.04
7.16	6

Chapter 8

8.6	16
8.9	−25
8.14	0.5 V
8.15	−5 V
8.23	40 kHz
8.26	(a) 31.3, 32 GΩ, 3.1 mΩ
	(b) −6.83, 12 kΩ, 680 μΩ

(c) 46.3, 22 GΩ, 4.6 mΩ
(d) 1, 1 TΩ, 100 μΩ

Chapter 9

9.20   12, 49, 23, 1.375
9.21   111000, 10000100, 1000011, 101.101
9.22   42179, 52037, 135, 1023
9.23   CDE4, 2D6, 22C4
9.24   1010010011000111
9.25   2CA5
9.26   100000, 11011, 1001101, 111

Chapter 11

11.2   0.1 Hz
11.3   40 ms
11.4   5 V
11.5   20 A
11.6   62.8 rad/s
11.7   25 Hz
11.8   5 V, 10 V, 250 Hz, 1571 rad/s
11.11 75 Hz, 25 V
11.13 6.37 V
11.14 7.85 A
11.17 2 W
11.18 4 W
11.19 6.66 V
11.20 5 V
11.21 1 W
11.22 2 mΩ
11.23 200 kΩ
11.24 11 percent (high)
11.25 11.1 V
11.32 5.3 V
11.34 60°, B leads A

Chapter 12

12.2   50 C
12.6   100 V, 50 V, 500 μV, −2.35 V
12.7   200 W, 1.25 W, 2.5 W, 117 μW
12.8   16 mΩ
12.9   150 Ω, 33.3 Ω, 42 Ω
12.10 5 kΩ

12.15 6 V, 375 V, 60 V
12.18 1.8 V
12.20 −446 mA
12.22 2.5 mA
12.24 470 mV
12.26 1.08 V
12.28 −62 mA
12.30 5 V

Chapter 13

13.6   45.5 V
13.7   20 μF
13.10 66 pF
13.11 13 nF
13.14 16.7 MV/m
13.17 67 mC/m^2
13.18 150 μF, 3.75 mF, 5.9 nF, 39.3 μF
13.21 100 ms
13.22 10 μF
13.26 562 mJ
13.27 15.6 mJ

Chapter 14

14.3   0.48 A/m
14.7   3000 ampere-turns, 3333 A/m, 4.19 mT, 1.68 μWb
14.9   3000 A/Wb
14.15 3 mH
14.17 36.3 μH
14.18 20.9 mH
14.19 1.43 H, 35 mH, 10.9 μH, 250 mH
14.23 2 s
14.29 49 mJ
14.34 50 V

Chapter 15

15.1   100 rad/s, 15 V
15.2   39.8 Hz, 17.7 V
15.14 12.6 Ω
15.15 200 kΩ
15.16 1.18 A r.m.s.
15.17 5 mV peak
15.21 28.0 V, ∠14.6°

15.26 $58.6\angle-65°$
15.28 $1000 + j0$
15.29 $0 - j159$
15.30 $0 + j6.28$
15.31 $80 + j124, 40 - j40$
15.32 $36\angle56°, 36e^{j56°}$
15.33 $19.1 + j16.1, 25e^{-j40°}$

Chapter 16

16.1 1 W
16.4 700 VA, 0.5, 350 W
16.6 1.97 A, 197 VA, 0.786, 155 W, 121 var
16.7 500 VA, 400 W, 300 var, 2 A
16.9 12.7 μF
16.10 4.5 μF

Chapter 17

17.1 15.9 Ω, 2 Ω
17.2 39.8 Hz
17.3 1571 rad/s
17.5 495 μs
17.8 15 Hz, 100 kHz, 8 kHz, 10 MHz, 3 Hz, 50 kHz
17.11 200 μs

Chapter 18

18.8 40 ms
18.9 576 μs

Chapter 19

19.13 9.1 V

Chapter 20

20.24 2 MΩ, 4 kΩ, 12, 2.9 Hz

Chapter 21

21.10 930 μA, 4.8 V
21.12 1.5 mA, 6.15 V, 3.9
21.16 5.2 mA, 5.2 V, 1

Chapter 22

22.19 15 W

Chapter 23

23.2 2.9 V
23.12 1000 rpm
23.16 18,000 rpm

Chapter 24

24.3 65 Hz
24.5 159 Hz
24.11 0.00004 percent

Chapter 26

26.2 8 kHz
26.6 4096

Index